Sasquatch
THE APES AMONG US

John Green

hancock

house

WPK 07 09

ISBN 0-88839-123-4 (Paperback Trade Edition)
Copyright 1978 Cheam Publishing Ltd.
First Paper Edition 1981
Second Paper Edition 2006

3 1116 02122 4867

Cataloging in Publication Data

Green, John, 1927 –
Sasquatch

ISBN 0-88839-123-4
1. Sasquatch. 1. Title
QL89.2.S2G74   001.9'44   C78-002016-2

Printed in Indonesia — TK PRINTING
Cover design: Rick Groenheyde

Published simultaneously in Canada and the United States by

HANCOCK HOUSE PUBLISHERS LTD.
19313 Zero Avenue, Surrey, B.C. Canada V3S 9R9
(604) 538-1114 Fax (604) 538-2262

HANCOCK HOUSE PUBLISHERS
1431 Harrison Avenue, Blaine, WA, U.S.A. 98230-5005
(604) 538-1114 Fax (604) 538-2262

Website: www.hancockhouse.com
Email: sales@hancockhouse.com

Contents

Maps

Locations of sighting and track reports are plotted on the maps as accurately as possible, but in many cases no precise location is known, and in others there are too many reports from a locality to place them all correctly.

The map on the facing page shows the number of reports from each state and province in the author's file as of November, 1977. Locations of reports are shown only for those states and provinces not covered by more detailed maps.

Maps drawn by Donald MacCallum, Harrison Hot Springs, British Columbia.

7

Grateful acknowledgement is made for permission to reprint copyright material:

From In Search of the Red Ape, by John Mackinnon. Copyright by John MacKinnon.
  Reprinted by permission of Holt, Rinehart and Winston, publishers, New York, and
  Collins Publishers, London.
From True North, by Elliott Merrick, Asheville, North Carolina.
From pp. 228-229 of The Long Walk: A Gamble for Life by Slovomir Rawicz. Copyright
1956 by Slavomir Rawicz. Used by permission of Harper & Row Publishers, Inc. and Harold
From The Apes, by Vernon Reynolds. Copyright 1967 by Vernon Reynolds. Reprinted by
  permission of the publishers, E.P. Dutton.
From Abominable Snowman, Legend Come to Life, by Ivan Sanderson. By permission of the
  publishers, Chilton Book Co.
From Wildlife Sketches Near and Far, by Bruce Wright. By permission of the publishers,
  ˙ University of New Brunswick Press.
From In the Shadow of Man, by Jane Goodall. By permission of the publishers, Houghton
  Mifflin Company, Boston, and Collins Publishers, London.
From Wild Men in the Middle Ages, by Richard Bernheimer. Copyright 1952 by the
  President and Fellows of Harvard College. By permission of the publishers, Harvard
  University Press.
From The Gentle Giants, by Geoffrey Bourne. Copyright by Geoffrey Bourne. By
  permission of the publisher,G.P. Putnam's Sons.
From Smokey and the Fouke Monster, by J.E. (Smokey) Crabtree. By permission of Carlton
  Press Inc., New York.

Foreword

It is tempting to speculate that somewhere in the dim past (one to two million years ago) one of the offshoots of man's evolutionary tree may have travelled from his point of origin in China or India up the mountain causeway which extends from Szechuan through Manchuria into the eastern part of Siberia. This causeway is covered with montane forest—the same type of forest which is present on the other side of the Bering Strait in Alaska and passes down the western side of the North and South American continents to Tierra del Fuego at the bottom of South America. There have been periods over the last two million years when the ice of the North Pole advanced and retreated and there must have been quite long periods when there was a complete sheet of ice across the Bering Strait connecting Siberia with Alaska.

Most probable candidate for the above script is Gigantopithecus, the remains of which were found in China and were estimated to be half a million to a million years old. The "abominable snowman" of Tibet and the sasquatch of North America may in fact be specimens of Gigantopithecus survived to the present day. The possibility of this northward migration and crossing into the North American continent would also explain the alleged presence of sasquatch-like creatures in the Soviet Union—the almas. In fact the Russians believe their almas are actually the remains of a race of dawn men who somehow or other managed to survive into the 20th Century. In the last two weeks have come reports that large naked but hair-covered manlike creatures are seen regularly by villagers in the eastern mountain parts of the Soviet Union.

As with the abominable snowman of Tibet, the existence of the North American sasquatch at the present time remains in doubt. That the sasquatch did occur in the northwest of the U.S.A. and in British Columbia in Canada in the past is supported by the fact that the Indians of these areas have old legends which tell of creatures like the abominable snowman or sasquatch, tall hairy creatures walking in a bipedal fashion which have been known in that part of the world for generations. Records of encounters between these creatures and the white man go back to 1811—but the question at issue remains--does a similar creature exist on the North American continent today? Claims to have seen sasquatches or creatures resembling them have been made from many areas of the country—from the far southeast in Florida and in the Midwest states. There have also been claims of sasquatch-like creatures along the western side of South America as well as alleged sightings in California, Washington State and Oregon in the U.S. The sasquatch comes into the category of monsters and the human race has a paranoid affection or a least a horrified interest in monsters, probably as a relic of a racial memory of the days when Neanderthal man played that role for the developing Cro-Magnon race.

Because of this quirk of human nature, the desire for notoriety which accompanies a sighting of these creatures expresses itself in quite a lot of people, and trivial bits of information or observation may be magnified out of all proportion. Then we have the lunatic fringe and we have the deliberate hoaxers who prey on such mysteries. The subject needs calm and thoughtful study. It needs an unemotional analysis of evidence and an unbiased approach to the subject. I believe that John Green in the present book has attained these objectives and has produced a very worthwhile compendium. His historical material is of great interest and his accounts and analysis of sasquatch sightings in the U.S. are factual and unemotional. There is no doubt that for anyone interested in this corner of human knowledge, or fable as some people prefer to describe it, Mr. Green's book is essential reading. Anyone who picks it up will find it easy and fascinating to read and will be richer in their understanding of this controversial subject.

February 12, 1978

Geoffrey H. Bourne
Yerkes Primate Research Centre
Emory University
Atlanta, Georgia

10

Author's Preface

I do not "believe in" sasquatches. Consequently, this is not a book for "believers." Nor is it a thriller about monsters. It is the story of an investigation, and one not yet successfully concluded.

The first 30 years of my life progressed quite satisfactorily without any need for monsters, weird animals or mysteries of any sort. My profession was newspaper work, my other interests were my family, politics and sailboats. I was not then, and am not now, attracted by fantasy or science fiction; and if I do have some curiosity about reports of such things as marine monsters, flying saucers, extra-sensory perception and telepathy, I am quite content to await possible further developments with no thought of personal involvement. My exposure to the sasquatch problem came about by accident and developed into serious interest through coincidence as much as anything else. If there had not been good tracks to see very shortly after I first became involved, and following that financial support for further investigation, I probably would have let the matter drop. Subsequent experience has shown that neither footprints nor finances are usually so readily come by.

Once well into it, I have found the whole matter far too interesting to get away from, but it has nothing to do with monsters. Although sasquatches are described as being very big, there is nothing monstrous about their behavior. A few incidents have undoubtedly been spine-chilling for the people involved, but I have found no one who tells of a sasquatch actually doing anything to hurt them. In most cases it was just there, not engaged in any activity more exciting that standing still or walking along.

The fascination of the subject, for me, involves the very thing that I am most inclined to complain about—that the scientific world ignores it. The material available should be of great interest to more than one branch of science. Footprints, film, and eye-witness accounts all have to have something behind them, either an animal or a hoax. If an animal, then it is one of the most interesting animals imaginable, both because of its physical resemblance to man and because it has remained unknown for so long. If the whole thing results from a hoax it is a gigantic one, operating all over the world,

11

century after century, and sorting it all out should be of great value to those who study the ways of human beings.

There are only the two alternatives that could provide an explanation for everything, since the footprints are real and must be either made by real animals or manufactured by men, but there is a third possibility with regard to the sighting reports. If there is really nothing to see, then they must involve some form of self-delusion or hallucination, often affecting more than one person at a time, or some human compulsion to tell wild stories in spite of unpleasant consequences. Such phenomena should be of interest also.

Instead, the amount of participation in the investigation by scientists or academics of any variety is very small, and the participation of scientific institutions non-existent. The field is left open to laymen, and a fertile field it has proved to be. There are people to be found in virtually every state and province in North America who describe encounters with upright, hair-covered animals unknown to science. There are written records, from a wide variety of sources, of similar reports far into the past. There are massive, mysterious footprints being found all over the continent. Laymen have all this to themselves because science has its eyes and ears tight shut, refusing to be disturbed by what it knows should not be. And laymen have built up a body of evidence for the existence of something unknown, be it a creature or a conspiracy, that cannot be ignored much longer.

One of the few scientists who has paid attention to the subject is Dr. John Napier, a British physical anthropologist of world-wide reputation who formerly headed the primate program at the Smithsonian Institute in Washington. In the preface to the 1976 edition of his book *Bigfoot, the Yeti and Sasquatch in Myth and Reality*, he had this to say:

> One is forced to conclude that a man-like life-form of gigantic proportions is living at the present time in the wild areas of the north-western United States and British Columbia. If I have given the impression that this conclusion is—to me—profoundly disturbing, then I have made my point. That such a creature should be alive and kicking in our midst, unrecognized and unclassifiable, is a profound blow to the credibility of modern anthropology.

My conclusion is the same as Dr. Napier's, that there is indeed such a creature in North America. The subject has long since grown beyond what can be written in a single volume, but in this book I am attempting to give some impression of the scope of the available information, and to present some of the best of the evidence and some of the more interesting stories of the investigation.

12

-1-

The Ape in the Night

You are sitting alone in the house at night when you hear a slight noise outside, and turning your head you are confronted by a bestial black face more like a gorilla's than anything else, staring in the window at you. It's a horror movie come true, the most terrifying thing you have ever experienced, yet somehow you survive it and after a while the face is gone.

The reason you survived is simple. You were never in any danger in the first place. An animal looked in out of the darkness to see what was inside your lighted room, but it wasn't trying to get in to eat you, and it felt no hostility towards you. It was just curious and for some reason abandoned its usual caution. In reality you were granted a rare privilege, a chance for a close look at one of the most interesting creatures on earth and one that very few people ever see. It's too bad you weren't in shape to appreciate it.

The face in your window belonged to a sasquatch, which is either a legend—in which case it is odd that you were able to see it—or the largest of the apes; an ape that lives in North America and that walks upright like a man.

If a moose or a buffalo looked in your window you would probably get quite a shock too, but by no means of the same magnitude. It would be an interesting and unexpected experience, and if the thing stayed there for any length of time you would probably find yourself wondering whether you could move closer for a better look without scaring it away. The only thing different about seeing a sasquatch is

13

that it is a good deal more unusual, but there may really be a chance that one will someday look in your window. There are people who say it has happened to them. Rod Pullar, a lifelong woodsman from Bellingham, Washington, has spent years looking for sasquatches and is certain that he has come close enough to hear them and smell them, but he has never seen one. On the other hand he can point out a house on Texas Street, now in the middle of a built-up area, where a friend of his tells of opening her curtains at 4 a.m. and seeing a huge flat face covered with oyster-white hair, "very sad looking" with downward-sloping eyes. She closed the curtains and told herself she couldn't have seen any such thing, but when she opened the curtains again it was still there. That was in September, 1969, and even then the house was by no means out in the woods, although a wooded hillside came down close to it.

I have talked at length with a lady who tells of two such experiences at her home seven miles northwest of Bremerton, Washington. At 2 a.m. on December 15, 1971, she saw from her kitchen window a dark-haired, erect animal about eight feet tall standing about five feet away, apparently looking into the house. She called her three children and they also saw it. The following February, in the daytime, she saw another one, seven or eight feet tall, covered with red brown hair, standing at the edge of the woods near her house, about 25 feet away. That time she was outside in the garden, so she went in the house. In neither case was there the slightest indication that she was in any danger.

Another woman that I interviewed told of stepping outside her house by the Nooksack River near Bellingham shortly before midnight because her dog was barking, and seeing a huge dark brown creature, covered with hair and with eyes that reflected her flashlight, but with a human look to its face. It was sitting on an up-ended oil drum in the back yard next door, its hands resting on the edges of the drum with its arms straight, pushing its shoulders up. As she shone her flashlight on it, it just kept sitting there, so she went back into her house. That was in September, 1967.

An incident near Fort Bragg, California, in Feb. 1962, involved a great deal more action, but again no harm to anyone. Here is how Bud Jenkins, one of the men involved, told the story in a taped interview made by Chuck Edmonds, of Ashland, Oregon, a short time after the event:

> My brother-in-law heard the dogs barking and he got up and went out to see what they were barking about. We have a fence between the house and the barnyard made of six-foot pickets and he saw something standing by that fence looking over the fence towards the dogs, which he thought was a

bear, and he came back in and woke me up and told me to come out and he would show me the largest bear I would ever see. It was standing upright.

I got up and went out with him and we didn't see it, so I said, "Well, wait a minute and I'll go in the house and get a flashlight and a gun."

So I came back in the house and my brother-in-law walked to the other side of the house then, to look back in the back yard, and as he stepped out from the corner of the house to look back there this creature stepped over this little two-foot fence we have out here right towards him, and he let out a scream and stepped backwards and as he stepped backwards he fell, so he came into the house on his hands and knees going like mad.

My wife was at this time holding the screen door open for him to come in. I heard the commotion and I ran to the inside door which we have here before you step onto the porch, and as he came through the door I saw this large creature going by the window, but I could see neither its lower body nor its head, all I could see was the upper part of its body through the window there.

When he came in my wife tried to close the door and they got it within about two to four inches of closing and they couldn't close it. Something was holding it open. My wife hollered at me and said, "Hurry and get the gun, it's coming through the door!"

Of course by that time I was standing right behind her here in this door leading onto the porch, and I said, "Well, let it through and I'll get it."

At that time the pressure went off the door and she pushed the door to and threw the lock on it, and I walked to the window and put my hand up to the window and looked out, so that I could see out into the yard, because it was still dark, and it was raining, and this creature was standing upright, and I would judge it to be about eight feet tall and it walked away from the house, back out to this little fence we have, and stepped over the little fence and walked past my car and out towards the main road . . .

I would judge it to weigh about 400 pounds and it walked upright at all times that I saw. It never went down on all fours at all, it stayed upright, and it had a very bad odor. The odor lingered on here for minutes after the creature was gone. And it left a hand print there by the door on the side of the house which was eleven and a half inches from the base

15

of the palm to the end of the finger. It didn't act harmful really. It acted more curious than harmful, but it certainly gave us quite a start . . .

It was twenty minutes before my brother-in-law could hold a cup of coffee in his hand steady enough to drink it. Of course he stepped right to the creature and met it right face to face.

Frightening though the experience had been, the brother-in-law, Robert Hatfield, agreed that the creature's attitude was one of curiosity, not menace. Sometimes they don't even seem to be curious. Mrs. Louise Baxter, of Skamania, Washington, had already had one look at a sasquatch in November, 1969, when a dark grey, shaggy biped, "very very big", walked across the Lewis and Clark Highway near North Bonneville, Washington, in front of her car at 10 o'clock at night. The following August she was driving on the same highway in the evening when she heard a noise that made her think she might have a flat tire. It was still daylight, so she pulled off the road and got out to have a look. Here is her story of what happened after that:

I kicked the tire, which was okay, and then bent over to see if possibly something was stuck under the fender to make the noise.

I suddenly felt as if I was being watched and without straightening up I looked towards the wooded area beside the road and looked straight into the face of the biggest creature I have ever seen except the one the time nearly a year before.

The creature was coconut brown and shaggy and dirty looking. It had one huge fist up to its mouth. The mouth was partly open and I saw a row of large square white teeth. The head was big and seemed to set right onto the shoulders. The ears were not visible due to the long hair about the head. It seemed the hair was about two inches long on its head.

It had a jutted chin and receding forehead. The nose and upper lip were less hairy and the nose was wide with big nostrils.

The eyes were the most outstanding as they were amber color and seemed to glow like an animal's eyes at night when car lights catch them.

It seemed contented there and seemed to be eating as the left fist was up toward the mouth as though it had something in it.

I screamed or hollered but whether I made any noise I can't tell I was so terrified. I know it didn't move while I looked. I don't remember how I got back in the car or how I

16

started it. As I pulled out I could see it still standing there, all 10 or 12 feet of him.

In the usual order of things such incidents are not reported to the police, or if someone does call them there is nothing for them to see when they get there. In the fall of 1975, however, there was an incident with a different twist. It was a policeman, answering a call about a prowler, who told of encountering the sasquatch.

The man concerned was the sergeant of the police force on an Indian reserve in Washington. Throughout the early part of October he and other officers had answered several calls concerning some creature that made loud noises in the vicinity of houses, and there had been four reports of hairy bipeds being seen. On one occasion the captain of the police force had shot with his revolver at an animal he saw in the bush beside a field that appeared to be about six feet tall and hair-covered. The descriptions given indicated that more than one creature was being seen, although not more than one at a time.

On October 23, at 7:30 p.m., the sergeant was called to a house where something had been heard pounding on the back wall. The woman who lived there had gone next door to her son's home, and there was no prowler in sight, but something had apparently torn some plastic that covered a back doorway and there was a broken window. At 2:20 a.m. the same night something was again reported behind the house, and when the sergeant arrived, along with several other people, their spotlight quickly picked up what looked like a very large ape standing in the back yard. While someone else held the light on it, the sergeant walked up to within 35 feet of the animal, which crouched down but made no attempt to run. There they stayed, for "many minutes", while the sergeant wondered what to do next. He had a shotgun loaded with buckshot, but he was not sure if the thing was some kind of human, and if it wasn't, he was not sure how much buckshot it could take. He noted afterwards in his report that it was black in color, would stand about seven to eight feet tall and appeared to have no neck. It was covered with short hair, except on the face. He could see no ears. The eyes were small. It appeared to have four teeth larger than the others, two upper and two lower. Its nose was flat. He could see the nostrils. By that time there were seven people there, all of whom could see the thing, although only two others had approached close to it. Then there were noises heard off in the dark on both sides, and the man with the spotlight swung it off to the right and called that there was "another one over there." At that point the sergeant decided to return to his patrol car.

17

-2-

Wild Men Long Ago

The stories told in the preceding chapter are a small sample of a phenomenon that has been going on for a long time, and is going on today. Hundreds of people are known who are still alive and who appear sane, sober and serious, who will tell of personal encounters with a type of creature which was generally unknown just a couple of decades ago, and which those who had heard of it understood to be just an Indian legend.

During the past 20 years, and particularly the last 10, North Americans have had considerable exposure to the subject. There have been a dozen books, numerous movies and television programs, and countless newspaper and magazine articles. To many people it may seem as if this has sprung from nothing—that the subject itself is new—but in fact the hairy biped has a long history which was familiar to our forefathers, but which in recent centuries has been suppressed. You can find suggestions of such creatures a long time back. In the Babylonian epic of Gilgamesh there is a hairy beast-man, Enkidu, and that is about as far back as literature goes. Greek mythology is full of hairy bipeds, albeit they often had hooves instead of feet, and our word "panic", so the dictionary tells us, is a living reminder of the unreasoning fear one of them could cause in mere humans.

In the earliest Old-English epic poem, the hero, Beowulf, is called on to subdue not only the manlike, man-eating monster Grendel, but also Grendel's mother. Richard Bernheimer, in *Wild Men in the Middle Ages* documents the fact that the wild man held such a

18

place in Medieval thought that he can be found carved, hair and all, on churches and cathedrals in England, France and Spain. He is represented in some of the earliest books ever printed and on some of the earliest playing cards. He appears in woodcuts and in etchings, and was commonly portrayed in dances and plays. King Charles VI of France was one of five men in shaggy wild-man costumes that caught fire from a torch during a dance in 1392. The other four died. The king was saved by his aunt, who smothered his flaming costume with the train of her robe, but he went out of his head from the shock.

Bernheimer's book, which was published by Harvard University Press in 1952, opens with the following description:

It is a hairy man curiously compounded of human and animal traits, without, however, sinking to the level of an ape. It exhibits upon its naked human anatomy a growth of fur, leaving bare only its face, feet and hands, at times its knees and elbows or the breasts of the female of the species . . . we find him in the clipped verses of French Arthurian romance, in the epics of German minstrel singers, and in the writings of Cervantes and Spenser. The whole vast field of late medieval secular art is his playground, from prints and paintings by Albrecht Durer and Peter Brueghel to the love caskets and tapestries which medieval swains presented to their ladies, or the tools of chivalry, such as saddles and ornamented weapons. His place in medieval daily life was assured by the appearance of his image on stove tiles, candlesticks, and drinking cups, and, on a larger scale, on house signs, chimneys and the projecting beams of frame houses . . .

It was agreed that he shunned human contact, settling, if possible, in the most remote and inaccessible parts of the forest, and making his bed in crevices, caves or the deep shadow of overhanging branches. In this remote and lonely sylvan home he eked out a living without benefit of metallurgy or even the simplest agricultural lore, reduced to the plain fare of berries and acorns or the raw flesh of animals. At all times he had to be ready to defend his life, for the inner forests teemed with savage beasts, real and imaginary, which were wont to attack him . . .

The conditions of the wild man's life thus required much more than the usual human strength, if he were to maintain himself without tools and almost without weapons in a raw and hostile environment . . .

It should be noted that while the author never for a moment

considered the possibility that anything real might exist or ever have existed to account for this preoccupation with the wild man, the people of the times about which he writes took exactly the opposite view. They did not prepare all that material to provide subject matter for the scholars of a later age to analyze. For them it was completely serious information. In fact, Bernheimer notes, folklore regarding the wild man did not end with the Middle Ages. It exists today in the mountains of central Europe.

It is in the mountainous regions and particularly in the Alps that the wild man has survived most vigorously, partly because such areas offer an excellent defense for the archaic modes of thinking against modern depredations, partly because there is a kinship between the raw grandeur of the mountains and the indomitable strength of the wild man. Ideas which had long since died out in the culturally more-advanced milieu of the plains could retain their hold over men's minds in isolated and retarded mountain valleys.

There is actually nothing noteworthy in the continued existence of such stories in the central European mountains, because they exist also in virtually every other mountainous region in the modern world. Nor is there anything unusual in the prominence of the wild men in the culture of Europe of a few hundred years ago. Information of that sort is accepted matter-of-factly in most cultures. The unusual thing is that it became suppressed in European culture after that time—a suppression that continues today in societies with cultural roots in Europe.

In North America there are no written records available before the arrival of the white man, and Indian information passed on to anthropologists about such things as wild hairy men has invariably been recorded as legend, even though the informants may not have meant it that way. The name I use for the hairy giants, "Sasquatch", is of Indian origin, but I do not claim any great knowledge of Indian material. I am aware that throughout Canada and the United States there were Indian stories of humanlike creatures, a good proportion of which were both giant and hairy. Some were considered to be eaters of human flesh. Some were supposed to have lived at a former time, others to be still in existence. They were often credited with attributes that we would consider supernatural.

There is a basic problem about such information, which was explained very well by Wayne Suttles, of Portland State University, in an article printed in *Northwest Anthropological Research Notes*, Volume 6, Number 1.

It is certainly true that we anthropologists have generally dumped sasquatch-like beings into a category 'supernaturals' and let it go at that. We may have done this because we are professionally interested more in native culture than in the facts of zoology, but I think it is more because we are operating with too simple a version of the Western dichotomy. In fact, if we were true to our earlier, Boasian objective of describing the native culture as seen by the participants, we ought not to categorize so freely the creatures our informants tell us about . . .

The sasquatch is one of many creatures the Lower Fraser people believe (or used to believe) exist (or once existed) in the wilderness around them. Most of these creatures can, from Indians' descriptions of them, be matched with animals known to Europeans. A few, however, cannot. Since we Europeans, scientifically trained or not, operate with a dichotomy real/mythical or natural/supernatural, we are inclined to place these creatures that are not part of our "real" world into our category "mythical" or "supernatural". As Green has pointed out, most of us have done this with the sasquatch and we may be wrong.

But I believe we would also be wrong to imagine that the Indians have (or had) the same dichotomy and that they would simply draw the line differently, putting the sasquatch in the category "real animals" and leaving other, to our minds more fanciful, creatures in the category "mythical animals" or "supernatural beings". In fact, I see little evidence for any such native dichotomy at all . . . a distinction between "real" and "mythical" or "natural" and "supernatural" beings just is not there. Thus a description of Coast Salish culture that is truly "emic"—that is, organized by native categories—should describe whales and bears, sasquatches and two-headed serpents all under the same heading as part of the "real" world of the Coast Salish . . .

I think that what Dr. Suttles has to say is correct. The question to which I have devoted 20 years of inquiry is precisely the one that the Indian does not ask, whether the sasquatch fits into the category of "real" or "mythical". If there are no such categories in the Indian culture, then information from that culture can be of only limited assistance.

To give a single example from the many stories available in ethnographic literature, the following is an excerpt from the *Archaeological Report of 1924-25* in the Province of Ontario. It is a small part of a paper by T. F. McIlwraith, assistant professor of anthropology, University of Toronto, entitled *Certain Beliefs of the*

21

A Kwakiutl totem pole at Alert Bay, British Columbia. The figure at the base represents the Dsonoqua, the cannibal woman.

Top and bottom views of a broken stone carving of a foot, found near Lillooet, B.C., and now in the Vancouver city museum. Missing the big toe and heel, what remains is a close match for part of an average-sized sasquatch track cast.

Bella Coola Indians Concerning Animals. Bella Coola is at the head of Burke Channel on the central coast of British Columbia.

Another animal which is rendered powerless by the same charm (a bullet smeared with the hunter's blood) is the Boqs. This beast somewhat resembles a man, its hands especially, and the region around the eyes being distinctly human. It walks on its hind legs, in a stooping posture, its long arms swinging below its knees: in height it is rather less than the average man. With the exception of its face, the entire body is covered with long hair, the growth being especially profuse on the chest, which is large, corresponding to the great strength of the animal.

The following story, which is regarded as an historical incident, illustrates the attitude of the Bella Coola towards these animals. Not many years ago a certain Qaktlis was encamped with his wife and child on a fiord near one of the haunts of boqs. He heard a number of the creatures in the forest behind him and seized his gun, at the same time calling out to them to go away. Instead, the breaking of branches and beating upon tree-trunks came nearer. Becoming alarmed, he called out once more: "Go away, or you shall feel my power."

They still approached and Qaktlis fired in the direction of the sounds. The answer was a wild commotion in the forest, roars, grunts, pounding, and the breaking of branches. The hunter, now thoroughly alarmed, told his wife and child to embark in the canoe while he covered their retreat with his gun. He followed them without molestation, and anchored his craft not far from shore. The boqs could be heard plainly as they rushed to and fro on the beach, but only the vague outlines of their forms were visible in the darkness. Presently, though there was no wind, the canoe began to roll as if in a heavy sea. Qaktlis decided to flee, but before he had gone far his paddle struck the bottom, in spite of the fact that he was in mid-channel. Looking up, he saw that the mountains were higher than usual; the boqs had, by their supernatural power, raised the whole area so that the water had been almost entirely drained away. They are the only supernatural beings with this power. Qaktlis jumped overboard into water which reached only to his knees, and towed his canoe several miles, the boqs following him along the shore. One of the features which the Bella Coola consider most surprising about this incident is that he did not tread upon the fish which must have been driven into the shallows.

23

This is not the only occasion on which boqs have appeared in that vicinity. Within the lifetime of the father of an informant, a chief was returning with some friends along the coast. As the canoe shot around the tip of a promontory, they saw a boqs gathering shell-fish. The paddlers backed behind some rocks from where they could watch without being seen. The creature acted as if frightened; it kept looking backwards, then hurriedly scraped up some clams with its fore-paws, dashed off with these into the forest, and came back for more. The chief decided to attack the animal. A frontal approach was impossible owing to lack of cover, so he landed and crept steadily through the forest, armed with his Hudson Bay Company's musket. Presently he stumbled upon a heap of clams which the animal had collected. He waited until it returned with another load, then raised his musket and fired. Instead of killing the boqs, its supernatural power was so great that the hunter's musket burst in his hands, though he himself was not injured. The boqs shrieked and whistled as if in anger, and at once hordes of its mates came dashing out through the forest. The frightened chief rushed out on the beach and called to his comrades to save him. They brought the canoe close to the shore so that he could clamber aboard, and then paddled away unharmed.

The Bella Coola believe that boqs, unlike most supernatural animals, have not abandoned the country since the coming of the white man. One man was most insistent that they still lived on the outer coast, and promised to point one out if a visit were made to that spot. The man in question refused to camp at the place where, he affirms, boqs are common. Another informant stated that though he had never actually seen one of the monsters, a horde of them surrounded his camp in the upper Bella Coola valley for a week, and every night roared and beat upon trees and branches.

Many such references are to be found in anthropological reports, but there may be a lot more information that has not been published anywhere. To quote again from Wayne Suttles' article:

If there is a real animal, shouldn't there be better descriptions in the ethnographic literature? Not necessarily. Anthropologists do not consciously suppress information, but they sometimes do not know what to do with it. There are ethnographies of peoples whom I know to have traditions of sasquatchlike beings that make no mention of such traditions; I suspect that these omissions occur not because

24

the writers have never heard the traditions but because they did not know how to categorize them.

Information from Indian sources is also to be found in journals and published recollections of early explorers and settlers. The oldest reference that I have to Indian traditions on the Pacific Coast is from a book written in Spanish by Jose Mariano Mozino, a naturalist who accompanied Juan Francisco de la Bodega y Quadra on his voyage from Mexico to the coast of what is now British Columbia in 1792. I have two different translations of a particular passage from his book *Noticias de Nutka*, referring to the beliefs of the Indian people of Nootka Sound on Vancouver Island in what is now British Columbia Here is one version:

One does not know what to say about a Matlox, inhabitant of the mountainous country, of whom all have an unspeakable terror. They figure that it has a monstrous body, all covered with black animal hair; the head like a human; but the eye teeth very large, sharp and strong, like those of the bear; the arms very large and the toes and fingers armed with large curved nails.

His howls fell to the ground those who hear them, and he smashes into a thousand pieces the unfortunate on whom a blow of his hand falls.

In *Told by the Pioneers*, a volume of interviews with old-timers in Washington State during the 1930's, P. H. Roundtree states:

P. O. and A. J. Roundtree took up donation claims on the Pe Ell Prairie. Before the white people came to this country, a big Skookum, or hairy man, came and drove all the Indians away that were living on the Pe Ell Prairie and the Indians never went back there to live until after the Roundtree boys took up claims there, and went there to live.

Paul Kane, a famous painter of Indians, in his book, *Wanderings of an Artist*, quotes the following journal entry made while travelling what is now southwestern Washington, March 26, 1847:

When we arrived at the mouth of the Kattle-poutal River twenty-six miles from Fort Vancouver I stopped to make a sketch of the volcano, Mount St. Helen's, distant, I suppose, about thirty or forty miles. This mountain has never been visited by either whites or Indians, the latter assert that it is inhabited by a race of beings of a different species, who are cannibals, and whom they hold in great dread. They also say that there is a lake at its base with a very extraordinary kind of fish in it, with a head more resembling that of a bear than any other animal. These superstitions are taken from the statement of a man who, they say, went to the mountain with another and escaped the fate of his companion, who was eaten by the "Skookums" or evil genii. I offered a considerable bribe to any other Indian who would accompany me in its exploration, but could not find one hardy enough to venture.

I am not at liberty to quote it, but I have seen a reference to a race of giants contained in the diary of Elkanah Walker, one of the original members of the Whitman mission in Oregon Territory. He wrote in 1840 of accounts by Spokane Indians of such creatures that stole salmon from their nets. The thieves were identified by their tracks, a foot and a half long, and by their intolerable smell.

The first newspaper reports in North America describing encounters by white settlers with hairy wild men did not appear on the west coast, since there were no newspapers there that early. There is no way of knowing when the very first report appeared, or how many there may have been. A few dedicated people have spent some time looking for them, and have found three or four, while others have been stumbled on accidentally. Some of the very early stories describe small creatures rather than giants. The first of which I have a copy was not the first to appear, since it refers to an earlier one. It was found by Gordon Strasenburgh in the Library of Congress, in the August 27, 1838, edition of the Dorchester County, Maryland, *Aurora*, but had been published originally by the Montrose, Pennsylvania, *Spectator*. The report, in its entirety, follows:

STRANGE ANIMAL, OR FOOD FOR THE MARVELLOUS

Something like a year ago, there was considerable talk about a strange animal, said to have been seen in the southwestern part of Bridgewater. Although the individual who described the animal persisted in declaring that he had seen it, and was at first considerably frightened by it, the story was heard and looked upon more as food for the marvellous, than as having any foundation in fact.

He represented the animal as we have it through a third person, as having the appearance of a child seven or eight years old though somewhat slimmer and covered entirely with hair. He saw it, while picking berries, walking towards him erect and whistling like a person. After recovering from his fright, he is said to have pursued it, but it ran off with such speed, whistling as it went, that he could not catch it. He said it ran like the 'devil', and continued to call it after that name.

The same or similar looking animal was seen in Silver Lake township about two weeks since, by a boy some sixteen years

26

old. We had the story from the father of the boy, in his absence, and afterward from the boy himself. The boy was sent to work in the backwoods near the New York state line. — He took with him a gun, and was told by his father to shoot anything he might see except persons or cattle.

After working a while, he heard some person, a little brother as he supposed, coming toward him whistling quite merrily. It came within a few rods of him and stopped.

He said it looked like a human being, covered with black hair, about the size of his brother, who was six or seven years old. His gun was some little distance off, and he was very much frightened. He, however, got his gun and shot at the animal, but trembled so that he could not hold it still.

The strange animal, just as his gun "went off," stepped behind a tree, and then ran off, whistling as before. The father said the boy came home very much frightened, and that a number of times during the afternoon, when thinking about the animal he had seen he would, to use the man's own words, "burst out crying."

Making due allowance for frights and consequent exaggeration, an animal of singular appearance has doubtless been seen. What it is, or whence it came, is of course yet a mystery. From the description, if an ourangoutan were known to be in the country we might think this to be it. As no such animal is known (without vouching for the correctness of the story) we shall leave the reader to conjecture, or guess for himself, what it is. For the sake of a name, however, we will call the "strange animal" The Whistling Wild Boy of the Woods.

Why is not this story as good as that copied into the Volunteer of a week before last, relative to the wild boy of Indiana? We acknowledge that the story has excited somewhat our propensity for the marvellous and we give it, as much as anything, to gratify the same propensity in others.

There isn't the slightest possibility of establishing now whether such a report was serious or not. Newspaper hoaxes were certainly not unknown in those times and the last paragraph of the story suggests the possibility that it was poking fun at, or trying to top, what the writer considered to be a whopper published in another paper. Even if there were some way of knowing that the story was in fact told by the boy, that would not establish that it was true. The only fact we are dealing with is that as far back as 1838 there were published reports of hairy humanlike creatures in North America.

I am not aware that anyone has found an Indiana report published earlier than this story, but considering the lack of rapid communication at the time—the first telegraph line had yet to be laid—there is a possibility that the Indiana story referred to has been found. Robert Sabaroff, of Tarzana, California, sent me a photo copy of the Philadelphia *Courier*, December 28, 1839, which contains a story copied from the Michigan City, Indiana, *Gazette*. It just might have been kicking around since the year before. Certainly it is about a wild boy in Indiana.

The story describes the boy as four feet high, with a light coat of chesnut colored hair. It says that he has often been chased but runs very fast, making frightful and hideous yells. The sightings had been made among the sand hills around Fish Lake, and the wild boy had been seen in summer running along the lake shore and also plunging into the lake and swimming very rapidly "all the time whining most piteously."

Whether that is the original Indiana story or not, it is established that there were stories, published virtually simultaneously, of small hair-covered bipeds being seen in Indiana and Pennsylvania. The next report that I have involves a wild giant. It was sent to me by Loren Coleman and is in the files of papers from Galveston and New Orleans, in May, 1851, but originates with the Memphis *Enquirer*. It runs as follows:

A WILD MAN OF THE WOODS

The Memphis Enquirer gives an account of a wild man recently discovered in Arkansas. It appears that during March last, Mr. Hamilton of Greene County, Ar., while out hunting with an acquaintance, observed a drove of cattle in a state of apparent alarm, evidently pursued by some dreaded enemy. Halting for the purpose, they soon discovered, as the animals fled by them, that they were followed by an animal bearing the unmistakable likeness of humanity. He was of gigantic stature, the body being covered with hair and the head with long locks that fairly enveloped the neck and shoulders. The 'wild man' after looking at them deliberately for a short time, turned and ran away with great speed, leaping twelve to fourteen feet at a time. His footprints measured thirteen inches each.

This singular creature, the Enquirer says, has long been known traditionally in St. Francis, Greene and Poinsett Counties, Ark., sportsmen and hunters having described him seventeen years since. A planter, indeed, saw him very recently but withheld his information lest he should not be credited, until the account of Mr. Hamilton and his friend placed the existence of the animal beyond cavil.

A great deal of interest is felt in the matter, by the inhabitants of that region, and various conjectures have been ventured in regard to him. The most generally entertained idea appears to be that he was a survivor of the earthquake disaster which desolated that region in 1811. Thrown helpless upon the wilderness by that disaster, it is probable that he grew up in his savage state, until now he bears only the outward resemblance of humanity.

So well authenticated have now become the accounts of this creature, that an expedition is organized in Memphis, by Col. David C. Cross and Dr. Sullivan, to scout for him.

Going back to about the same period is a brief reference in an interview with Agnes Louise (Ducheney) Eliot published in *Told by the Pioneers*. Although she refers in the interview to "Grandpa Ducheney" the person she is talking about is her father. The story is not dated, but she notes that General Grant had stayed with the Ducheneys, which would have to have been sometime from 1851 to 1853, and that Grandpa Ducheney died when his children were still young:

> Grandpa Ducheney firmly believed the story of the huge apes near St. Helens Mountain. He went there to hunt once and one of these apemen beckoned to him. He just turned and ran and ran until he reached home.

Rocque Ducheney came west from Montreal as a trader with the Hudson's Bay Company. The men in charge of that company's posts kept journals that span two centuries, but most of their time was spent at their forts, trading with Indians who brought in their furs. In the American west the "mountain men" did their own trapping and for a brief period spread out through most of the valleys in the western mountains, but they left few records of what they had seen. The only account from that period that I have is written by Theodore Roosevelt in his book *Wilderness Hunter*, published in 1892. No date is given for the incident, but it must have been about the middle of the century. The story was told as follows:

> Frontiersmen are not, as a rule, apt to be very superstitious. They lead lives too hard and practical, and have too little imagination in things spiritual and supernatural. I have heard but few ghost-stories while living on the frontier, and those few were of a perfectly commonplace and conventional type.
>
> But I once listened to a goblin-story which rather impressed me. It was told by a grizzled, weatherbeaten old

mountain hunter, named Bauman, who was born and had passed all of his life on the frontier. He must have believed what he said, for he could hardly repress a shudder at certain points of the tale; but he was of German ancestry, and in childhood had doubtless been saturated with all kinds of ghost and goblin lore, so that many fearsome superstitions were latent in his mind; besides, he knew well the stories told by the Indian medicine-men in their winter camps, of the snow-walkers, and the spectres, and the formless evil beings that haunt the forest depths, and dog and waylay the lonely wanderer who after nightfall passes through the regions where they lurk; and it may be that when overcome by the horror of the fate that befell his friend, and when oppressed by the awful dread of the unknown, he grew to attribute, both at the time and still more in remembrance, weird and elfin traits to what was merely some abnormally wicked and cunning wild beast; but whether this was so or not, no man can say.

When the event occurred Bauman was still a young man, and was trapping with a partner among the mountains dividing the forks of the Salmon from the head of Wisdom River. Not having had much luck, he and his partner determined to go up into a particularly wild and lonely pass through which ran a small stream said to contain many beaver. The pass had an evil reputation because the year before a solitary hunter who had wandered into it was there slain, seemingly by a wild beast, the halfeaten remains being afterwards found by some mining prospectors who had passed his camp only the night before.

The memory of this event, however, weighed very lightly with the two trappers, who were as adventurous and hardy as others of their kind. They took their two lean mountain ponies to the foot of the pass where they left them in an open beaver meadow, the rocky timber-clad ground being from there onward impracticable for horses. They then struck out on foot through the vast, gloomy forest, and in about four hours reached a little open glade where they concluded to camp, as signs of game were plenty.

There was still an hour or two of daylight left, and after building a brush lean-to and throwing down and opening their packs, they started upstream. The country was very dense and hard to travel through, as there was much down timber, although here and there the sombre woodland was broken by small glades of mountain grass. At dusk they again reached camp. The glade in which it was pitched was not many yards wide, the tall, close-set pines and firs rising round it like a wall. On one side was a little stream, beyond which rose the steep mountain slope, covered with the unbroken growth of evergreen forest.

They were surprised to find that during their absence something, apparently a bear, had visited camp, and had rummaged about among their things, scattering the contents of their packs, and in sheer wantonness destroying their lean-to. The footprints of the beast were quite plain, but at first they paid no particular heed to them, busying themselves with rebuilding the lean-to, laying out their beds and stores and lighting the fire.

While Bauman was making ready supper, it being already dark, his companion began to examine the tracks more closely, and soon took a brand from the fire to follow them up, where the intruder had walked along a game trail after leaving the camp. When the brand flickered out, he returned and took another, repeating his inspection of the footprints very closely. Coming back to the fire, he stood by it a minute or two, peering out into the darkness, and suddenly remarked, "Bauman, that bear has been walking on two legs." Bauman laughed at this, but his partner insisted that he was right, and upon again examining the tracks with a torch, they certainly did seem to be made by but two paws or feet. However, it was too dark to make sure. After discussing whether the footprints could possibly be those of a human being, and coming to the conclusion that they could not be, the two men rolled up in their blankets, and went to sleep under the lean-to.

At midnight Bauman was awakened by some noise, and sat up in his blankets. As he did so his nostrils were struck by a strong, wild-beast odor, and he caught the loom of a great body in the darkness at the mouth of the lean-to. Grasping his rifle, he fired at the vague, threatening shadow, but must have missed, for immediately afterwards he heard the smashing of the underwood as the thing, whatever it was, rushed off into the impenetrable blackness of the forest and the night.

After this the two men slept but little, sitting up by the rekindled fire, but they heard nothing more. In the morning they started out to look at the few traps they had set the previous evening and put out new ones. By an unspoken agreement they kept together all day, and returned to camp towards evening.

On nearing it they saw, hardly to their astonishment, that the lean-to had again been torn down. The visitor of the preceding day had returned, and in wanton malice had tossed about their camp kit and bedding, and destroyed the shanty. The ground was marked up by its tracks, and on leaving the camp it had gone along the soft earth by the brook, where the footprints were as plain as if on snow, and, after a careful scrutiny of the trail, it certainly did seem as if, whatever the thing was, it had walked off on but two legs.

The men, thoroughly uneasy, gathered a great heap of dead logs and kept up a roaring fire throughout the night, one or the other sitting on guard most of the time. About midnight the thing came down through the forest opposite, across the brook, and stayed there on the hillside for nearly an hour. They could hear the branches crackle as it moved about, and several times it uttered a harsh, grating, long-drawn moan, a peculiarly sinister sound. Yet it did not venture near the fire.

In the morning the two trappers, after discussing the strange events of the last 36 hours, decided that they would shoulder their packs and leave the valley that afternoon. They were the more ready to do this because in spite of seeing a good deal of game sign they had caught very little fur. However it was necessary first to go along the line of their traps and gather them, and this they started out to do. All the morning they kept together, picking up trap after trap, each one empty. On first leaving camp they had the disagreeable sensation of being followed. In the dense spruce thickets they occasionally heard a branch snap after they had passed; and now and then there were slight rustling noises among the small pines to one side of them.

At noon they were back within a couple of miles of camp. In the high, bright sunlight their fears seemed absurd to the two armed men, accustomed as they were, through long years of lonely wandering in the wilderness, to face every kind of danger from man, brute or element. There were still three beaver traps to collect from a little pond in a wide ravine near by. Bauman volunteered to gather these and bring them in, while his companion went ahead to camp and made ready the packs.

On reaching the pond Bauman found three beavers in the traps, one of which had been pulled loose and carried into a beaver house. He took several hours in securing and preparing the beaver, and when he started homewards he marked, with some uneasiness, how low the sun was getting. As he hurried toward camp, under the tall trees, the silence and desolation of the forest weighed on him. His feet made no sound on the pine needles and the slanting sun-rays, striking through among the straight trunks, made a gray twilight in which objects at a distance glimmered indistinctly. There was nothing to break the gloomy stillness which, when there is no breeze, always broods over these sombre primeval forests.

At last he came to the edge of the little glade where the camp lay, and shouted as he approached it, but got no answer. The camp fire had gone out, though the thin blue smoke was still curling upwards.

Near it lay the packs wrapped and arranged. At first Bauman could see nobody; nor did he receive an answer to his call. Stepping forward he again shouted, and as he did so his eye fell on the body of his friend, stretched beside the trunk of a great fallen spruce. Rushing towards it the horrified trapper found that the body was still warm, but that the neck was broken, while there were four great fang marks in the throat.

The footprints of the unknown beast-creature, printed deep in the soft soil, told the whole story.

The unfortunate man, having finished his packing, had sat down on the spruce log with his face to the fire, and his back to the dense woods, to wait for his companion. While thus waiting, his monstrous assailant, which must have been lurking in the woods, waiting for a chance to catch one of the adventurers unprepared, came silently up from behind, walking with long noiseless steps and seemingly still on two legs. Evidently unheard, it reached the man, and broke his neck by wrenching his head back with its fore paws, while it buried its teeth in his throat. It had not eaten the body, but apparently had romped and gambolled around it in uncouth, ferocious glee, occasionally rolling over and over it; and had then fled back into the soundless depths of the woods.

Bauman, utterly unnerved, and believing that the creature with which he had to deal was something either half human or half devil, some great goblin-beast, abandoned everything but his rifle and struck off at speed down the pass, not halting until he reached the beaver meadows where the hobbled ponies were still grazing. Mounting, he rode onwards through the night, until beyond reach of pursuit.

In writing that story, Theodore Roosevelt became not only one of the first to publish what we would consider a sasquatch sighting report, but also one of the first to try to explain the story away. Apparently he was not aware of the existence of similar reports. The location he gives would be in the Bitterroot Mountains on the Idaho-Montana border, one of the areas where sighting reports are most common at the present time. Fortunately there are no reports from there or elsewhere to indicate that the malevolent disposition of this individual has been passed on to any of its modern descendants. There is no modern story that describes the killing or injuring of a human by one of these creatures, deliberately or otherwise.

It is easy to think of such century-old reports as coming down from a time when people didn't yet know very much about the animal life of the planet and would readily accept stories about imaginary beasts or fail to identify real ones properly. In fact, nothing could be much further from the truth. During the 18th and 19th centuries the exploration of the world of nature probably excited a greater interest among both scientists and the public at large than has been the case since. Herbert Wendt in *In Search of Adam* notes that systematic classification of species was begun in the early 1700's, and that by 1766 Count Buffon had published the 20 volumes of his *Natural History* and had concluded that "everything seems to have been shaped in accordance with an original and general structural scheme which can be traced very far back." Buffon had heard that there were a smaller and a larger ape in Africa, known to him as "Jacko" and "Pongo", and had seen a gibbon.

There had been chimpanzees on display in Holland and England in the 17th century, and an orangutan was exhibited and later dissected in Holland by 1776. France, England and Russia all sent scientific expeditions to remote parts of the world to observe the passage of Venus across the sun on June 3, 1796, and those expeditions included naturalists. Among the results were the discovery of the duckbilled platypus and the lemurs. From then on it became customary for naturalists to be included on voyages of exploration. Discoveries of new animals proceeded steadily and the names of leading men in the field of zoology were probably household words to a much greater extent than they are today. Major excavations of the fossilized bones of extinct animals began just a little later.

The existence of an enormous ape in Africa had been reported in the early 1800's, but no physical remains of the "Pongo" were obtained for study until 1847, when bones of the lowland gorilla were studied. The mountain gorilla remained unknown until 1902. Vernon

34

Reynolds, in *The Apes* describes the reaction to the gorilla's debut in 1847:

> The discovery of the gorilla made an enormous impact on both public and scientists, in Europe and in the United States, much more than the discovery of the other apes, and perhaps more than the discovery of any other animal. There were several reasons for this. First, the discovery came late, when the era of modern science was already well underway, so that the techniques already existed for evaluating the new find and placing it among the apes, as a close relative of man. Second, the huge gorilla had "monster appeal" for an imaginative public. Here was a new monster, and one the scientists approved of! Third, although the gorilla was described before the publication of Darwin's works, these, and the furious "Am I ape or man?" argument they triggered off helped to keep the gorilla in the forefront of people's minds.

Charles Darwin came along fairly late among the voyaging naturalists, and saw no formerly-unexplored lands, but he was able to discern more than others in what he did see, and after publication of his *On the Origin of Species* in 1859, followed by *The Descent of Man* in 1871, debate about the animal world and man's place in it became one of the great preoccupations of European and American society. The latter part of the 19th century also saw the discovery of early apes and ape men such as Oreopithecus in 1872 and Homo erectus (Pithecanthropus) in 1892, and it was in the same period that the great discoveries of fossils of huge dinosaurs, ancient mammals and giant flying reptiles were being made in the American west. It was in this atmosphere that Thomas Huxley stated, in 1863:

> The question of all questions for humanity, the problem which lies behind all others and is more interesting than any of them is that of the determination of man's place in Nature and his relation to the cosmos. Whence our race came, what sort of limits are set to our power over Nature and Nature's power over us, to what goal we are striving, are the problems which present themselves afresh with undiminished interest, to every human being born on earth.

Plainly it was not a time when thinking people were unaware of the importance of manlike beasts, or uninformed as to what creatures were known to exist. That at least one of the first Europeans to penetrate the forests of the Pacific slope was very well aware of the discoveries being made elsewhere is on record. David Thompson, who has been called "the greatest land geographer the world has ever known," wrote concerning his travels from 1792 to 1812 in the western area of what is now the northern United States and southern Canada:

> From the numerous remains in Siberia and parts of Europe of the Elephant, Rhinocerous, and the other large Animals, especially near the Rivers, and in their banks, of those countries, I was led to expect to find the remains of those Animals in the Great Plains and the Rivers that flow through them: but all my steady researches, and all my enquiries led to nothing.

One thing that Thompson did find was an extraordinary set of footprints. The story appears under the entry for January 7, 1811, in *David Thompson's Narrative.* At the time he was attempting to cross the Rocky Mountains via the valley of the Athabaska River and the Yellowhead Pass, and was in the vicinity of the present site of Jasper, Alberta—the first white men known to have followed what is now the route of the Canadian National Railway and the Yellowhead Highway. He wrote:

> Continuing our journey in the afternoon we came on the track of a large animal, the snow about six inches deep on the ice; I measured it: four large toes each of four inches in length to each a short claw; the ball of the foot sunk three inches lower than the toes, the hinder part of the foot did not mark well, the length fourteen inches, by eight inches in breadth, walking from north to south, and having passed about six hours. We were in no humour to follow him: The Men and Indians would have it to be a young mammoth and I held it to be the track of a large old grizzled Bear; yet the shortness of the nails, the ball of the foot, and its great size was not that of a Bear, otherwise that of a very large old Bear, his claws worn away; this the Indians would not allow.

Thompson referred to the incident again in his account of the return journey in the autumn:

> I now recur to what I have already noticed in the early part of last winter, when proceding up the Athabaska River to cross the Mountains, in company with . . . Men and four hunters, on one of the channels of the River we came to the track of a large animal, which measured fourteen inches in length by eight inches in breadth by a tape line. As the snow was about six inches in depth the track was well defined, and we could see it for a full one hundred yards from us, this animal was proceeding from north to south. We did not attempt to follow it, we had not time for it, and the Hunters, eager as they are to follow and shoot every animal made no attempt to follow this beast, for what could the balls of our

fowling guns do against such an animal. Report from old times had made the head branches of this River, and the Mountains in the vicinity the abode of one, or more, very large animals, to which I never appeared to give credence; for these reports appeared to arise from that fondness for the marvellous so common to mankind: but the sight of the track of that large beast staggered me, and I often thought of it, yet never could bring myself to believe such an animal existed, but thought it might be the track of some monster Bear.

I have seen Thompson's account referred to as the first report of a sasquatch track, but that can hardly be the case. Obviously Thompson measured and studied the track closely, so he would not have been mistaken in his observation of claw marks, which no sasquatch track would show. On the other hand, what he describes will not do for a bear track either. It is too large, there are not enough toes, and while a huge bear could have made tracks of such proportions with his hind feet, the shorter tracks of his front feet would also have been present. What made the tracks will presumably never be known, but there can be no doubt that they were something completely out of the ordinary, for a woodsman of Thompson's experience to be "staggered" by them. Neither can there be any doubt about their size, since they were measured with a tape. David Thompson was not a man to make mistakes. He did his work as a partner in a private fur company, and his feat of mapping more than a million and a half square miles of western North America remained generally unknown, in fact the man himself was forgotten. His rediscovery resulted from a Canadian government survey in the Rocky Mountains in 1880. The chief surveyor, Dr. J. B. Tyrell, was so astonished at the accuracy of the old maps with which he had been supplied that he investigated their origin and brought to light Thompson's journals and survey notes and the unpublished *Narrative*.

After the California gold rush got going in 1849, the dispersion of white miners and then settlers throughout the Pacific slope became very rapid, but I know of no sasquatch reports in that period. All I have is a letter written to *True* magazine after it printed an article on California's "Bigfoot" in December, 1959. John M. Weeks, Providence, R.I., wrote as follows:

My grandfather prospected for gold in the eighteen fifties throughout the region described as being the home of the Snowman. Upon grandfather's return to the East he told stories of seeing hairy giants in the vicinity of Mount Shasta. These monsters had long arms but short legs. One of them picked up a 20-foot section of a sluiceway and smashed it to bits against a tree.

When grandfather told us these stories, we didn't believe him at all. Now, after reading your article, it turns out he wasn't as big a liar as we youngsters thought he was.

From 1869 on, the reports became too numerous to quote them all. The reporter doing the "100 Years Ago in Kansas" column for the Osage City, Kansas, *Free Press* in early June, 1969, apparently found one in the files of the Junction City *Union*, quoting from a letter to a St. Louis paper from Crawford County, dated August 15, 1869. It read as follows:

> We of the Arcadia Valley, in the southern part of Crawford County, are having a new sensation which may lead to some new disclosures in natural history, if investigated as it should be. It is nothing less than the discovery of a wild man or gorilla, or 'what is it.'
>
> It has, at different times, been seen by almost every inhabitant of the valley, and it occasionally has been seen in the adjoining counties of Missouri, but it seems to make its home in this vicinity. Several times it has approached the cabins of the settlers, much to the terror of the women and children, especially if the men happen to be absent working in the fields. In one instance it approached the house of one of our old citizens, but was driven away with clubs by one of the men.
>
> It has so near a resemblance to the human form that the men are unwilling to shoot it. It is difficult to give a description of this wild man or animal. It has a stooping gait, very long arms with immense hands or claws; it has a hairy face and those who have been near it describe it as having a most ferocious expression of countenance; generally walks on its hind legs but sometimes on all fours. The beast or 'what is it?' is as cowardly as it is ugly and it is next to impossible to get near enough to obtain a good view of it.
>
> The settlers, not knowing what to call it, have christened it 'Old Sheff'. Since its appearance our fences are often found down, allowing the stock free range in our corn fields. I suppose Old Sheff is only following his inclination, as it may be easier for him to pull them down than to climb over. However, as it is, curses loud and deep are heaped upon its head by the settlers. The settlers are divided in opinion as to whether it belongs to the human family or not. Probably it will be found to be a gorilla or large orangutan that has escaped from some menagerie in the settlements east of here.

At one time over sixty of the citizens turned out to hunt it down, but it escaped; but probably owing to the fright that it received it kept out of sight for several days, and just as the settlers were congratulating themselves that they were rid of an intolerable nuisance, Old Sheff came back again, seemingly as savage as ever.

If this meets the eye of any showman who has lost one of his collection of beasts, he may know where to find it. At present it is the terror of all women and children in the valley. It cannot be caught and nobody is willing to shoot it.

That is all that has come to hand about Old Sheff, but there must have been other stories in circulation about the same time, witness this article from the Lansing, Michigan, *Republican* on August 4, 1870, copied from the New York *Tribune*:

THE LATEST "WILD MAN"

The wild man has now turned up as far away as Nevada. Like the sea-serpent, he appears from time to time, in different parts of the continent; and the stories which the country papers get up about him are always very wonderful and awful.

The last appearance of the wild man or rather "the object", as he, or she, or it is called, was in a desolate region of Northern Nevada, where an intense state of excitement has been roused about it. A large party, armed and equipped, lately started in pursuit of "it," and one night a splendid view was obtained of the object which, it was concluded, had once been a white man, but was now covered with a coat of fine, long hair, carried a huge club in the right hand, and in the left a rabbit. The moment it caught sight of the party, as the moon shone out, it dashed past the camp "with a scream like a roar of a lion," brandished the huge club and attacked the horses in a perfect frenzy of madness.

The savage bloodhounds which the party had brought along refused to pursue the object; and so the party hastily raised a log rampart for self-defense; but, instead of making attack, the object merely uttered the most terrible cries through the night, and in the morning had disappeared. It was evident, however, from the footprints, that the object would require a "pair of No. 9 shoes," and this is all we know. The party could have shot it on first seeing it, but failed to do so.

From a careful reading of the accounts, we should judge that this wild man or object made tracks immediately after seeing the party, for Salt Lake, to keep company with the sea-serpent which the Mormons who lately saw it there say is a half a mile long, and looks dreadfully terrible.

The big-city newspaperman who wrote that article obviously wanted to make it clear that he wasn't about to be taken in by any of the "wonderful and awful" stuff those country papers were printing. He may, of course, have been quite right to take that attitude in this case.

Out in California, a single small article resulted in a series of reports from two different places. The original item was printed on September 19, 1870, in the San Joaquin *Republican*, in a column of "local Brevities", as follows:

WILDMAN — We learn from good authority that a wild man has been seen at Crow Canyon, near Mount Diablo. Several attempts have been made to capture him, but as yet have proved unsuccessful. His tracks measure thirteen inches.

That little item stirred the Oakland *Daily Transcript* to respond on September 27:

An item appeared in the San Joaquin Republican the other day stating that a wild man had been seen in some part of San Joaquin county, and we afterward noticed the statement copied into several other papers, with brief comments indicating disbelief in the report. We must confess to a want of credulity on our part also as to the exact correctness of the item at the time, but we were yesterday placed in possession of certain information which leads us to believe that there may be some foundation for the report. As our columns are somewhat crowded this morning, we will give the reports as we received them and as briefly as possible.

F. J. Hildreth and Samuel De Groot, of Washington Corners, in this county, while out hunting on Orias Timbers Creek in Stanislaus county about three weeks ago, discovered footprints along the bank of the creek resembling the impressions of a human being's feet. Mr. Hildreth, who gave us this information, states that the tracks were like those of a human being with the exception that the impressions of the toes were much larger.

Hildreth afterward became separated from his companion, and upon proceeding some distance up the creek, saw a few yards ahead of him what he believed to be gorillas. If the description Mr. Hildreth has given us of these animals is true, he is certainly warranted in believing them to be of that species of animal. Mr. De Groot also reports that he saw the same objects and is positive that they are gorillas. The appearance of these strange animals in that neighborhood is notorious and that they are gorillas is firmly believed by a great many people in that vicinity.

A number of old hunters have started out to capture them, and we are promised whatever further facts may occur as soon as the party returns. The above we gathered from various parties, and whether true or not, there are many persons in the neighborhood of the Washington Corners who firmly believe that the animals referred to are veritable gorillas.

That story was reprinted on October 1 by the Antioch *Ledger*, and as a result the Ledger received the following letter from a correspondent in Grayson, California.

I saw in your paper a short time since an item concerning the 'gorilla' which was said to have been seen in Crow Canyon and shortly after in the mountains at Orestimba Creek. You sneered at the idea of there being any such a 'critter' in these hills, and, were I not better informed, I should sneer too, or else conclude that one of your recent prospecting party had got lost in the wilderness, and did not have sense enough to find his way back to Terry's.

I positively assure you that this gorilla or wildman, or whatever you choose to call it is no myth. I know that it exists, and that there are at least two of them, having seen them both at once not a year ago. Their existence has been reported at times for the past twenty years, and I have heard it said that in early days an ourang-outang escaped from a ship on the southern coast; but the creature I have seen is not that animal, and if it is, where did he get his mate? Import her as the web-foot did their wives?

Last fall I was hunting in the mountains about 20 miles south of here, and camped five or six days in one place, as I have done every season for the past fifteen years. Several times I returned to camp, after a hunt, and saw that the ashes and charred sticks from the fireplace had been scattered about. An old hunter notices such things, and very soon gets curious to know the cause. Although my bedding and traps and little stores were not disturbed, as I could see, I was anxious to learn who or what it was that so regularly visited my camp, for clearly the half burnt sticks and cinders could not scatter themselves about.

I saw no tracks near the camp, as the hard ground covered with leaves would show none. So I started in a circle around the place, and three hundred yards off, in damp sand, I struck the track of a man's foot, as I supposed—bare and of immense size. Now I was curious, sure, and I resolved to lay for the barefooted visitor. I accordingly took a position on a hillside, about sixty or seventy feet from the fire, and, securely hid in the brush, I waited and watched. Two hours and more I sat there and wondered if the owner of the feet would come again, and whether he imagined what an interest he had created in my enquiring mind, and finally what possessed him to be prowling about there with no shoes on.

The fireplace was on my right, and the spot where I saw the track was on my left, hid by the bushes. It was in this direction that my attention was mostly directed, thinking the visitor would appear there, and besides, it was easier to sit and face that way. Suddenly I was surprised by a shrill whistle, such as boys produce with two fingers under their tongues, and turning quickly, I ejaculated, 'Good God!' as I saw the object of my solicitude standing beside my fire, erect, and looking suspiciously around. It was the image of a man, but it could not have been human.

I was never so benumbed with astonishment before. The creature, whatever it was, stood fully five feet high, and disproportionately broad and square at the fore shoulders, with arms of great length. The legs were very short and the body long. The head was small compared with the rest of the creature, and appeared to be set upon his shoulders without a neck. The whole was covered with dark brown and cinnamon colored hair, quite long on some parts, that on the head standing in a shock and growing close down to the eyes, like a Digger Indian's.

As I looked he threw his head back and whistled again, and then stooped and grabbed a stick from the fire. This he swung round, until the fire on the end had gone out, when he repeated the manoeuver. I was dumb, almost, and could only look. Fifteen minutes I sat and watched him as he whistled and scattered my fire about. I could easily have put a bullet through his head, but why should I kill him? Having amused himself, apparently, as he desired, with my fire, he started to go, and, having gone a short distance he returned, and was joined by another—a female, unmistakably—when both turned and walked past me, within twenty yards of where I sat, and disappeared in the brush.

I could not have had a better opportunity for observing them, as they were unconscious of my presence. Their only object in visiting my camp seemed to be to amuse themselves with swinging lighted sticks around. I have told this story many times since then, and it has often raised an incredulous smile; but I have met one person who has seen the mysterious creatures, and a dozen of whom have come across their tracks at various places between here and Pacheco Pass.

That story was in turn reprinted by the Butte *Record*, where Jim McClarin came upon it. Research to trace the series back to its San Joaquin origin was done by Joyce Kearney. Mount Diablo is just east of Oakland. Orestimba Creek is about 50 miles to the southeast. Twenty miles south of Grayson indicates the same vicinity as Orestimba Creek.

At about this period there were three reports of the capture of small hairy bipeds, which are dealt with in other chapters, plus the only report I have from the Maritime Provinces of Canada. J. Kenneth Johnson wrote to researcher Loren Coleman quoting his father, Rev. A. Fulton Johnson, about a creature sometimes seen at the edge of the woods near Moncton, New Brunswick. Reverend Johnson said it was very squatty, with long arms and with its entire body covered with hair.

On March 9, 1876, the San Diego, California, *Union* printed the following story under the heading:

A WILD MAN IN THE MOUNTAINS

The following strange story is sent us by a correspondent at Warner's Ranch in this county. We know the writer to be a perfectly reliable person and believe his statement, singular though it may seem, to be fully entitled to credence:

WARNER'S RANCH — March 5 — About ten days ago Mr. Turner Helm and myself were in the mountains about ten miles east of Warner's Ranch, on a prospecting tour, looking for the extension of a quartz lode which had been found by some parties sometime before. When we were separated, about half a mile apart—the wind blowing very hard at the time—Mr. Helm, who was walking along looking down at the ground, suddenly heard someone whistle.

Looking up he saw 'something' sitting on a large boulder, about fifteen or twenty paces from him. He supposed it to be some kind of an animal, and immediately came down on it with his needle gun. The object instantly rose to its feet and proved to be a man. This man appeared to be covered all over with coarse black hair, seemingly two or three inches long, like the hair of a bear; his beard and the hair of his head were long and thick; he was a man of about medium

size, and rather fine features — not at all like those of an
Indian, but more like an American or Spaniard.

They stood gazing at each other for a few moments, when
Mr. Helm spoke to the singular creature, first in English and
then Spanish and then Indian, but the man remained silent.
He then advanced towards Mr. Helm, who not knowing
what his intentions might be, again came down on him with
the gun to keep him at a distance. The man at once stopped,
as though he knew there was danger.

Mr. Helm called to me, but the wind was blowing so hard
that I did not hear him. The wild man then turned and
went over the hill and was soon out of sight; before Mr.
Helm could come to me he had made good his escape. We
had frequently before seen this man's tracks in that part of
the mountains, but had supposed them to be the tracks of an
Indian. I did not see this strange inhabitant of the mountains
myself; but Mr. Helm is known to be a man of unquestioned
veracity, and I have no doubt of the entire truth of his
statement.

L.T.H.

Colorado is one of the states where sasquatch reports are all but
unknown. I do not have a single report from there with the minimum
details of who, what, when and where. There is, however, a story in
Treasure Tales of the Rockies, by Percy Eberhart, stating that several
miners in the vicinity of Grizzly Peak described seeing "The Lake
Creek Monster," a creature resembling a man except for the
extraordinary length of its arms and the long shaggy hair covering its
body. Tim Church, a sasquatch researcher from Rapid City, South
Dakota, states that the original account was published in the
Leadville *Chronicle* about 1881. He also sent me the oldest Oregon
report that I have, taken from the Carson City, Nevada, *Morning
Appeal*, December 31, 1885.

WILD MAN IN THE MOUNTAINS

Much excitement has been created in the neighborhood of
Lebanon, Oregon, recently over the discovery of a wild man
in the mountains above that place, who is supposed to be the
long lost John Mackentire. About four years ago Mackentire,
of Lebanon, while out hunting in the mountains east of
Albany with another man, mysteriously disappeared and no
definite trace of him has ever yet been found. A few days ago
a Mr. Fitzgerald and others, while hunting in the vicinity of
the butte known as Bald Peter, situated in the Cascades,
several miles above any settlement saw a man resembling the

44

long-lost man, entirely destitute of clothing, who had grown as hairy as an animal, and was a complete wild man.

He was eating the raw flesh of a deer when first seen, and they approached within a few yards before he saw them and fled. Isaac Banty saw this man in the same locality about two years ago. It is believed by many that the unfortunate man who was lost became deranged and has managed to find means of subsistence while wandering about in the mountains, probably finding shelter in some cave. A party of men is being organized to go in search of the man.

There are two more Oregon reports before the turn of the century, but I do not have either one from anything approaching an original source. Ivan Sanderson, in *Abominable Snowmen: Legend Come to Life*, tells of enormous human footprints being seen around a mining camp 50 miles inland in the Chetko River area and two men standing guard being smashed to death in the night. No source for the story is given, nor is there anything specifically pinning the violence on the sasquatch. I have a copy of another story, written by Dale Vincent in 1947, which tells of two miners prospecting above the Sixes River in 1899 seeing their camp gear thrown off a cliff by a big, powerful erect creature, "neither man nor beast", covered with a yellow fuzz over its body. They shot at it but it fled. Again no source is given for the information, and the clipping does not even show what paper it is from. Both those locations are in the southwest corner of Oregon, and there are two more specific reports from the Sixes River within the following five years, both published in the Myrtle Point *Enterprise* and copied in other papers. The first appeared in the Roseburg, Oregon, *Daily Review* on December 24, 1900:

A KANGAROO MAN — The Sixes mining district in Curry county has for the past 30 years gloried in the exclusive possession of a "kangaroo man". Recently while Wm. Page and Johnnie McCullock, who are mining there, went out hunting McCulloch saw the strange animal-man come down to a stream to drink. In calling Page's attention to the strange being it became frightened, and with cat-like agility, which has always been a leading characteristic, with a few bounds was out of sight.

The appearance of this animal is almost enough to terrorize the rugged mountain sides themselves. He is described as having the appearance of a man—a very good looking man —is nine feet in height with low forehead, hair hanging down near his eyes, and his body covered with a prolific growth of hair which nature has provided for his protection. Its hands reach almost to the ground and when its tracks were measured its feet were found to be 18 inches in length with five well formed toes. Whether this is a devil, some strange animal or a wild man is what Messrs. Page and McCulloch would like to know, says the Myrtle Point Enterprise.

The second story was reprinted in the *Lane County Leader*, Cottage Grove, Oregon, on April 7, 1904:

SIXES WILD MAN AGAIN

Visits the Cabins of Miners and Frightens the Prospectors.

At repeated intervals during the past ten years thrilling stories have come from the rugged Sixes mining district in Coos County, Oregon, near Myrtle Point, regarding a wild man or a queer and terrible monster which walks erect and which has been seen by scores of miners and prospectors. The latest freaks of the wild man is related as follows in the last issue of the Myrtle Point Enterprise:

The appearance again of the "Wild Man" of the Sixes has thrown some of the miners into a state of excitement and fear. A report says the wild man has been seen three times since the 10th of last month. The first appearance occurred on "Thompson Flat." Wm. Ward and a young man by the name of Burlison were sitting by the fire of their cabin one night when they heard something walking around the cabin which resembled a man walking and when it came to the corner of the cabin it took hold of the corner and gave the building a vigorous shake and kept up a frightful noise all the time—the same that has so many times warned the venturesome miners of the approach of the hairy man and caused them to flee in abject fear.

Mr. Ward walked to the cabin door and could see the monster plainly as it walked away, and took a shot at it with his rifle, but the bullet went wild of its mark. The last appearance of the animal was at the Harrison cabin only a few days ago. Mr. Ward was at the Harrison cabin this time and again figures in the excitement. About five o'clock in the morning the wild man gave the door of the cabin a vigorous shaking which aroused Ward and one of the Harrison boys who took their guns and started in to do the intruder. Ward fired at the man and he answered by sending a four pound rock at Ward's head but his aim was a little too high. He then disappeared in the brush.

Many of the miners avow that the "wild man" is a reality. They have seen him and know whereof they speak. They say he is something after the fashion of a gorilla and unlike anything else that has ever been known; and not only that but he can throw rocks with wonderful force and accuracy. He is about seven feet high, has broad hands and feet and his body is covered by a prolific growth of hair. In short, he looks like the very devil.

In 1904 we are apparently approaching close to modern conditions. The same page of the Lane County *Leader* contains the following account of a current labor dispute in Roseburg:

ANOTHER STRIKE

The "lady stenographers" of Roseburg have gone on a strike, formed a trust, and through intimidation or persuasion induced the local lawyers to sign the following agreement:

"We, the attorneys, counselors and solicitors of Roseburg County Bar, who can afford to hire and maintain lady stenographers, hereby promise and agree to and with one another to have and to hold said stenographers for five and one half days of each and every week, and to release, quitclaim and surrender said lady stenographer of one half of one day of each and every week for the other fellow to have and to hold. The one-half day thus set apart and dedicated to the other fellow shall begin each and every Saturday at 12 M., during the life of this agreement.

The year 1904 ended with the first specific report I know of from Vancouver Island. On December 14, both the Victoria *Colonist* and the Vancouver *Province* reported the sighting of a wild man by four men from Qualicum. The story in the Colonist reads, in part, as follows:

A. R. Crump, J. Kincaid, T. Hutchins and W. Buss, four sober-minded settlers of Qualicum, are the new witnesses and there is not the slightest deviation or variation in detail in the stories they tell with an earnestness that defies ridicule. They were out hunting in the vicinity of Horne Lake, which lies mid-way between Great Central Lake and Comox Lake, in an uninhabited and little explored section of the interior of Vancouver Island, when they came upon the uncouth being whom they describe as a living, breathing and intensely interesting modern Mowgli. The wild man was apparently young, with long matted hair and a beard, and covered with a profusion of hair all over the body. He ran like a deer through the seemingly impenetrable tangle of undergrowth, and pursuit was utterly impossible.

The story in the *Province* notes that the wild man had appeared frequently in the vicinity of Horne Lake in the past few years, but that the reports had been made by only one or two people and were generally doubted. Of the most recent sighting, the paper stated that the wild man had been distinctly seen by four reliable people, and their testimony should be accepted without any doubt. On March 8, 1907, the *Province* had another story:

A monkey-like wild man who appears on the beach at night, who howls in an unearthly fashion between intervals of exertion at clam digging, has been the cause of de-populating an Indian village, according to reports by officers of the steamer Capilano, which reached port last night from the north.

The Capilano on her trip north put in to Bishop's Cove where there is a small Indian settlement. As soon as the steamer appeared in sight the inhabitants put off from the shore in canoes and clambered on board the Capilano in a state of terror over what they called a monkey covered with long hair and standing about five feet high which came out on the beach at night to dig clams and howl.

The Indians say that they had tried to shoot it but failed, which added to their superstitious fears. The officers of the vessel heard some animals howling along the shore at night but are not prepared to swear that it was the voice of the midnight visitor who has so frightened the Indians.

What can be said about these old stories? Not too much beyond the fact that they exist. There are no witnesses to talk to and no one can be sure if any one of them is true or false. In most cases we don't even know whether the people named ever existed. The stories are important in one way, however. If there really is a species of hairy bipeds living in North America there should have been people who encountered them in former years and reports of some of those encounters should be on record. If there were no such stories it would be a strong indication that there could be no such creatures. Many people assume that to be the case. There is no history or tradition telling them that reports of hairy men are a part of their heritage, so they take it for granted that such things have not been reported in the past. If there had been such sightings they feel sure they would know about it. Those assumptions have been proven to be mistaken. The old stories do exist, and people did not know about them.

-3-

The Centennial Sasquatch Hunt

It seems certain that there have always been Indian traditions of wild giants virtually everywhere in North America, but the only place where such information got through to the rest of the population seems to have been in British Columbia, and that was the work of just one man, J. W. Burns. Mr. Burns spent many years as a teacher on the Chehalis Indian Reserve beside the Harrison River about 60 miles east of Vancouver. He wrote numerous newspaper stories about the encounters his Indian friends had with the hairy giants, including an article in a major national magazine in 1927. While those stories certainly did not convince the non-Indian society that such creatures actually existed, they did make the name "Sasquatch" a household word in that corner of the world.

Growing up in Vancouver, I knew of stories of the sasquatch from childhood, although I certainly did not take them seriously, or have any reason to think that anyone else did. In 1954 I bought the newspaper at Agassiz, which serves the Harrison area, and the following year wrote my first sasquatch story, an April Fool's Day special about a beautiful guest at the Harrison Hot Springs Hotel being carried off into the mountains by a hairy monster. At no time did I have the slightest idea that there was anyone in the community, or anywhere else, who didn't think the whole business was a joke. In 1956 Rene Dahinden came to my office looking for information that might assist him in actually finding a sasquatch. An immigrant from Switzerland, he had been told about them while working on a farm in Alberta, and swallowed the whole thing. His visit made a good story, but I felt rather sorry for him.

After that came one of those chance events that sometimes have effects out of all proportion to their real importance. The village council at Harrison Hot Springs was pondering what to do with a small grant available from the government of British Columbia to help finance some local project commemorating the 100th birthday of the province in 1958. Since the village population was small, the amount was only a few hundred dollars. Someone suggested spending it to put up a statue of a sasquatch. Then someone else proposed using it instead to finance a sasquatch hunt. Because sasquatch hunts were not exactly the kind of thing the grant was meant for, the councillors voted to ask the provincial Centennial Committee for permission to use the money in that way. In doing so, they accidentally crossed the wires of a detonator.

I didn't know it then, but it is clear now that almost everyone is interested in monsters, only most people don't like to admit it. A monster story by itself meets considerable resistance and usually doesn't circulate very far. But let some official organization show an interest in a monster, and the situation reverses. When it has somehow been made respectable, a monster story attracts tremendous public response. A recent example of this was the television special put on in 1974 about the Loch Ness Monster and the hairy giants of the Himalayas and North America. Plenty of programs have been done on the same subjects before without creating any kind of a stir, but that one was different. It was sponsored by the Smithsonian Institution. The result was that it had 60 million viewers in the United States, the most to watch any program put on during that week and the most ever to watch a television documentary. Here's how the Los Angeles *TV Guide* told the story in its January 25-31 issue in 1975:

AS WE SEE IT

The job of Network Executives, we're told, is to Give the Public What It Wants. They do this with the help of Nielsen Ratings, Audience Testing, Track Records, Demographic Surveys, Three-Hour Lunches and other Invaluable Data. The result is the Mystique of Network Infallibility, which is powerful enough to make a witch doctor turn in his chicken entrails.

On November 25, the Mystique met its match — mugged by Bigfoot, the Abominable Snowman and the Loch Ness Monster. Here is how it happened.

Smithsonian Institution specials are produced in the desperate hope of making science more appealing than escapist potboilers. One gets the impression that the networks carry such shows solely to appease the FCC, Congress and

50

others who might raise the old "vast wasteland" cry again. But to make sure that documentaries don't unduly damage the ratings averages, they are sent to Death Valley—a time slot where the ratings are already down.

And so it came to pass on November 25 that a Smithsonian Institution special about three legendary monsters pre-empted *Gunsmoke*, whose ratings on CBS had been slow on the draw. The Smithsonian folk bravely hoped that their show wouldn't do much worse. Then the ratings came out— And lo! Smithsonian's name led all the rest. A humble documentary had beaten all other network shows that week, including all of the Top-10 regulars.

Smithsonian's astounding triumph conjures up two pleasing if entirely imaginary visions. One is of CBS executives frantically scanning lists of rejected documentaries. The other is of the Smithsonian Institution itself, where dozens of scientists and dusty, professorial types caper joyously among their learned journals, toss their mortarboards in the air and chant, like a bunch of kids at a football game, "We're No. 1! We're No. 1!"

The effect of what Harrison Hot Springs did was every bit as explosive and as unexpected. The story that a government, even the government of a tiny village, wanted to search for a monster was news all around the world. It also started a lot of talk around Agassiz and Harrison Hot Springs, and I was quickly exposed to the fact that there were local people who took the sasquatch very seriously indeed, and not all of them were Indians. The man who impressed me the most was Esse Tyfting, head custodian at the high school in Agassiz. I already knew him, and knew that he had an excellent reputation in the community. He told me that he had seen enormous footprints at Ruby Creek, only a dozen miles away. It had been 16 years ago, and he didn't make a habit of talking about it, but the picture of those footprints was still so fresh in his mind that he could sketch the outline of one without hesitation and, as I was later able to establish, with great accuracy.

At the time of the incident concerned he was a railroad maintenance worker at Ruby Creek. One of the other men on the crew was an Indian named George Chapman, who lived in an isolated house near the bank of the Fraser River a couple of miles north of Ruby Creek. One afternoon Jeannie Chapman, George's wife, came hurrying down the track with her two small children, shouting, "The Sasquatch is after me!"

She had obviously been upset by something, and the men went to the Chapman house. There they found the huge tracks crossing the field to the house, going around to a shed from which a barrel of fish had been dumped out, then down to the river and back across the field to the mountains. Mr. Tyfting remembered the measurements of the tracks, 16 inches long, four inches across the heel and eight inches across the ball. More impressive even than the size, where the tracks crossed the potato patch the potatoes were crushed in the ground, and where they came to the railroad it was plain that the animal had stepped over a fence more than four feet high without breaking stride. He estimated the creature's weight at 800 to 1,000 pounds.

Mrs. Chapman had refused to return to the house the sasquatch came to, but she still lived in the vicinity and Rene and I had the opportunity to discuss the incident with her on two occasions. She was still upset by the recollection, but all that had actually happened was that one of her children had run in the house saying there was "a big cow coming out of the woods" and she had looked out to see approaching the house a manlike creature about eight feet tall, all covered with dark hair. She didn't spend much time looking at it, but it had a flat face and she knew it was nothing like a bear. Terrified, she led the children out the door, keeping the building between them and the creature, and fled. For an Indian to see a sasquatch was believed to be very bad luck, in fact the observer was in danger of dying. Mrs. Chapman survived, but her relatives told me she never got over the experience.

The story had been publicized at the time and had been investigated by Joe Dunn, a deputy sheriff from Bellingham, Washington, who was personally interested in the subject. Esse Tyfting was with him when he measured and cast the tracks. Mr. Dunn had since died, but one of his sons showed me a report he had written that confirmed what I had already been told. The footprint cast had long since been broken, but there was a tracing of the outline of a print which verified the accuracy of Mr. Tyfting's recollection. Not realizing that it would mean nothing to the scientists, I went to the trouble of having several participants in this incident interviewed by a magistrate before whom they made sworn statements concerning the tracks, which all of them did without hesitation.

The Ruby Creek incident is still one of the most interesting and best documented in the sasquatch file, and two more reports that have never been surpassed with regard to observation of detail also came to light in the explosion of publicity regarding the Harrison sasquatch hunt. One of them, Albert Ostman's story of being kidnapped by a sasquatch, is given in another chapter. The other was told by the late William Roe, who then lived in Cloverdale, B.C. I did not ever meet

Mr. Roe, who moved to Alberta shortly after his story was first published in one of the Vancouver newspapers, but I had heard him in a taped radio interview, and in the period when I was assembling sworn statements I wrote and asked him if he would put an account of his experience in sworn form. He took the trouble of going to the legal department of the city of Edmonton and forwarded to me a story and a supporting statutory declaration. He also had his daughter do a drawing at his direction. Here is his sworn story:

Ever since I was a small boy back in the forest of Michigan, I have studied the lives and habits of wild animals. Later, when I supported my family in Northern Alberta by hunting and trapping, I spent many hours just observing the wild things. They fascinated me. But the most incredible experience I ever had with a wild creature occurred near a little town called Tete Jaune Cache, British Columbia, about eighty miles west of Jasper, Alberta.

I had been working on the highway near Tete Jaune Cache for about two years. In October, 1955, I decided to climb five miles up Mica Mountain to an old deserted mine, just for something to do. I came in sight of the mine about three o'clock in the afternoon after an easy climb. I had just come out of a patch of low brush into a clearing when I saw what I thought was a grizzly bear, in the bush on the other side. I had shot a grizzly near that spot the year before. This one was only about 75 yards away, but I didn't want to shoot it, for I had no way of getting it out. So I sat down on a small rock and watched, my rifle in my hands.

I could see part of the animal's head and the top of one shoulder. A moment later it raised up and stepped out into the opening. Then I saw it was not a bear.

This, to the best of my recollection, is what the creature looked like and how it acted as it came across the clearing directly toward me. My first impression was of a huge man, about six feet tall, almost three feet wide, and probably weighing somewhere near three hundred pounds. It was covered from head to foot with dark brown silver-tipped hair. But as it came closer I saw by its breasts that it was female.

And yet, its torso was not curved like a female's. Its broad frame was straight from shoulder to hip. Its arms were much thicker than a man's arms, and longer, reaching almost to its knees. Its feet were broader proportionately than a man's, about five inches wide at the front and tapering to much thinner heels. When it walked it placed the heel of its foot

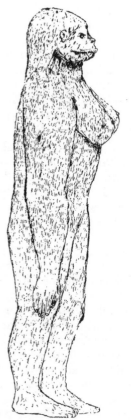

The sasquatch from Mica Mountain, drawn by William Roe's daughter under his direction.

The Chapman house at Ruby Creek, abandoned after the sasquatch came there.

down first, and I could see the grey-brown skin or hide on the soles of its feet.

It came to the edge of the bush I was hiding in, within twenty feet of me, and squatted down on its haunches. Reaching out its hands it pulled the branches of bushes toward it and stripped the leaves with its teeth. Its lips curled flexibly around the leaves as it ate. I was close enough to see that its teeth were white and even.

The shape of this creature's head somewhat resembled a Negro's. The head was higher at the back than at the front. The nose was broad and flat. The lips and chin protruded farther than its nose. But the hair that covered it, leaving bare only the parts of its face around the mouth, nose and ears, made it resemble an animal as much as a human. None of this hair, even on the back of its head, was longer than an inch, and that on its face was much shorter. Its ears were shaped like a human's ears. But its eyes were small and black like a bear's. And its neck also was unhuman. Thicker and shorter than any man's I had ever seen.

As I watched this creature, I wondered if some movie company was making a film at this place and that what I saw was an actor, made up to look partly human and partly animal. But as I observed it more, I decided it would be impossible to fake such a specimen. Anyway, I learned later there was no such company near that area. Nor, in fact, did anyone live up Mica Mountain, according to the people who lived in Tete Jaune Cache.

Finally the wild thing must have got my scent, for it looked directly at me through an opening in the brush. A look of amazement crossed its face. It looked so comical at the moment I had to grin. Still in a crouched position, it backed up three or four short steps, then straightened up to its full height and started to walk rapidly back the way it had come. For a moment it watched me over its shoulder as it went, not exactly afraid, but as though it wanted no contact with anything strange.

The thought came to me that if I shot it, I would possibly have a specimen of great interest to scientists the world over. I had heard stories of the Sasquatch, the giant hairy Indians that live in the legends of British Columbia Indians, and also many claim, are still in fact alive today. Maybe this was a Sasquatch, I told myself.

I levelled my rifle. The creature was still walking rapidly away, again turning its head to look in my direction. I lowered the rifle. Although I have called the creature "it", I felt now that it was a human being and I knew I would never forgive myself if I killed it.

Just as it came to the other patch of brush it threw its head back and made a peculiar noise that seemed to be half laugh and half language, and which I can only describe as a kind of a whinny. Then it walked from the small brush into a stand of lodgepole pine.

I stepped out into the opening and looked across a small ridge just beyond the pine to see if I could see it again. It came out on the ridge a couple of hundred yards away from me, tipped its head back again, and again emitted the only sound I had heard it make, but what this half-laugh, half-language was meant to convey, I do not know. It disappeared then, and I never saw it again.

I wanted to find out if it lived on vegetation entirely or ate meat as well, so I went down and looked for signs. I found it in five different places, and although I examined it thoroughly, could find no hair or shells of bugs or insects. So I believe it was strictly a vegetarian.

I found one place where it had slept for a couple of nights under a tree. Now, the nights were cool up the mountain, at this time of year especially, and yet it had not used a fire. I found no sign that it possessed even the simplest of tools. Nor a single companion while in this place.

Whether this was a Sasquatch I do not know. It will always remain a mystery to me, unless another one is found.

I hereby declare the above statement to be in every part true, to the best of my powers of observation and recollection.

William Roe

In a later letter he added a few details. The nails were not like a bear's claws, but short and heavy like a man's fingernails. There were no bulging muscles, but the animal was as deep as it was wide, and if it was seven feet tall then it weighed close to 500 pounds. The picture shows, in profile, a massively-built creature with an extremely long and heavy arm, huge thighs and breasts, but little indication of buttocks. It has a short neck and a low head, very long from front to back, with tiny ears closer to the front than to the back. The jaws project beyond the small nose, and a thin lower lip sticks out farther yet. I did not know anything about Mr. Roe at that time, but when I made a trip across Canada some years after his death, zoologists in two different cities told me that they had corresponded with him about buffalo. Both assured me that he was an expert observer of wildlife.

J. W. Burns is unquestionably responsible for the public awareness of the sasquatch in British Columbia decades before similar stories gained circulation elsewhere, and his writings may also have had something to do with the fact that there are more stories from British Columbia in the early years of this century than from anywhere else. However, there were other people who took an active interest in the subject quite independently. One of those was Bruce McKelvie, a journalist and historian. Another was Charles Cates, who at the time Rene and I talked to him was the mayor of North Vancouver. Both men died in the late 1950's, which was before any of us began to accumulate systematic records, so most of what they knew is lost. Bruce McKelvie had a file of several old stories, and he told us that he knew a man who had killed a sasquatch, but he was committed not to repeat the story. Charles Cates had grown up as a neighbour of the Indians on a North Vancouver reserve. He told us that a friend of his had been one of a group of people camped on Anvil Island in Howe Sound when a sasquatch put its head into their tent at night. Both men knew of an incident involving a man named Mike King, although they disagreed as to when it had taken place. Here is Mr. McKelvie's version:

> One of the most outstanding timber cruisers who ever operated in British Columbia, Mike King was a fine type of man with an enviable reputation for reliability. He told of being in the Campbell River locality of Vancouver Island. He was alone, for his Indian packers refused to accompany him into that particular area, being afraid of the "monkey men" of the forest. It was late afternoon when he saw the "man-beast" bending over a water hole, washing some roots, which he placed in two neat piles. When appraised of Mr. King's presence, the creature gave a startled cry and started up a hillside. He stopped at some distance and looked back at the timber cruiser, while Mike King kept "it" covered with his rifle. Mr. King described the sasquatch as being "covered with reddish-brown hair, and his arms were peculiarly long and were used freely in climbing and in bush running; while the trail showed a distinct human foot, but with phenomonally long and spreading toes.

I do not have very many reports from Vancouver Island, and most of those I have are old. I expect a lot more would be known if there had been active research there in recent years. Alex Oakes, of Coombes, Vancouver Island, told me he had heard of several people who had seen sasquatches in the 1940's and '50's. He himself had seen one in the early 1940's as it ran across the road in front of his car not far from his home. It was running at high speed, and hurdled the fences on either side of the road. What particularly struck him was the way the hair streamed out behind its shoulders as it ran and leaped. It was about seven feet tall, he estimated, and was the color of a brown bear. The hair on its shoulders was about six inches long.

After a conversation we had in 1957, Charles Cates wrote me a brief letter including the following observations:

The story always seemed to me to have some basis in truth. Some of my reasons are as follows: Practically all the Indian peoples of British Columbia have a name for these creatures. In the Squamish language it sounds somewhat like "Smy-a-likh". Their stories always seem to have the same description as to their physical appearance, and also the Indians did not use the existence of these creatures to tell any fantastic tales about them. They merely refer to them by name, adding that they are quite shy and will do no harm, the reference always being quite casual. Also, I knew two or three old men who said that in their youth they had seen the creatures in various parts of the coast, as apparently they were quite plentiful years ago.

Another story that came to light at that period, and that was backed by a sworn statutory declaration, was that of the late Charles Flood, then a resident of New Westminster, as follows:

I, Charles Flood of New Westminster (formerly of Hope) declare the following story to be true:

I am 73 years of age and spent most of my life prospecting minerals in the local mountains to the south of Hope, toward the American boundary and the Chilliwack Lake area.

In 1915, Donald McRae and Green Hicks of Agassiz, B. C. and myself, from Hope, were prospecting at Green Drop Lake twenty-five miles south of Hope, and explored an area over an unknown divide, on the way back to Hope, near the Holy Cross Mountains.

Green Hicks, a half-breed Indian, told McRae and me a story, he claimed he had seen alligators at what he called Alligator Lake, and wild humans at what he called Cougar Lake. Out of curiosity we went with him; he had been there a week previous looking for a fur trap line. Sure enough, we saw his alligators, but they were black, twice the size of lizards in a small mud lake.

A mile further up was Cougar Lake. Several years before a fire swept over many square miles of mountains which resulted in large areas of mountain huckleberry growth.

While we were travelling through the dense berry growth, Green Hicks suddenly stopped us and drew our attention to a large, light brown creature about eight feet high, standing on its hind legs (standing upright) pulling the berry bushes with one hand or paw toward him and putting berries in his mouth with the other hand, or paw.

I stood still wondering and McRae and Green Hicks were arguing. Hicks said "it is a wild man" and McRae said "it is a bear." The creature heard us and suddenly disappeared in the brush around 200 yards away. As far as I am concerned the strange creature looked more like a human being, we seen several black and brown bears on the trip, but that "thing" looked altogether different. Huge brown bear are known to be in Alaska, but have never been seen in southern British Columbia.

I never have seen anything like this creature, before or after this incident in 1915, in all my days of hunting and prospecting in British Columbia.

Charles Flood

Reports of such creatures being seen eating something have been rare, but there is another early British Columbia account of a sasquatch eating berries, although it did not come to light until 1968. The man concerned, a retired insurance broker, wrote that about 35 years earlier he had been on a weekend trip to the head of Pitt Lake, just east of Vancouver, and that he and another man climbed about 1500 feet to the edge of a small plateau, looking for interesting rocks. As they were sitting eating their lunch:

A movement behind a thicket some quarter mile away caught my eye and I said, "Cartie, there is something down there." He looked, then asked for the field glasses. We both thought it was a black bear feeding on berries, then he exclaimed, "Here, look at its face!"

Through the glasses it was quite plain — a human face on a fur-clad body.

"What the hell," I said. "He must be a hermit or something of the kind but look at the size of him."

Cartie replied, "Wait until he leaves and let us go down and look at the tracks."

So we waited. I don't think the creature saw us, though he or she may have sensed us, as presently it went away across the plateau and vanished among the rocks. We went down after a considerable interval and examined the tracks, which were quite distinct. Cartie looked pretty grim and said, "Let's go back. What you have just seen is a Sasquatch; don't mention this to anyone, not even to your wife. No one will believe you, you will just be laughed at and you will have a miserable time of it. Just forget the whole thing and keep quiet."

So I did, and I have, until now.

In a later letter this man told me that the plateau was stony, so that the tracks were not really good, but they seemed to be a large human track with no claw marks, and were definitely not bear tracks. He and his friend were a bit apprehensive and did not search very long. The berries the creature was eating were a type of Oregon grape.

When the thing left, he said, it "shuffled off on two feet, without haste."

One of the most fascinating stories I have ever heard is told by Burns Yeomans, of Deroche, British Columbia. Here is an abridged transcript of an interview I had with him about 1965:

How long have you lived in this part of the country?
I was born in this part of the country.
How old are you now?
Sixty-four.
Have you spent much time in the bush?
Quite a lot of it.
What sort of work would you be doing?
Prospecting a lot.
The occasion we're interested in, what year would that be?
I'm not sure, 1939 or 1940, one or the other.
What time of year?
August.
Were you alone?
No. Actually there were five of us up there, but there were only two of us at the top of that hill where we could see the other side.
What were you doing up there then?
Looking for the molybdenite that's supposed to be in that country.
Where did you start from? You were up Harrison Lake?
Yes. We were up on Silver Creek, the headwaters of Silver Creek.
About how far?
Well, I imagine about 14 miles.
And then where did you go?
We went east up that mountain . . . it's straight east from that mine.

(There was a cabin that had been built the year before near a molybdinum prospect.)

Do you know how high it would be?

I don't . . . but it was in August and there was still snow on the top of that ridge.

When you got to the peak you could see down on the other side?

On the other side down in a big valley there.

What was the valley like?

Well, it's fairly open.

Any idea how high it would be?

No. I imagine it would be 5,000 feet anyway.

The mountain would be a good deal higher than that?

Yes.

What did you see in the valley?

We saw these animals. I don't know what they were but there were four or five of them, wrestling just like men, down in this valley.

What color were they?

They were about three quarters of a mile away, and we didn't have any glasses with us, but they looked to me to be black. They were a dark color anyway.

Could you tell whether they had hair?

Yeah, well I don't know. The color of them, it looked like hair.

It didn't look like skin?

No.

It didn't look like clothes?

No.

They were wrestling like people?

Just like men would do.

Having fun, you'd say?

Yes.

Did you see them stand up on two legs?

Yes they were. I can't recall them getting around any other way but on two legs. As far as I remember of it they were on two legs all the time.

Is there any possibility that they could have been bears?

Well, I don't know, but if there's any sasquatches in this country, I seen them. I didn't think they were bears.

Could you give us an idea how long you were there?

About half an hour.

You were sitting up there for about half an hour, watching

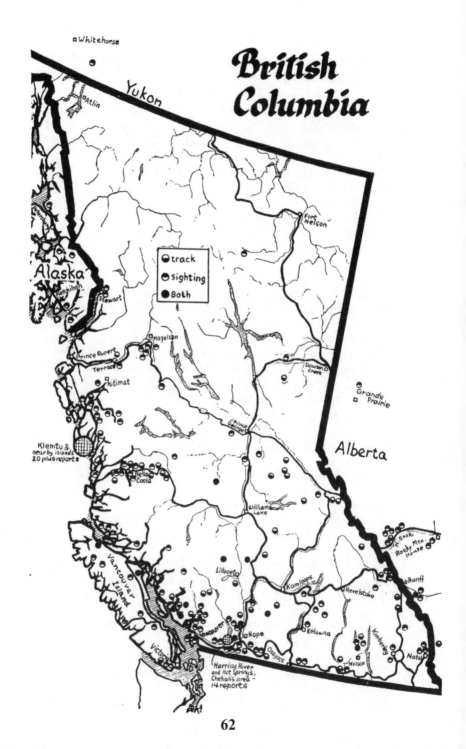

British Columbia

track
sighting
Both

Whitehorse

Yukon

Atlin

Alaska

Ketchikan

Stewart

Hazelton

Prince Rupert

Terrace

Kitimat

Fort Nelson

Dawson Creek

Grande Prairie

Prince George

Klemtu & near by islands 20 plus reports

Bella Coola

Williams Lake

Alberta

R. Sask.
Rocky Mtn. House

Vancouver Island

Lillooet

Kamloops

Revelstoke

Banff

Victoria

Vancouver

Hope

Kelowna

Kimberley

Natal

Osoyoos

Nelson

Harrison River and Hot Springs, Chehalis area - 14 reports

four or five black creatures that looked like men wrestling?
They kept wrestling all that time?

Seemed to, yes.

Any sign of them getting hurt at all?

No, it didn't seem to. One would throw the other down, he'd jump right up on his hind legs again.

Did they grab each other round the neck?

Well I don't remember that. I don't remember just how they done it, but it seemed like a wrestling match.

Did you have an impression when you were watching them of what size they were?

I'd say they were seven feet high anyway.

Bigger than a man?

Bigger than a man, yeah. I'd say they'd weigh 400 pounds or more.

You think they'd be built heavier than a man?

Yeah . . . mind you with hair on . . . but they looked to me they'd weigh that much anyway.

Had you ever been there before?

No.

Did you ever go again?

No, I've never been back there.

Ever seen anything else like this anywhere else?

Never anywhere else in this country. I've been all over these mountains.

Talking to old-timers is one of the most interesting aspects of the sasquatch investigation, although in recent years I am beginning to find that some of them aren't such an awful lot older than I am, and that the old times they recall are times I can remember too. Mrs. Jane Patterson came to a ranch near Bridesville, B. C., just north of the U. S. border, in 1936. She thinks it was the following year that she went alone to an old abandoned house on the ranch, because she had been told there was some rhubarb growing in the old garden there. She ducked under a tree branch and straightened up to find herself facing a creature that was sitting about 10 feet away with its back against another tree. Her story is told with many a merry chuckle.

I said, "Oh, there you are." I don't know what I was speaking to it for. It was sitting down. The house had sort of blown over so it leaned against a tree, and here it was sitting by this tree, you know, like you'd sit on the floor, and it had hands like that (on its knees). It was taller than me though, I would think. It looked just like a monkey to me . . .

It just blinked its eyes. I suppose it was as surprised as I was. It wasn't really a tan, but it wasn't a brown. Maybe you'd call it a light brown.

It never moved except to blink its eyes. I thought it was a big monkey.

She said it had hair all over, but not thick. Its eyes were like slits. She backed away from it, and headed home.

After I got quite a ways back, the other side of the house from it, I began walking a little faster.

I wanted to go back, and I told my husband about it, and he said he hadn't lost no monkey, he wasn't going up there. I talked him into it about three days later . . . he said he hadn't lost no monkey and he didn't want to see no monkey, but anyway he did it. That was three days after, of course.

I couldn't help reflecting that the late Mr. Patterson should have been a scientist. He would have been a natural.

Oh, yes, Harrison Hot Springs never did have its sasquatch hunt. The B. C. committee turned down the idea, and the council bought a furnace for the community hall instead.

-4-

Bigfoot at Bluff Creek

"Bigfoot" is a name that someone on a road construction crew tagged on an unknown something that was making huge footprints at night in the loose dirt of the right of way. If he had known at the time that he was providing a common name for a whole species of animals all over the United States, he might have given a little more thought to it and saved a lot of confusion. When people first hear of something called "Bigfoot" they naturally assume that there is only one.

In a way it is sheer fluke that people ever heard of Bigfoot at all. The 16-inch footprints that showed up on the Bluff Creek forest road in the northwest corner of California in 1958 were by no means the first ever seen. Similar prints had been seen from time to time in many places without the world in general becoming aware of them. The two things that were different about Bigfoot and Bluff Creek were that someone went to a newspaper with the story, and that tracks kept appearing there fairly frequently for some time after the news was out.

The Bluff Creek valley runs approximately north from the Klamath River, starting about four miles northeast of the tiny town of Weitchpec. In 1958, Highway 96 along the Klamath was little more than a one-lane road, unpaved, but there were several forest access roads running north and west from it, mainly high up on the ridges. The new road up Bluff Creek was not penetrating any unknown territory, but it entered an area completely wild, completely buried in closed-canopy forest, and completely uninhabited. The Bluff Creek road was to be different from the others, wider, with better grades, staying more in the bottom of the valley. It was to provide access to the timber in the area, and was eventually to link up with Highway 199. It later turned out that the hillside was unstable and after a series of huge slides much of the road was eventually abandoned.

At the end of August, 1958, when the construction crew was roughing out a roadbed some 20 miles north of the Klamath, big tracks started appearing in the dirt overnight. They usually came down from the hillside and crossed the road, going towards the creek. Sometimes they followed the road for a distance, and passed close to parked earth-moving equipment. The prints were like those of a flat-footed human, but huge—16 inches long, seven inches wide at the ball of the foot and only two inches narrower at the heel. Average stride was over four feet. All were made by the same individual. Most of the prints would be obliterated when the machines started to work in the morning, but every few days a new set would be there. After about a month of this a bulldozer operator, Jerry Crew, went to see a friend who was a taxidermist and got some casting plaster and instructions on how to use it. A few days later he appeared at the office of the *Humboldt Times* in Eureka and had his picture taken with a cast of a track. The picture, and an accompanying story, was sent all over the continent by the Associated Press, and "Bigfoot" was born.

After seeing Jerry Crew's picture, which was printed in a Vancouver newspaper, three of us decided to head for California. Just at the end it turned out to be quite a trip. Taking what appeared on the map to be the quickest route over from the coast to Weitchpec we found ourselves wandering around on dirt tracks in the mountains west of the Klamath River, lost and almost out of gas. Finally we flagged down a pickup truck only to be told by the driver that the bridge we were looking for had been washed out the preceding spring. He was half way through an explanation of how we could get across the Klamath another way when we heard a heavy motor roaring somewhere below us on the hillside. The man in the pickup suddenly remembered that he was driving the lead car for a wide load, shouted a warning, and shot off up the hill. We were left to negotiate a winding hill, backwards, with a flatbed truck carrying a bulldozer threatening to run right over us, and showing no sign of slowing down. Our driver finally managed to pull far enough off the road to let the truck by, but then we were left with no option but to follow in its wake hoping that the man in the pilot car would find some way to signal to us where we were to turn off. We couldn't try to pass the truck because we were completely blinded by dust if we got anywhere near it. I forget how we got out of that. Perhaps the truck stopped to let us by. In any event we managed to coast down the hill and limp into a gas station—but there was almost a quarter inch of dust on everything in the car, including us.

On that trip there were only a few old tracks to be seen, all the most recent ones had been wiped out just before we got there. Still, those that remained were very impressive. Looking at them was quite an experience. I realized that in spite of having undertaken a 2,000-mile trip just to see them for myself, deep down I had never expected that there would be anything to see. Fortunately my wife was with me. She might have found it more difficult to be so understanding over the years had she not seen those tracks herself. We spent one evening with Jerry Crew, and on the way home went to Anderson, California, to call on Bob Titmus, Jerry's taxidermist friend, who in the meantime had been to Bluff Creek himself and had seen much better tracks than those we saw. He told us that from time to time hunters had told him of seeing such tracks, but he had always been sure they were made by bears. Having now seen some himself, he couldn't say what had made them, but certainly not a bear. Nor did he think that the tracks could have been faked.

I found that there was no public tradition in that part of California of the sasquatch or its equivalent, and people were thinking up silly explanations for the tracks about a crazy man who had run away to the woods. It turned out later that the Indians did have similar stories to those in B. C. but the non-Indian community had never been exposed to them. There was no doubt that the same thing was involved in both areas. I had a tracing of a track from Ruby Creek, in British Columbia, that matched the Bluff Creek tracks so closely there was not a half inch difference at any point.

Calling on Bob Titmus turned out to be the most important thing we did on the trip. A couple of weeks later he wrote to say that he and a friend named Ed Patrick had found fresh tracks on a sandbar in the bed of Bluff Creek and they were not the same as those up on the road. With Gus Milliken, from Yale, B. C., I made the trip south again and we spent a couple of days with Bob camping near the end of the Bluff Creek Road. By then it was November and road work was about over for the season. Bob had destroyed four of the best tracks by casting them, but there were several left on the sandbar, and I could find more on an adjoining gravel bar by feeling for them, although they were covered with fallen leaves. The new track was an inch shorter than the original "Bigfoot", and considerably less like a human track. The original track could conceivably have been that of an enormous human with a very wide foot and fallen arches. The new track had shorter toes, and what appeared to be the ball of the foot had a deep split on the outer edge and part way across it,

67

making a sort of double ball. What was really impressive, however, was the depth of the tracks.

The few old tracks I had seen by the road were in the loose dirt on the shoulder, where human prints would also make a considerable impression. The sand on the bar was hard-packed and damp. When I walked around on it my boots showed a complete heel print, but of no measurable depth, and the soles made only a scuff mark. The big tracks had an average depth of close to an inch, and various heel and toe impressions exceeded an inch and a quarter. I had to jump off a log about two feet high and land on the point of one heel to make a hole as deep.

We could not think of any way a man could have made the tracks without the use of some sort of specialized heavy equipment, and there was no apparent way that such equipment, assuming that it existed, could have reached the sandbar. Both sides of the valley were steep and covered with heavy underbrush. Taking a machine down without leaving evidence of its passage seemed out of the question. At the downstream end of the sandbar there was a tangled log jam, and close upstream the creek ran through a small canyon, with a straight rock wall on each side dropping into a deep pool. About the only answer would have been to fly the machine in with a large helicopter, but that could not have been done secretly because at the time the tracks were made there were construction workers living in a camp just a few hundred yards away.

We were faced with the basic problem that makes the whole matter so fascinating. Something made the tracks. They were there, and they were real. At some time just a few days before something had been on that sandbar making those marks in the ground, and whatever had done it was as real as the tracks. If it couldn't have been a man, then there had to be an animal that was heavy almost beyond imagining, and that walked upright like a man. It was years before I could accept that proposition emotionally, but on a logical basis it was inescapable. Something had made the tracks, and it had to be either a man with a machine or an animal. Both explanations were ridiculous, but one of them had to be true. There was a possibility that somebody a lot smarter than we were had a way of making such tracks in such a place, and that possibility still exists, but in the intervening years no such person has come forward, nor has one been caught in the act. Animals capable of making such tracks, on the other hand, are reported about a hundred times a year.

By the time of my second trip there were several stories circulating in the Bluff Creek area of a huge apelike creature being seen on the road. I have never talked to any of the people who reported those sightings, but I had a good chuckle over a remark attributed to one of them. He is supposed to have said that the thing he saw crossing the road was only a man in some sort of fur suit that had a droopy look around the seat. Asked whether he had found any tracks, he explained that he had not ventured to get out of his truck, because "that guy was eight feet tall."

Rene Dahinden was unable to go to Bluff Creek in 1958 because he was not yet a Canadian citizen and did not have a U. S. visa. In the fall of 1959 he had his citizenship, and we went south for a look around. I don't recall whether we knew there would be tracks there, but there were—a new set of the 15-inch size on the sandbar, again found by Bob Titmus. In the meantime, Ivan Sanderson had read an article of mine about the sasquatch and had called to see me. On that trip, in some manner the details of which I no longer recall, negotiations started involving Bob, Ivan, Rene, myself and Tom Slick, a San Antonio millionaire who had financed several expeditions to look for the abominable snowman in Nepal. We made another trip to California to meet with Tom a couple of weeks later, and "The Pacific Northwest Expedition" was organized.

Looking back, it should have been obvious that things were going to be a little strange. We were looking for a financial backer so we could spend enough time in the area to intercept Bigfoot or his smaller cohort on one of their trips through. Tom was accustomed to being in charge of whatever and whoever he paid for. The title of the "expedition" was his idea, and he was to be the "leader" of it. As a result there was a protracted negotiating session in the lobby of a motel in Willow Creek. One of the most active participants was a man who had accompanied Tom. We found out afterwards that he was not an associate, but had met Tom by appointment to talk about selling him a mine or something. An elderly lady knitting in one corner of the lobby also put in a few words from time to time.

We all believed that it would only take a few months at most to achieve some results, and considering how frequently the tracks had been turning up that probably wasn't an unreasonable assumption. It has been my opinion ever since that anyone can be excused for getting money from someone when they are new to the game, but anyone who can go on raising money after they have learned how poor the odds really are has got to be mighty careless with the truth. The Pacific Northwest Expedition didn't disturb Bigfoot or his friend any, but it taught all of us some useful things, one of which was that the average sasquatch hunter is so pig-headed that two of them together are pretty sure to have a falling out before long. It's fairly obvious when you think about it. People who will go hunting for an animal that is rejected by the world of science and almost everybody else are bound to be people who don't pay much attention to any

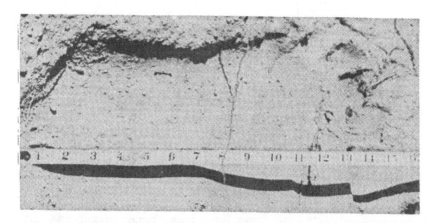

The oldest track photo that I know of, this was taken on a utility right of way between Eureka and Cottonwood in northern California about 1947, more than 10 years before the first excitement over "Bigfoot."

Original personnel of the "Pacific Northwest Expedition," from left, Ed Patrick, Tom Slick, Rene Dahinden, Kirk Johnson (co-sponsor), Bob Titmus and Tom's secretary, Jerri Walsh.

opinion but their own, and who expect not only to have an opinion but to act on it.

There were all sorts of things to argue about. Slick insisted that the thing should be shot, but I think all the rest of us had serious doubts about that idea at that time. We all had different ideas about how we should be hunting it, and where. The amount of financial support provided was woefully inadequate, especially when Tom assigned a share of the funds to a couple of local characters who showed up with a cock-and-bull story about tracking Bigfoot half way across the county. That was the first real intimation we had of Tom's strange failing as a judge of people. It must have been different in the business world, but when it came to sasquatch hunting he was chronically doubtful about the people who really wanted to find the thing, but an easy mark for any con artist. Maybe it was because the con artists didn't argue with him. Part of the trouble was that he expected someone to get Bigfoot almost anytime, and he wanted whoever it was to be on his payroll.

Tom favored hunting with hounds, so much of the time was spent driving the dirt roads looking for tracks to put the hounds on. Rene, who was pretty disgusted with the whole proceeding anyway, packed it in after a few days of riding around in a station wagon while a hired hunter and his brand-new bride nuzzled each other in the front seat. If Rene felt like doing any nuzzling, well, he was sharing the back seat with the hounds. Bob Titmus was the "deputy leader" but he had his business to look after and couldn't spend much time in the field. Ed Patrick was active a good deal of the time, and I managed to spend one month there, in the spring of 1960, at which time I saw some more good tracks. Those were only about 14 inches long and were narrower than the others. As I recall they also had the split in the ball of the foot. I think it may have been the track of the creature in the Patterson movie, but one of the conditions under which we were financed was that all pictures and other material went to Tom Slick in San Antonio, and unless those things should turn up some day we will never know for sure.

Those tracks were punched deep into a big pile of loose dirt heaped beside the Cedar Camp road, a few miles east of Bluff Creek. As if they weren't noticeable enough by themselves, we had camped near them and had covered some of them with plastic because they were being rained on. While we were there a forestry worker drove up the road past them, and on the way back down he stopped and had coffee with us. We didn't tell him what we were doing but he probably had an idea, for he made occasion to tell us that he had been patrolling those roads for many, many years, and there was no truth to the stories about big tracks because if there were any he

71

would have seen them. Then he drove on down the road, passing the tracks for the second time. I wouldn't be surprised if he has never seen any yet. There were lots of other people like him.

We tried a lot of things. Smelly things, edible and otherwise, didn't attract Bigfoot. Neither did shiny things hung in trees, or screams and other noises in the night. He avoided camera trip lines. We didn't have any more-sophisticated equipment. One man who was with us for a while insisted that if we stationed men with horses on all the ridges the actions of the horses would tell the horsemen where the sasquatches were moving in the valleys. Maybe he had something there, but none of us knew any horses. Between arguments we had a lot of fun, and nobody got hurt. We solemnly shipped off quantities of suspected hair and droppings for analysis, some of which remained unidentified, but all that was found of Bigfoot and his friends was tracks, and not very many of those. No one was maintaining the roads in the winter and spring, so we all became experts at manoeuvering our cars through tight places, and digging them out of old snow drifts. One huge boulder that had come down on a steep part of the Cedar Camp road had to be passed on the outside with one side of the car scraping the rock and the outside wheels leaving only a partial track on the edge of the sharp embankment. Tom used to fly out to join his "expedition" whenever he could, and once I drove him past that boulder without giving it a second thought. He was sitting on the outside, and he didn't say anything, then or for a while afterwards.

Tom got a tremendous kick out of Bigfoot hunting. Once he and I made a short sweep around camp while Bob was cooking supper. We had just driven in, and there wasn't much chance of any animal hanging around, but Tom clutched his rifle at the ready and said to me, "Boy! We're hunting the biggest game in the world."

Another time, Bob tells me, Tom was supposed to leave after a weekend in the bush to attend a directors' meeting of one of his companies. I suppose it was Slick Airways. He finally decided he wasn't going to go, and he had Bob drive him to the nearest phone, which was in a booth in Hoopa, about 30 miles away. The booth lacked a door and as Bob sat in the car he could hear Tom telling someone to go ahead and chair the meeting, that he couldn't possibly get away, and they didn't need him and why didn't they just go ahead and buy those 707's. Then he got back in the car and headed for the hills to hunt Bigfoot.

The following year both Bob and I had shifted our interest to an area on the coast of British Columbia, where we also had a deal with Tom, but one providing us with a veto over the other people he might feel an urge to hire. I wasn't back in California until 1963, by which time Tom was dead. Bob, who had returned to visit his parents, had been shown tracks of three different sizes at Hyampom, about 60 miles south of Bluff Creek site, and I drove down to have a look. By the time I arrived Bob had made some casts, but sightseers had trampled on nearly all the tracks, leaving only a couple of 16-inch ones. The other tracks, one 17 inches long and very wide, the other 15 inches, with widely splayed toes, were both new to all of us, but the 16-inch could well have been the original Bigfoot. One of the men who had found the tracks was a local game guide named Syl McCoy, whom I met again on my next trip south.

Early in 1967, after being out of touch with what was going on in northern California for several years, Rene and I drove down there in the course of a trip we were making to contact other people active in sasquatch investigations. We knew that Roger Patterson had cast a track near Bluff Creek about 1964, but we did not know that for parts of 1963 and 1964 Bigfoot had been as active as in the fall of 1958. That was when logging operations were going on in the area opened up by the Bluff Creek road, and apparently the creature was interested, because the loggers saw his tracks often. In fact, his presence made it hard to keep men on the job. Pat Graves, who was in the area sometimes as a road contractor and sometimes inspecting road work, told us he had seen the tracks several times far out in the bush while walking lines where new roads were to be built. He also was one of several witnesses we talked to who told of finding tracks indicating that Bigfoot had picked up a trailer loaded with culverts, turned it upside down and dropped it. That was done in the daytime and several men heard the crash.

At other times he had thrown big culverts, too heavy for men to lift without machinery, from the road into the creek, without any damage to the bushes on the slope below the road. There was nothing to prove it, but our impression was that these were the 16-inch tracks, not the 15-inch. There had been floods all through that country in the fall of 1964, and heavy rains had caused some tremendous landslides. The 16-inch track had not been seen since, and there was speculation that Bigfoot had been buried in a slide.

In Willow Creek we stopped at the forestry station to enquire if anyone there knew anything about recent track reports. I was told that one of the staff had pictures of tracks, and that gentleman was called out to show them to us. His first words were: "Why should I show them to him? He took them."

It was Syl McCoy, now working for the forestry department. He had no information for us then, but late that August he phoned my home to say that some tracks had turned up above Bluff Creek on Onion Mountain. At that time I had made contact with Harold

McCullough, who trains tracking dogs at Coquitlam, B. C. He and I and Dale Moffit drove south with his best dog, a white Alsatian named White Lady. The tracks were there alright, the familiar 15-inch print and a 13-inch one that was new to me. They ambled around where a water trailer was parked at the junction of two roads. The tracks turned out to be too old to interest the dog, but they were good enough to cast. It was not the first time those two tracks had been seen together. Bob Titmus had followed them twice in an area on the other side of the Klamath from Bluff Creek in the summer of 1960. On that trip we met Al Hodgson, owner of the Willow Creek Variety Store, who had a display of track casts and photographs made during the years when I had been away from California. They were all of the 15-inch track except for one extremely deep cast made in a mudhole, which was only 12 inches long. Another person we talked to was Mrs. Bud Ryerson, wife of the contractor who owned the water trailer.

The first morning after I got home I was awakened by a phone call from Bud Ryerson, who had just driven out to Blue Creek Mountain, where his crew was working, and had found fresh tracks running for hundreds of yards in the loose dust on the new road. He was calling on a radiophone, which half northern California could hear, and all he said was that what I was looking for was there. I knew what he meant, and it turned out that a lot of other people did too. I got on the phone trying to round up scientists, a tracking dog and money to charter a plane. The Vancouver *Sun* came through with $500—the only time anything like that has ever happened to me—and Rene and I got away early in the afternoon with Dale Moffit and White Lady but no scientist. In the meantime I had made a call to Al Hodgson, whom I met only briefly a couple of days before, and asked him to have someone meet us at the Orleans airstrip with transportation, food and American money.

Landing at Orleans was interesting. The airstrip runs at right angles to the valley, which is rather narrow, and the pilot shouted in my ear that he would have to side-slip to get low enough to touch down. Side-slipping turned out to involve a sharp drop to one side and then to the other, like a falling leaf. Knowing that it was going to happen, I found it interesting if slightly disconcerting. Not knowing it was going to happen, Rene and Dale got a lot more out of it than that.

At Orleans we found waiting for us a school teacher cum bear hunter, S. C. Buttram, with a jeep, a stock of groceries and $100 cash. There aren't too many people who would come through in a pinch like that for a new acquaintance, and we said a silent 'thankyou' to Al Hodgson. Syl McCoy had meanwhile been sent to

74

some other part of the country to fight a fire. We drove immediately to meet Bud Ryerson, but by the time we got to where the tracks were it was already night. At the first sniff the dog turned as rigid as if she had been given an electric shock, but no one cared to try following the tracks off into the woods in the dark. At dawn the dog's reaction was entirely different. Exhibiting no sign of excitement, she puttered around sniffing the leaves of little bushes at the edge of an old jeep road where the larger of the two creatures had apparently gone. The ground there was too hard to show anything, and Dale told us that underneath the leaves was the last place where a trace of scent would hang. Lady went a few hundred yards down the road, but showed no sign of having anything definite to follow.

The main purpose of the trip having failed, there were a lot more things we could have done, and we picked away at them, but not very efficiently. The evening before Dale had found that he had left the dog's tracking harness in the plane. I had some phone calls to make anyway, so Rene and I drove to Orleans. The pilot, having found nothing to pass the time in the tiny town, elected to come back with us, and we stopped to show him the old tracks on Onion Mountain.

I don't remember what he said on first seeing those giant tracks outlined by flashlight with black woods all around us, but he said plenty after Rene let out a scream just behind him. The next time we stopped there we found new tracks, only 12 inches long but with an enormous stride and sinking in far deeper than our own tracks. I thought they might have been made by some barefoot person taking high leaping strides, and tried to duplicate them that way, but sharp little stones in the gravel put an immediate end to the experiment. Whether the scream had anything to do with the new tracks appearing I don't know. Since then a hearty scream just before hitting the sack has been tried many times in hopes of attracting visitors in the night, but without result. I don't know what the sasquatches think of it, but it gets interesting results with people.

After the letdown when the dog wouldn't track, Rene and I started feeling the effects of lack of sleep, especially when the heat passed 100 in the shade. We didn't exactly waste the rest of the day, but I have wished ever since that we had been more painstaking and systematic in examining and recording the tracks. There were an awful lot of them, covering two sections of road totalling about 600 yards right at the peak of Blue Creek Mountain, which is the high point of a fairly level ridge just under 5,000 feet in elevation. Bud Ryerson said that when he first saw the tracks there were plainly two small sets and one large one. During the day before we arrived traffic on the road had wiped out the tracks on about a quarter of the road surface, and we

Bruce Berryman, left, Bob Titmus and Syl McCoy with the casts they made from footprints found near Hyampom, California in 1963, demonstrating that two feet can sometimes make a yard.

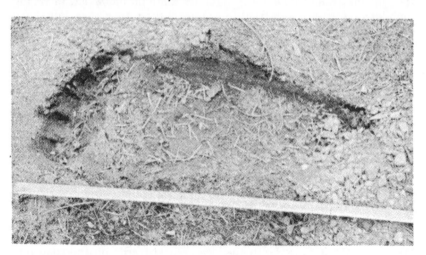

A 13" track driven deep into a sand and gravel bar on Bluff Creek.

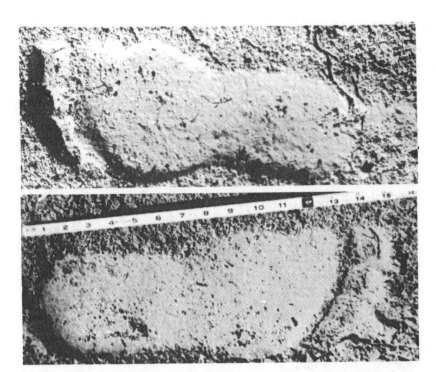

Don Abbott and Blue Creek Mountain tracks.

could no longer find a place where there was a double set of small tracks.

Two creatures had come up on the road at the same point, out of the valley of Pecwan Creek, west of Bluff Creek, and had wandered north. They had not walked side by side, and weren't necessarily together at all. At some points one set or the other would go off the side of the road onto hard-packed ground, then return after a while. Once the larger creature apparently did a loop off the road and came back on it farther back than where he left it, making a double set of big tracks. Near the abandoned forestry lookout at the peak of the mountain both sizes went off the road for a considerable distance. The smaller tracks finally left the road, on the east side, a considerable distance before the point where the large tracks went off to the same side down the old jeep road.

Just after they first came on the road the large tracks went over to where the men had left some small tractor parts in a box, and those were scattered out on the road, which was what had first drawn Bud Ryerson's attention to the tracks. Deep dust stirred up by the road equipment was the only material the tracks would show in, and it was so soft that human tracks would sink in it to almost equal depth. There had been a sprinkle of big drops of rain that evening, however, and all the tracks were literally cast in a delicate skin of dried mud. Some of them would have shown perfect detail, but traffic had filled them with a layer of dust and they were so fragile that there was no way to remove the dust without damaging the tracks. Perhaps half of them had been wiped out in the wheel ruts, yet there were still 590 tracks left.

We took a lot of pictures and made some casts, but we did not cast the best tracks because we had word that the British Columbia Museum was sending an anthropologist to see them. It took him two days to arrive, and despite the co-operation of Bud Ryerson in trying to preserve the tracks—even at the cost of interfering with his road project—traffic and curious visitors were constantly wiping out some and putting more dust in the others. In the meantime we heard that similar sizes of tracks had been reported made the same night down in the bed of Bluff Creek several miles to the northwest. They were on a sandbar where some loggers had a trailer camp, but the night they were made the loggers were all away for the weekend. When they got back and found the tracks they moved the trailers away, destroying most of the tracks in the process. There were a few left, however, and they were definitely made by the same individuals. Even the smaller track sank an inch deep in the damp sand, where we did not sink in at all.

When Don Abbott arrived from the provincial museum he chose to try to lift a track out of the ground by impregnating it with glue. It proved difficult to soak the glue in without eroding the track, and then the glue showed no sign of wanting to set, so we left it overnight. Next day Don persuaded some zoologists from Humboldt State University to come and look at the tracks, but waiting for them proved a fatal mistake. Before we got back the road crew had decided we were finished with the tracks and they had graded the road, wiping out almost all of them, including those we had tried to solidify with glue. Not only had we made only a few casts of second-rate tracks, we had promised casts to a number of people who had come up to make their own, persuading them to wait until the scientists had been there. As a result we didn't have much left for ourselves. The museum does have one good cast showing one track of each size in a single piece of plaster, but things being as they are between museums and sasquatches, they don't display it. Dr. Clifford Carl, the director of the B. C. Museum, was almost solely responsible for the interest they did show, and his death not long afterwards pretty well put an end to it. Don Abbott was entirely convinced, when he was looking at them, that the tracks were those of an unknown primate. He is a cultural anthropologist, however, and he was not able to convince the zoologists.

The Blue Creek Mountain tracks eliminated one of the possible ways that it seemed tracks might be manufactured by humans. Even though many tracks gave every appearance of having been made by pressure, even to the point of making cracks in the ground around them, we had wondered if an extremely skillful person could dig and shape the tracks, removing the material instead of compressing it. The tiny layer of mud on the dust at Blue Creek Mountain ruled that out. It had sprinkled early in the evening, and in such hot weather the dampness could hardly have lasted more than an hour or so. There was no dew at all the following night. To have shaped a thousand tracks by hand in that brief period and in the dark would be impossible many times over. It did not appear that they could have been made with false feet, either, since they showed evidence of flexibility and muscular control, including one 15-inch track that appeared to have as sharp an outline as the others, yet was narrower and showed only four toes.

It probably will never be possible to rule out entirely the possibility that tracks are faked. There have been a lot of attempts to imitate them and some have fooled people for a time. There are almost certainly other fakes that have not been detected. To go from that to faking all the tracks, however, is a colossal step. Tracks have been reported for centuries, and from all over North America as well as other parts of the world. In the years before the subject was widely

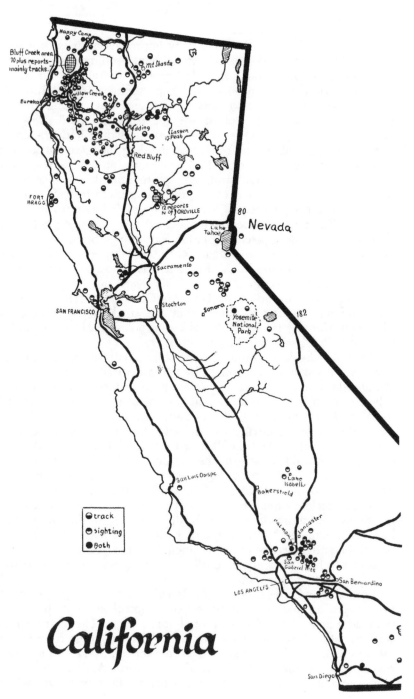

Happy Camp

Bluff Creek area
70 plus reports—
mainly tracks.

Mt Shasta

Willow Creek

Eureka

Redding

Lassen
Peak

Red Bluff

FORT
BRAGG

12 reports
N. of OROVILLE

80 Nevada

Lake
Tahoe

Sacramento

Stockton

SAN FRANCISCO

Sonora

Yosemite
National
Park

182

San Luis Obispo

Lake
Isabella

Bakersfield

Palmdale Lancaster

San
Gabriel Mts

LOS ANGELES

San Bernardino

San Diego

track
sighting
Both

California

publicized there would probably have had to be some kind of continuing undercover organization to have achieved those results. It is not reasonable to argue that such an effort could go on without anyone ever being caught in the act or coming forward to brag of his exploits. In the height of the Bigfoot excitement a television program offered a reward of $1,000 for anyone who could show how to make the tracks. That was a more impressive sum of money then than it is now, and all sorts of people applied, but no one collected. It would also be unreasonable that there should be an animal making the tracks without ever being caught in the act, but that is not the case. Many hundreds of people have reported seeing just such an animal, and one has been photographed in the act of making a fine set of tracks. The problem is to produce the animal in the flesh.

People still cling to the idea that bears make the tracks, and certainly they are the only possible candidate among North American animals, but the differences are specific and unmistakable. A bear has long claws which always show beautifully in a track of any depth. Sasquatch tracks never show claw marks. A bear's hind foot is wider than a sasquatch track in proportion to its length and more pointed in the heel. Its front foot has a totally different shape, with the sole far wider than it is long. A bear's toes are fairly uniform in size. The one extending farthest forward is the middle one, and the outside one is well forward of the inside one. Only two of the many five-toed sasquatch tracks I am familiar with have toes of uniform size. All the others have a "big toe" on the inside of the foot. In every track I know of but one, the longest toe is the inside toe. There is one exception that has toes all the same size with the middle one farthest forward, but even it doesn't look anything like a bear track. The middle toe is farthest forward because the others stick out toward the sides.

You can get the approximate shape of a sasquatch track where a bear's hind foot has stepped ahead of his front foot with the hind heel obliterating some of the front toe imprints. It doesn't cover them all, because it isn't wide enough, and the back of a bear's front print is concave, bearing no resemblance to a sasquatch heel. In any event it takes a fair-sized grizzly to make an overlapping print even 13 inches long, and there are not supposed to be any grizzlies in California.

Another popular explanation for giant tracks is that they were originally smaller and have melted out. That could only apply to tracks in snow, and even then it won't work because melting does not change the length of the stride, while it does destroy all the detail. I have seen huge melted-out tracks that could once have been those of humans or some other smaller creature. A person who was perfectly certain that sasquatches exist might take them for sasquatch tracks—I know that Indians often do—and they might be right, but no researcher would claim anything concerning them.

There is one thing really strange about the "Bigfoot" situation at Bluff Creek, and that is the fact that so many tracks showed up in the one locality. Tracks are generally seen so seldom that one has to assume that if these creatures are real they must be in the habit of avoiding places where their tracks will show. For one individual to get careless about this would perhaps be understandable, but in Bluff Creek two different individuals started doing it almost at the same time, and several others have made contributions on a less frequent scale. I used to think that the explanation was the kind of soil in the Bluff Creek valley combined with the weather conditions there— that there was so little gravel available and so much dry weather that it was not worth while to gravel the logging roads, so that they would always be either dusty or muddy, and would show a track. There is some truth in that compared to the situation in coastal British Columbia, where logging roads are always heavily gravelled, but the conditions in Bluff Creek are not as different as I once thought, and certainly not so unique as to account for so many more tracks being found there than anywhere else. Besides, a lot of the tracks were in the creekbed. There is a partial explanation in the fact that many more man-hours have been spent looking for tracks in the Bluff Creek area than anywhere else, perhaps more than everywhere else put together. But in 1958, 1963 and 1964 nobody had to look for the tracks, they were showing up all over.

Why did they do it then, and why have they stopped since? There are some possible explanations. The new road may have entered an area where "Bigfoot" had always been carelessly leaving his tracks around. He may indeed have died, or moved elsewhere, after 1964, and the maker of the 15-inch tracks may have departed since 1967. Perhaps there are just so many Bigfoot hunters roaming those hills now that the sasquatches have left or learned caution. Whatever the truth about him may be, Bigfoot at Bluff Creek certainly left his mark on North America.

-5-

The Sasquatch Classics

Every field of investigation probably has its classic cases. Certainly the sasquatch investigation does. Dealing with more recent events the list would probably vary for each investigator, depending on what he or she personally has been involved in, but up to the time of the Patterson movie in 1967 I think almost everyone would agree that the main items would be "Jacko", Ape Canyon, Albert Ostman and "Bigfoot".

All of those reports have been the subject of speculation and scrutiny for nearly 20 years except for the Patterson movie, and it has undoubtedly been more thoroughly shaken down than all the others put together. I will deal with it in a later chapter, and I have already dealt with "Bigfoot". As to the other three classics, how do they look now?

"Jacko" has been around by far the longest, and has suffered the most from modern enquiry—although he has also picked up a certain amount of supporting testimony. So long after the event, (or non-event) it will certainly never be possible to settle the status of the story beyond dispute. The full story, published in the Victoria, B. C. *Daily Colonist* on July 4, 1884, ran as follows:

WHAT IS IT?

A Strange Creature Captured Above Yale
A British Columbia Gorilla

(Correspondence of the Colonist)
YALE, B.C., July 3rd, 1882[*sic*]

In the immediate vicinity of No. 4 tunnel, situated some

twenty miles above this village, are bluffs of rock which have hitherto been unsurmountable, but on Monday morning last were successfully scaled by Mr. Onderdonk's employees on the regular train from Lytton. Assisted by Mr. Costerton, the British Columbia Express Company's messenger, and a number of gentlemen from Lytton and points east of that place who, after considerable trouble and perilous climbing, succeeded in capturing a creature which may truly be called half man and half beast. "Jacko" as the creature has been called by his capturers, is something of the gorilla type standing four feet seven inches in height and weighing 127 pounds. He has long, black, strong hair and resembles a human being with one exception, his entire body, excepting his hands, (or paws) and feet are covered with glossy hair about an inch long. His fore arm is much longer than a man's fore arm, and he possesses extraordinary strength, as he will take hold of a stick and break it by wrenching or twisting it, which no man living could break in the same way.

Since his capture he is very reticent, only occasionally uttering a noise which is half bark and half growl. He is, however, becoming daily more attached to his keeper, Mr. George Tilbury, of this place, who proposes shortly starting for London, England, to exhibit him. His favorite food so far is berries, and he drinks fresh milk with evident relish. By advice of Dr. Hannington raw meats have been withheld from Jacko, as the doctor thinks it would have a tendency to make him savage. The mode of his capture was as follows:

Ned Austin, the engineer, on coming in sight of the bluff at the eastern end of the No. 4 tunnel saw what he supposed to be a man lying asleep in close proximity to the track, and as quick as thought blew the signal to apply the brakes. The brakes were instantly applied, and in a few seconds the train was brought to a standstill. At this moment the supposed man sprang up, and uttering a sharp quick bark began to climb the steep bluff. Conductor R. J. Craig and Express Messenger Costerton, followed by the baggageman and brakemen, jumped from the train and knowing they were some twenty minutes ahead of time immediately gave chase. After five minutes of perilous climbing the then supposed demented Indian was corralled on a projecting shelf of rock where he could neither ascend nor descend. The query now was how to capture him alive, which was quickly decided by Mr. Craig, who crawled on his hands and knees until he was about forty feet above the creature. Taking a small piece of loose rock he let it fall and it had the desired effect of rendering poor Jacko incapable of resistance for a time at least.

The bell rope was then brought up and Jacko was now lowered to terra firma. After firmly binding him and placing him in the baggage car "off brakes" was sounded and the train started for Yale. At the station a large crowd who had heard of the capture by telephone from Spuzzum Flat were assembled, each one anxious to have the first look at the monstrosity, but they were disappointed, as Jacko had been taken off at the machine shops and placed in charge of his present keeper.

The question naturally arises, how came the creature where it was first seen by Mr. Austin? From bruises about its head and body, and apparent soreness since its capture, it is supposed that Jacko ventured too near the edge of the bluff, slipped, fell and lay where found until the sound of the rushing train aroused him. Mr. Thos. White and Mr. Gouin, C. E., as well as Mr. Major, who kept a small store about half a mile west of the tunnel during the past two years, have mentioned having seen a curious creature at different points between Camps 13 and 17, but no attention was paid to their remarks as people came to the conclusion that they had either seen a bear or stray Indian dog. Who can unravel the mystery that now surrounds Jacko! Does he belong to a species hitherto unknown in this part of the continent, or is he really what the train men first thought he was, a crazy Indian!

Long before I ever saw that story it had been investigated by Bruce McKelvie, and he told me that the only things he had been able to determine were that every man mentioned in the story was a real person who could have been there, and that the files of the only other newspaper published in British Columbia at that time, at New Westminster, had been destroyed in a fire. Ironically, there had been a paper at Yale itself until a few days before, but the headquarters of railway construction had moved on from Yale to Kamloops, and the editor had moved his paper after it to be where the action was. When he resumed publication in Kamloops he printed very little Yale news and did not mention Jacko, but that is not surprising, since he had been ill and did not publish for many weeks.

Mr. McKelvie lived on Vancouver Island and did his research in the provincial archives at Victoria. He may well have been correct that there were no copies of the New Westminster paper available there. In any event I took his word for it. It was not until 1974 that I learned that the University of British Columbia, in Vancouver, did have microfilm copies. It turned out that there were actually two different papers, but although they were published much closer to Yale than the *Colonist* the first mention either of them made of Jacko was when the *Columbian* reprinted the article from the *Colonist* on July 5. Then on July 9 the *Mainland Guardian* hit the street with the following brief item:

THE "WHAT IS IT"

Is the subject of conversation in town this evening. How the story originated, and by whom, it is hard for one to conjecture. Absurdity is written on the face of it. The fact of the matter is, that no such animal was caught, and how the "Colonist" was duped in such a manner and by such a story, is strange: and stranger still, when the Columbian reproduced it in that paper. The "train" of circumstances connected with the discovery of "Jacko" and the disposal of same was and still is, a mystery.

Rex.

Yale, B. C., July 7, 1884.

Nothing more was said until July 12, when *The Columbian* printed the following:

THE WILD MAN — Last Tuesday it was reported that the wild man, said to have been captured at Yale, had been sent to this city and might be seen at the gaol. A rush of citizens instantly took place, and it is reported that not fewer than 200 impatiently begged admission into the skookum house. The only wild man visible was Mr. Moresby, governor of the gaol, who completely exhausted his patience answering enquiries from the sold visitors.

That is the sum total of the contemporary evidence, and in my estimation it doesn't look good for Jacko. The original story did not really have "absurdity written on the face of it" in anything except that it involved an animal of a type not known to exist. Other newspaper hoaxes I have read (and written) have not been nearly so restrained, but the joker who wrote in the *Colonist* may well have been of a subtler sort. Even for someone having every reason to believe that Jacko could have been a real animal, it is hard to imagine that the *Colonist* and the *Columbian* would have accepted the *Guardian*'s rebuke without reply if the story were factual. Still, Jacko has his defenders.

Some years ago I talked to an old man named August Castle who was said to have been at Yale all his life. He said that he was a small child in 1884 and lived in the Indian section of the community. He did not see Jacko, because it was the white men who had the animal, but he remembered the excitement about it. If Mr. Castle's age was

86

even approximately what it was reported to be at the time of his death he would actually have been a teenager in 1884, and one would think he would not have missed out on seeing the beast just because it was in another part of town, but of course there might have been any number of other reasons why he did not see it, or his age may have been exaggerated. Certainly 80 years is long enough for memory to blur.

There are also two indirect but independent accounts that refer to Jacko as a real animal. One was contained in a letter written to Dr. Grover Krantz at Washington State University in 1970 by Chilco Choate, a game guide at Clinton, B. C., as follows:

Received your letter re my grandfather's experience with the sasquatch which was captured near Yale, B. C. in the last century. The case you mentioned is the same to be sure.

I heard this story from my father (who is now getting on and his memory is getting a little vague most of the time). When he first told us the story many years ago it was still quite clear in his mind and this is how it went.

My grandfather was the B. & B. engineer (buildings & bridges) for the C.P.R. when it was being built west of Revelstoke. There is still a small train stop near Yale that was named after him. (Choate B.C.) After the C.P.R. was built he became a circuit judge for the County Court of Yale, although I don't know just how long he was actually a judge. Anyway, he was there when this "Ape" was brought in and kept at Yale. "Ape" was the word my Dad says his Dad called the captive. The Ape was kept in Yale until the owner loaded it crated onto a train heading east. The story goes that he was taking it to London, Eng. to set up a side show with it and make his fortune. This was the end of the story as nobody heard of either of them again. It was either my Grandfather's opinion or Dad's opinion that the Ape must have died on the trip and was probably disposed of in any way possible. Personally I would imagine it probably died at sea and would have simply been thrown overboard.

The other reference is in a recent letter to me from Mrs. Hilary Foskett, of Ucluelet, B. C.

Did I mention before that my mother, Adela Bastin, was educated in Yale at All Hallows in the West, a school run by Anglican nuns from All Hallows in Ditchingham, England? The school was of course a boarding school and accepted pupils from all over . . .

87

When the stories of the 'Yeti' and Sasquatch appeared in the press, Mother recalled stories of Jacko at Yale. She was probably eight or nine when she started school there and local inhabitants were still talking about the 'wild man', and the good sisters at the school took care in shepherding the pupils from school to chapel or church. In spite of this local 'fear' in her later years at the school, Mother climbed Mt. Leakey behind Yale with a group of local young people. Until well in her eighties she could recall Yale days in detail, but before her death at 93 a year ago, her memory was slipping. The Dr. Harrington referred to was well known to Mother and her sisters . . .

One objection that has been raised to the Jacko story is based on a misprint—that Mr. Gouin could not have been a C.B.E. because the Order of the British Empire did not exist in 1884. I don't know about that, but in the original story Mr. Gouin is only a C.E., presumably a civil engineer.

Jacko could not have been taken east from Yale by train, since the rails of the C.P.R. were still about 100 miles short of connecting in eastern B. C. at that period, but could have been taken west by train to go east, heading for the coast to connect by boat with the Northern Pacific in the United States, which had been completed the year before. Grover Krantz has pointed out that P. T. Barnum first exhibited "Jo-Jo the Dog-Faced Boy" in 1884, and that the posters used in that year describe a creature that could have been Jacko, but there are no photos. The photos of the 1885 version show a hairy-faced man who, Dr. Krantz says, does not fit the descriptions of the year before. He points out that Barnum would have paid more money for Jacko than anyone in England, and that if the animal had died before the 1885 season his death could have been concealed to avoid losing a valuable drawing card.

There is no way to reconcile all those bits of information. Whether there was an animal or not, some of them have to be wrong. Bearing in mind the vital interest in the whole question of man and ape at that period in history, I don't find it believable that there could have been that much attention attracted and then nothing further written about it, anywhere, if there really was an animal. Had it escaped immediately that would be different, but none of the stories suggest that. However there is no way to be certain of anything at this late date.

The Albert Ostman and Ape Canyon stories both are dated 1924, but only the latter came to public attention at that time. The story broke in the Portland *Oregonian*, July 13, 1924, as follows:

FIGHT WITH BIG APES
REPORTED BY MINERS

FABLED BEASTS ARE SAID TO
HAVE BOMBARDED CABIN

One of Animals, Said to Appear
Like Huge Gorilla, Is
Killed by Party

KELSO, Wash., July 12 — (Special) The strangest story to come from the Cascade mountains was brought to Kelso today by Marion Smith, his son Roy Smith, Fred Beck, Gabe Lefever and John Peterson, who encountered the fabled "mountain devils" or mountain gorillas of Mount St. Helens this week, shooting one of them and being attacked throughout the night by rock bombardments of the beasts.

The men had been prospecting a claim on the Muddy, a branch of the Lewis River about eight miles from Spirit Lake, 46 miles from Castle Rock. They declared that they saw four of the huge animals, which were about 400 pounds and walked erect. Smith and his companions declared that they had seen the tracks of the animals several times in the last six years and Indians have told of the "mountain devils" for 60 years, But none of the animals ever has been seen before.

Smith met with one of the animals and fired at it with a revolver, he said. Thursday Fred Beck, it is said, shot one, the body falling over a precipice. That night the animals bombarded the cabin where the men were stopping with showers of rocks, many of them large ones, knocking chunks out of the log cabin, according to the prospectors. Many of the rocks fell through a hole in the roof and two of the rocks struck Beck, one of them rendering him unconscious for nearly two hours.

The animals were said to have the appearance of huge gorillas. They are covered with long, black hair. Their ears are about four inches long and stick straight up. They have four toes, short and stubby. The tracks are 13 to 14 inches long. These tracks have been seen by forest rangers and prospectors for years.

The prospectors built a new cabin this year and it is believed it is close to the cave thought to be occupied by the animals. Mr. Smith believes he knows the location of the cave.

On the 14th the paper re-affirmed that rangers at the Spirit Lake ranger station had seen queer four-toed footprints too, and that trappers and other prospectors had reported giant footprints in recent years. It explained that the rocks which entered the cabin, supposedly 200 in all, came through a hole in the roof used in lieu of a chimney.

The following day Ranger Huffman from Spirit Lake was quoted pouring cold water on the story. He said that a party of 70 from the Y.M.C.A. had been camped within a mile of the scene at the time of the alleged attack and had noticed nothing. There was also a suggestion that boys from the camp had carried out the attack as a prank.

For the next two days the *Oregonian* reprinted stories by the Indian editor of the *Real American*, at Hoquiam, Washington, regarding the Seeahtic tribe of giant Indians, over seven feet tall with bodies hairy like those of bears. On July 19 and 20 the paper carried a long report by L. H. Gregory on an expedition to what is now known as Ape Canyon. He reported that the tracks were made with human knuckles and every one represented a right foot. Ranger Huffman was quoted as enlarging on his unfavorable comments, but adding:

> Old Man Smith, who started this ape stampede, absolutely believes it. If ever a man was "wild-eyed" it was Smith when he came down here from the cabin with the story of having been attacked by apes. Something happened up there, but I can't imagine what though. It wasn't apes. Another funny thing is that you can't shake the stories of the other men with Smith. Oh there's a mystery about it. The mystery to me is who put up the job on Smith and his companions and how in the world they did it.

It turned out that Ranger Huffman did not blame the boys from the Y.M.C.A. Camp after all. The camp was at Spirit Lake, a long way from the mine:

> They were all in camp that night and the day these fellows claim to have had their fight. Wait 'till you see the cabin.
> Boys could never have pulled the job.

Mr. Gregory's story dwells at length on the extreme difficulty of the climb down into the canyon to the cabin. It also makes clear that many parties of ape hunters were running all over the area before his group arrived. The rocks in the cabin, he reports, were found to be fresh-broken ones, presumably from the mine, rather than from the loose rocks to be found on the hillside near the cabin.

I have never found any indication that anyone tried to investigate the Ape Canyon incident after the initial "Gee whiz" and debunking reports, until the mid 1960's. By that time the only one of the miners left who would talk about the incident was Fred Beck. I spent only one evening with him, but I have heard tapes of interviews by other people. He had apparently told the story often enough so that he had a set pattern of things to say, and there was not much to be gained by further questions. To my understanding there was a difficulty in fitting all the elements of his story in the correct order, and I was not able to clear that up. He simply repeated what he had said before. However I was talking to him 40 years after the event.

As with all the old accounts, there is no prospect of being able to establish now the truth or otherwise of the basic story. One thing to bear in mind is that the important sightings took place in the daytime, not during the attack on the cabin. That being so, the Y.M.C.A. camp explanation would be irrelevent even if it had not been challenged. As a matter of fact it would simplify matters to be able to blame the night's events on some human agency, as the behavior described is unique. There is no other report of sasquatches engaging in any elaborate form of combined activity, let alone a concerted attack on a human habitation.

The comments about the footprints cannot be assessed because of a lack of information. It does not appear that any of the original party were on hand to confirm that the prints seen by L. H. Gregory were those attributed to the apes, and with ape hunters all over the place there would be a great liklihood both that the original footprints would have been trampled on and that attempts to imitate prints would have been made—not necessarily with any intention to deceive.

To give some of the original flavor, here is a transcript of an interview Roger Patterson recorded with Fred Beck in 1966:

> We was mining in there working our claims about two years before we had the contact with them. We seen tracks down on the Muddy. We didn't know what they were, thought that probably some big Indians in there fishing, barefooted. But we heard rumours people been up there. One fellow was telling about fishing, had a string of fish and laid them on the bank and when he looked around there was a great big, like a man, hairy, great big fella, had his fingers and was just smashing them along the rocks. He come out of there faster than he went in.
>
> And nobody believed him. But mind you my father-in-law that was with me, he believed him. He said that fella didn't lie, 'cause he knew him. He said that fella was scared when he came out of there, 'cause he wasn't no human.
>
> Then when we seen them tracks there we never thought so much about it. My father-in-law was telling about these people he knew, his experience years ago seeing them tracks

up there and one thing and another. And the Indians—there was lots of Indians around here them days—them old Indians was afraid to go back in there to Spirit Lake 'cause they was all afraid of the spirit.

When we went up there why we never expected nothing like that. We heard noises there, whistling, noise like a pounding on their chest.

Q — When was the first time that you actually seen one?

A — Well about two years after that we were seeing them. There was a spring a little way from our mining cabin. We had to go down into a gully to get water, and there's a ridge. The old man always carried his rifle with him when he went to the spring, because he said he seen tracks, awful suspicious tracks, he always carried his rifle. We went down, I went with him, went down to get some water, and we looked up on a ridge there, oh, about 100 yards away and we seen this peeking out of this tree, and I seen him first and I said, "Marion there's a, look at that on that tree."

He looked there and he just up and bang, bang, bang, I see the bark fly, and the bullets were all in just the place, just like that (gesturing?) right around the bark of the tree, just you know how . . .

Q — Yeah.

A — So we seen him running down this ridge then, and then he took a couple more shots at him. Marion when he first shot I rushed over there, it was hard going, he said:

"Don't run, don't run, Fred, don't run," he said, "he won't go far," he said, "I put three shots through that fool's head, he won't go far."

So we got up the ridge and looked down there he was goin', just jumpin', looked like it'd be twelve, fourteen feet a jump, runnin'. The old man took a couple more shots at him and the old man said:

"My God, I don't understand it, I don't understand it, how that fella can get away with them slugs in his head," he says. "I hit him with the other two shots, too."

Well he got away alright.

Q — Well now the tracks that you seen before, you would estimate at around how big?

A — Well, say 19 inches they measured. You see the two detectives up with us from Portland and they measured the tracks. It rained that night. We stayed at the lake. A regular cloudburst. We went up there and we had where we washed out dishes, right under the little seep hole, you know, springs,

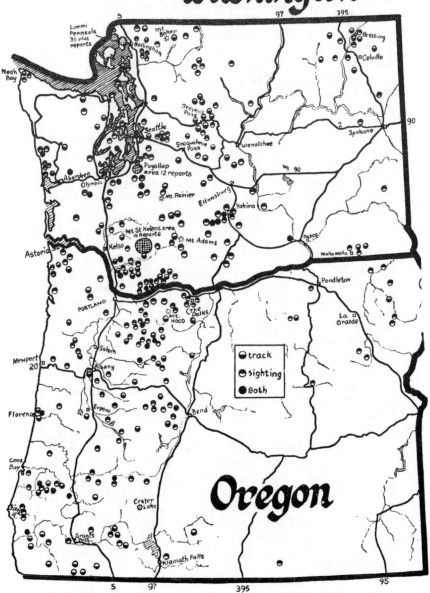

Washington

Lummi
Peninsula
30 plus
reports

Mt Baker

Bellingham

Bessburg

Colville

Neah
Bay

Stevens
Pass

Seattle

Snoqualmie
Pass

Wenatchee

Spokane

Aberdeen

Olympia

Puyallup
area 12 reports

Mt. Rainier

Ellensburg

Yakima

Mt St Helens area
16 Reports

Kelso

Mt Adams

Pasco

Walla Walla

Astoria

Pendleton

PORTLAND

Mt.
HOOD

The
Dalles

La
Grande

Salem

● track
◑ sighting
● Both

Newport
20

Albany

Florence

Eugene

Bend

Coos
Bay

Oregon

Crater
Lake

Grants
Pass

Klamath Falls

you might call them springs, melted snow, kind of a little gully there. We had cleaned our pots out, our beans and rice. It quit raining before we got there. When we went up there, there were tracks made since it rained, fresh tracks, about 19 inches long. And the detective said, "Well, these are tracks all right," but, he said, "who made them?"

I said, "Well, who did make them?" I said, "We was all together, and they've been made since the rain."

And they talked about it and they said it might have been a coon making them, or part human—all that stuff—a bear part human!

Q — When you finally did see . . .

A — When we seen 'em, you know, why we heard that noise—pounding and whistling, at night they come in there and we had a pile of shakes piled up there, big shakes. Our cabin was built out of logs. We didn't have rafters on it, we had good-sized pine logs, you know, for rafters, two-inch shakes, pine shakes. We had them rafters close apart, they was about a foot apart, 'cause he said he wanted to make a roof what'd hold the snow. We made one to hold the snow.

Them buggers attacked us, knocked the chinking out on my dad's, on my father-in-law's chest, and had an axe there, he grabbed the axe. And the old man grabbed the axe and turned it so it wouldn't go through the chinking hole between the logs and then he shot on it, right along the axe handle, and he let go of it. And then the fun started!

Well! I wanta tell you, pretty near all night long they were on that house, trying to get in, you know. We kept a shootin'. Get up on the house we'd shoot up through the ceiling at them. Couldn't see them up there, you could hear them up there. My God, they made a noise. Sounded like a bunch of horses were running around there.

Next day, we'd find tracks, anywhere there was any sand on the rocks, we found tracks of them.

Q — Were those the same tracks that you'd seen?

A — Yes, the same tracks, them tracks, but never measured them. They were big fellas. That was the only track we measured, was them the detectives measured.

Q — Now, did you explain to us a while back that there was at one time that you had shot one, when did this . .

A — Well, that was the next morning, I guess it was, if I remember that.

Q — After the attack?

A — Yeah, I was, we was going to go down to the tunnel

to get some tools out that we had, drills and things like that in the tunnel, we was going to clean out and go home. When we went down I took a rifle with me. We all carried rifles after that happened.

I, down the ridge there a couple of hundred yards, no it wasn't that far, why there was one of them fellas run out of a clump of brush and run down the gorge, and I shot him in the back, three shots, and I could hear the bullets hit him and I see the fur fly on his back. I shot for his heart. And he stopped and he just fell right over a precipice, and I heard him go doonk, zoop down into the canyon.

Q — You said he fell into Ape Canyon.

A — Yeah, and the sun come out in the afternoon, that water was really a torrent goes down there, it'd wash anything out fall in there. And that's the reason I don't know if they're human or not, cause I couldn't kill 'em. And I hit.

Q — Well how would you describe, Mr. Beck, as far as what they look like in their body and their head?

A — Well, they was tall, I dunno, they looked to me like they was eight foot tall, maybe taller, and they was built like a man, little in the waist, and big shoulders on, and chest, and their necks was kinda what they call bull necks, you know how they are.

Q — No neck at all, hardly.

A — That's it, and then their ears, turns out like ours do, and so big, you know, and hair all over, you couldn't tell nothin' about 'em.

Q — Did they have hair on their face, or could you, did you ever . . .

A — No, let's see, I don't believe they did . . . I believe they did have hair on their face.

Q — But not as much as . . .

A — No (cough) can't have whiskers.

Q — Sure. How about their nose?

A — I couldn't of but I was, uh, they seemed to have a kind of pug nose, flat nose, kind of flat.

Q — And their eyes?

A — All I can know is, we were excited, you know, you don't see very good detail when you're excited, but I know one thing, that they was no human.

Q — They did, though, walk upright. Did you ever see . .

A — I never seen one on the four.

Q — Their arms, probably, was they . . .

A — Arms down below the hips, long, I figured . . .

Q — Below their knees.

A — Yeah, their knees. Long arms. And big arms.

Q — What would you estimate maybe for weight?

A — Pretty heavy. I'd say they're six or eight hundred pounds. Like you know, estimated. And maybe more, I don't know. I couldn't tell ya the weight of them. The way they sunk down in the ground I'd have some idea about it. I'd say they weighed eight, nine hundred pounds, or more, 'cause it made a deep imprint in the dirt. There's so much rock up there, you know, only can see them in places where there's sand, you know.

Q — Sure. Well now after you had the attack, what happened then, the next morning?

A — Well, we come out, out of there. Come down, and my father-in-law he was so excited and scared. I told him, he promised never to tell nobody, 'cause I said it wouldn't do, people wouldn't believe it, don't tell anybody. He says, "I won't, I won't," but he did. Went down to the lake and the rangers down there knew him. He was so excited they found, took him in the other room and talked to him and he acknowledged what the trouble was. They said they believed him, because the old man had been a hunter, they knew him. All his life . . .hunting until no little thing would ever scare him, no animal or anything like that. Then he went to Kelso and told some of his friends down there. Then the newspaper reporter give us a merry time, day and night.

Q — Had they ever heard of anything, anybody before this?

A — I don't think so. They really didn't believe it.

Q — I see. Well, we sure thank you for this interview.

In my own discussion with Fred Beck, he explained more fully the incident of the chinking and the axe. The spaces between the logs in the cabin were wide, and large strips of split saplings were wedged in to close the holes. One of these was knocked out above the bunk where Marion Smith was lying and it hit him on the chest. Then a hairy hand and arm came through the hole, groped around, and got hold of an axe by the handle. As the axe was disappearing through the gap in the log wall, Smith managed to turn the head at right angles to the crack so that it could not be pulled through. Then he grabbed a gun and shot down the line of the handle, at which the axe was released.

As far as I could determine, that hairy arm was the only part of any of the attackers that the men saw that night. Their cabin had no windows. There must have been at least two creatures outside, as there was one heard on the roof at the same time that one was heard pounding on a wall. The impression is that there were more than two, but that could not be definitely established. Mr. Beck said nothing to me about a barrage of rocks coming into the cabin. He did mention hearing rocks land on the roof and roll off. At that time I had not seen the newspaper reports and did not know that he was reported to have been knocked out by a rock. What he did say was that at the height of the thumping on the roof one of the men was sitting in a chair rocking himself back and forth and calling out in singsong fashion:

You leave us alone, and we'll leave you alone; and we'll all go home in the morning.

Did all this really happen? I think so. To the people at that time and place, knowing nothing of such creatures except the old legends of mountain devils at Mount St. Helens, the miners' story was not believable. However if such creatures do exist, then certainly the most acceptable explanation for the miners having claimed to have seen them is that they did see them. There isn't a shadow of a suggestion as to why they would make up such a story and keep telling it all their lives.

If the story is true, then it has to be accepted that on this one occasion an attempt to shoot a sasquatch did result in a form of retaliation, and by more than one of the animals, although it is worth noting that no one was hurt. That would not mean that every detail of every version of the story would be true, of course.

In some ways this is the best of the old stories. There is no doubt that it involved real people and that it was told at the time when it is supposed to have happened. With Albert Ostman's story, from that same summer of 1924, there is no proof that he ever told it before 1957, at a time when the media were making a great fuss over the Harrison Hot Springs centennial sasquatch hunt. Its great strength is in the wealth of anatomical detail it contains that has since been confirmed over and over, but for which there was no known source at the time the story was first told.

As a matter of fact, it is not strictly true to say that the Ostman story was ever told at all. It was written. Although he was questioned about it repeatedly over a period of years, I don't know of any occasion on which Albert Ostman went through the whole story as a narrative except when he wrote it out in longhand in a scribbler. He did this after writing a letter to William Roe, whose story he had seen in a newspaper. Roe told him he was bringing a reporter to interview him, and in an attempt to bring back as best he could the recollection of events already 33 years in the past, Ostman gathered all the things

Two men with stories to tell, Fred Beck, top, with the gun he used at Ape Canyon, and Albert Ostman, talking to the author.

he could that reminded him of that period of his life, even finding a shopping list he had written out before one of his prospecting trips. He then tried to rebuild the atmosphere by writing down a complete account of the trip on which he encountered the sasquatch. When he was later asked to swear to the accuracy of the written account he stressed that he could do so only regarding the part directly concerning the sasquatch, some of the rest had been patched together to take him back into the past and might not be related to that particular trip.

I first heard of Albert Ostman through a newspaper story, and I thought he was just telling a wild yarn. Probably I would never have gone to see him, had I not met a radio newsman who had interviewed him, and who insisted that if Ostman was lying then he couldn't be sure he had ever interviewed anyone who was telling the truth. When I did go to see him he gave me the scribbler with the story in it. It is quite long, even in edited form, but it is certainly worth reading once, and anyone who has seen it before can skip it. Here is what he wrote.

I have always followed logging and construction work. This time I had worked over one year on a construction job, and thought a good vacation was in order. B. C. is famous for lost gold mines. One is supposed to be at the head of Toba Inlet—why not look for this mine and have a vacation at the same time? I took the Union Steamship boat to Lund, B.C. From there I hired an old Indian to take me to the head of Toba Inlet.

This old Indian was a very talkative old gentleman. He told me stories about gold brought out by a white man from this lost mine. This white man was a very heavy drinker— spent his money freely in saloons. But he had no trouble in getting more money. He would be away a few days, then come back with a bag of gold. But one time he went to his mine and never came back. Some people said a Sasquatch had killed him.

At that time I had never heard of Sasquatch. So I asked what kind of an animal he called a Sasquatch. The Indian said, "They have hair all over their bodies, but they are not animals. They are people. Big people living in the mountains. My uncle saw the tracks of one that were two feet long. One old Indian saw one over eight feet tall."

I told the Indian I didn't believe in their old fables about mountain giants. It might have been some thousands of years ago, but not nowadays.

The Indian said: "There may not be many, but they still exist."

We arrived at the head of the inlet about 4:00 p.m. I made camp at the mouth of a creek . . .The Indian had supper with me, and I told him to look out for me in about three weeks. I would be camping at the same spot when I came back . . .

Next morning I took my rifle with me, but left my equipment at the camp. I decided to look around for some deer trail to lead me up into the mountains. On the way up the inlet I had seen a pass in the mountain that I wanted to go through, to see what was on the other side.

I spent most of the forenoon looking for a trail but found none, except for a hogback running down to the beach. So I swamped out a trail from there, got back to my camp about 3:00 p.m. that afternoon, and made up my pack to be ready in the morning. My equipment consisted of one 30-30 Winchester rifle, I had a special home-made prospecting pick, axe on one end, pick on the other. I had a leather case for this pick which fastened to my belt, also my sheath knife.

The storekeeper at Lund was co-operative. He gave me some cans for my sugar, salt and matches to keep them dry. My grub consisted mostly of canned stuff, except for a side of bacon, a bag of beans, four pounds of prunes and six packets of macaroni, cheese, three pounds of pancake flour and six packets of Rye King hard tack, three rolls of snuff, one quart sealer of butter and two one-pound cans of milk. I had two boxes of shells for my rifle.

The storekeeper gave me a biscuit tin. I put a few things in that and cached it under a windfall, so I would have it when I came back here waiting for a boat to bring me out. My sleeping bag I rolled up and tied on top of my pack sack, together with my ground sheet, small frying pan, and one aluminum pot that held about a gallon. As my canned food was used, I would get plenty of empty cans to cook with.

The following morning I had an early breakfast, made up my pack, and started out up this hogback. My pack must have been at least eighty pounds, besides my rifle. After one hour, I had to rest. I kept resting and climbing all that morning. About 2:00 p.m. I came to a flat place below a rock bluff. There was a bunch of willow in one place. I made a wooden spade and started digging for water. About a foot down I got seepings of water, so I decided to camp here for the night, and scout around for the best way to get on from here.

I must have been up to near a thousand feet. There was a most beautiful view over the islands and the Strait—tugboats with log booms, and fishing boats going in all directions. A lovely spot. I spent the following day prospecting round. But no sign of minerals. I found a deer trail leading towards this pass that I had seen on my way up the inlet.

The following morning I started out early, while it was cool. It was steep climbing with my heavy pack. After a three hours climb, I was tired and stopped to rest. On the other side of a ravine from where I was resting was a yellow spot below some small trees. I moved over there and started digging for water.

I found a small spring and made a small trough from cedar bark and got a small amount of water, had my lunch and rested here 'till evening . . . I made it over the pass late that night.

Now I had downhill and good going, but I was hungry and tired, so I camped at the first bunch of trees I came to . . . I was trying to size up the terrain—what direction I would take from here. Towards west would lead to low land and some other inlet, so I decided to go in a northeast direction . . . had good going and slight down hill all day. I must have made 10 miles when I came to a small spring and a big black hemlock tree.

This was a lovely campsite, I spent two days here just resting and prospecting. The first night here I shot a small deer . . .

(Two days later) . . . I found an exceptionally good campsite. It was two good-sized cypress trees growing close together and near a rock wall with a nice spring just below these trees. I intended to make this my permanent camp. I cut lots of brush for my bed between these trees. I rigged up a pole from this rock wall to hang my packsack on, and I arranged some flat rocks for my fireplace for cooking. I had a really classy setup . . .

And that is when things began to happen.

I am a heavy sleeper, not much disturbs me after I go to sleep, especially on a good bed like I had now.

Next morning I noticed things had been disturbed during the night. But nothing missing I could see. I roasted my grouse on a stick for breakfast . . .

That night I filled up the magazine of my rifle. I still had

101

one full box of 20 shells and six shells in my coat pocket. That night I laid my rifle under the edge of my sleeping bag. I thought a porcupine had visited me the night before and porkies like leather, so I put my shoes in the bottom of my sleeping bag.

Next morning my pack sack had been emptied out. Some one had turned the sack upside down. It was still hanging on the pole from the shoulder straps as I had hung it up. Then I noticed one half-pound package of prunes was missing. Also my pancake flour was missing, but my salt bag was not touched. Porkies always look for salt, so I decided it must be something else than porkies. I looked for tracks but found none. I did not think it was a bear, they always tear up and make a mess of things. I kept close to camp these days in case this visitor would come back.

I climbed up on a big rock where I had a good view of the camp, but nothing showed up. I was hoping it would be a porky, so I would get a good porky stew. These visits had now been going on for three nights . . .

This night it was cloudy and looked like it might rain. I took special notice of how everything was arranged. I closed my pack sack, I did not undress, I only took off my shoes, put them in the bottom of my sleeping bag. I drove my prospecting pick into one of the cypress trees so I could reach it from my bed. I also put the rifle alongside me, inside my sleeping bag. I fully intended to stay awake all night to find out who my visitor was, but I must have fallen asleep.

I was awakened by something picking me up. I was half asleep and at first I did not remember where I was. As I began to get my wits together, I remembered I was on this prospecting trip, and in my sleeping bag.

My first thought was—it must be a snow slide, but there was no snow around my camp. Then it felt like I was tossed on horseback, but I could feel whoever it was, was walking.

I tried to reason out what kind of animal this could be. I tried to get at my sheath knife, and cut my way out, but I was in an almost sitting position, and the knife was under me. I could not get hold of it, but the rifle was in front of me, I had a good hold of that, and had no intention to let go of it. At times I could feel my packsack touching me, and could feel the cans in the sack touching my back.

After what seemed like an hour, I could feel we were going up a steep hill. I could feel myself rise for every step. What was carrying me was breathing hard and sometimes gave a slight cough. Now, I knew this must be one of the mountain Sasquatch giants the Indian told me about.

I was in a very uncomfortable position—unable to move. I was sitting on my feet, and one of the boots in the bottom of the bag was crossways with the hobnail sole up across my foot. It hurt me terribly, but I could not move.

It was very hot inside. It was lucky for me this fellow's hand was not big enough to close up the whole bag when he picked me up—there was a small opening at the top, otherwise I would have choked to death.

Now he was going downhill. I could feel myself touching the ground at times and at one time he dragged me behind him and I could feel he was below me. Then he seemed to get on level ground and was going at a trot for a long time. By this time, I had cramps in my legs, the pain was terrible. I was wishing he would get to his destination soon. I could not stand this type of transportation much longer.

Now he was going uphill again. It did not hurt me so bad. I tried to estimate distance and directions. As near as I could guess we were about three hours travelling. I had no idea when he started as I was asleep when he picked me up.

Finally he stopped and let me down. Then he dropped my packsack, I could hear the cans rattle. Then I heard chatter—some kind of talk I did not understand. The ground was sloping so when he let go of my sleeping bag, I rolled downhill. I got my head out, and got some air. I tried to straighten my legs and crawl out, but my legs were numb.

It was still dark, I could not see what my captors looked like. I tried to massage my legs to get some life in them, and get my shoes on. I could hear now it was at least four of them. they were standing around me, and continuously chattering. I had never heard of Sasquatch before the Indian told me about them. But I knew I was right among them.

But how to get away from them, that was another question. I got to see the outline of them now, as it began to get lighter, though the sky was cloudy, and it looked like rain, in fact there was a slight sprinkle.

I now had circulation in my legs, but my left foot was very sore on top where it had been resting on my hobnail boots. I got my boots out from the sleeping bag and tried to stand up. I found that I was wobbly on my feet, but I had a good hold of my rifle.

I asked, "What you fellows want with me?"

Only some more chatter.

It was getting lighter now, and I could see them quite clearly. I could make out forms of four people. Two big and two little ones. They were all covered with hair and no clothes on at all.

I could now make out mountains all around me. I looked at my watch. It was 4:25 a.m. It was getting lighter now and I could see the people clearly.

They look like a family, old man, old lady and two young ones, a boy and a girl. The boy and the girl seem to be scared of me. The old lady did not seem too pleased about what the old man dragged home. But the old man was waving his arms and telling them all what he had in mind. They all left me then.

I had my compass and my prospecting glass on strings around my neck. The compass in my lefthand shirt pocket and my glass in my right hand pocket. I tried to reason our location, and where I was. I could see now that I was in a small valley or basin about eight or ten acres, surrounded by high mountains, on the southeast side there was a V-shaped opening about eight feet wide at the bottom and about twenty feet high at the highest point—that must be the way I came in. But how will I get out? The old man was now sitting near this opening.

I moved my belongings up close to the west wall. There were two small cypress trees there, and this will do for a shelter for the time being. Until I find out what these people want with me, and how to get away from here. I emptied out my packsack to see what I had left in the line of food. All my canned meat and vegetables were intact and I had one can of coffee. Also three small cans of milk—two packages of Rye King hard tack and my butter sealer half full of butter. But my prunes and macaroni were missing. Also my full box of shells for my rifle. I had my sheath knife but my prospecting pick was missing and my can of matches. I only had my safety box full and that held only about a dozen matches. That did not worry me—I can always start a fire with my prospecting glass when the sun is shining, if I got dry wood. I wanted hot coffee, but I had no wood, also nothing around here that looked like wood. I had a good look over the valley from where I was—but the boy and girl were always watching me from behind some juniper bush. I decided there must be some water around here. The ground was leaning towards the opening in the wall. There must be water at the upper end of this valley, there is green grass and moss along the bottom.

All my utensils were left behind. I opened my coffee tin and emptied the coffee in a dishtowel and tied it with the metal strip from the can. I took my rifle and the can and went looking for water. Right at the head under a cliff there was a lovely spring that disappeared underground. I got a drink, and a full can of water. When I got back the young boy was looking over my belongings, but did not touch anything. On my way back I noticed where these people were sleeping. On the east side wall of this valley was a shelf in the mountain side, with overhanging rock, looking something like a big undercut in a big tree about 10 feet deep and 30 feet wide. The floor was covered with lots of dry moss, and they had some kind of blankets woven of narrow strips of cedar bark, packed with dry moss. They looked very practical and warm—with no need of washing.

The first day not much happened. I had to eat my food cold. The young fellow was coming nearer me, and seemed curious about me. My one snuff box was empty, so I rolled it toward him. When he saw it coming, he sprang up quick as a cat, and grabbed it. He went over to his sister and showed her. They found out how to open and close it—they spent a long time playing with it—then he trotted over to the old man and showed him. They had a long chatter.

Next morning, I made up my mind to leave this place—if I had to shoot my way out. I could not stay much longer, I had only enough grub to last me till I got back to Toba Inlet. I did not know the direction but I would go down hill and I would come out near civilization some place. I rolled up my sleeping bag, put that inside my pack sack—packed the few cans I had—swung the sack on my back, injected the shell in the barrel of my rifle and started for the opening in the wall. The old man got up, held up his hands as though he would push me back.

I pointed to the opening. I wanted to go out. But he stood there pushing towards me—and said something that sounded like "Soka, soka." I backed up to about sixty feet. I did not want to be too close, I thought, if I had to shoot my way out. A 30-30 might not have much effect on this fellow, it might make him mad. I only had six shells so I decided to wait. There must be a better way than killing him, in order to get out from here. I went back to my campsite to figure out some other way to get out.

I could make friends with the young fellow or the girl, they might help me. If I only could talk to them. Then I thought of a fellow who saved himself from a mad bull by blinding him with snuff in his eyes. But how will I get near enough to this fellow to put snuff in his eyes? So I decided next time I give the young fellow my snuff box to leave a few grains of snuff in it. He might give the old man a taste of it.

But the question is, in what direction will I go, if I should get out? I must have been near 25 miles northeast of Toba Inlet when I was kidnapped. This fellow must have travelled at least 25 miles in the three hours he carried me. If he went west we would be near salt water—same thing if he went south—therefore he must have gone northeast. If I then keep going south and over two mountains, I must hit salt water someplace between Lund and Vancouver.

The following day I did not see the old lady till about 4:00 p.m. She came home with her arms full of grass and twigs and of all kinds of spruce and hemlock as well as some kind of nuts that grow in the ground. I have seen lots of them on Vancouver Island. The young fellow went up the mountain to the east every day, he could climb better than a mountain goat. He picked some kind of grass with long sweet roots. He gave me some one day—they tasted very sweet. I gave him another snuff box with about a teaspoon of snuff in it. He tasted it, then went to the old man—he licked it with his tongue. They had a long chat. I made a dipper from a milk can. I made many dippers—you can use them for pots too—you cut two slits near the top of any can—then cut a limb from any small tree — cut down back of the limb — down the stem of the tree — then taper the part you cut from the stem. Then cut a hole in the tapered part, slide the tapered part in the slit you have made in the can, and you have a good handle on your can. I threw one over to the young fellow, that was playing near my camp, he picked it up and looked at it then he went to the old man and showed it to him. They had a long chatter. Then he came to me, pointed at the dipper then at his sister. I could see that he wanted one for her too. I had other peas and carrots, so I made one for his sister. He was standing only eight feet away from me. When I had made the dipper, I dipped it in water and drank from it, he was very pleased, almost smiled at me. Then I took a chew of snuff, smacked my lips, said that's good.

The young fellow pointed to the old man, said something

106

that sounded like "Ook". I got the idea that the old man liked snuff, and the young fellow wanted a box for the old man. I shook my head. I motioned with my hands for the old man to come to me. I do not think the young fellow understood what I meant. He went to his sister and gave her the dipper I made for her. They did not come near me again that day. I had now been here six days, but I was sure I was making progress. If only I could get the old man to come over to me, get him to eat a full box of snuff that would kill him for sure, and that way kill himself, I wouldn't be guilty of murder.

The old lady was a meek old thing. The young fellow was by this time quite friendly. The girl would not hurt anybody. Her chest was flat like a boy's—no development like young ladies. I am sure if I could get the old man out of the way I could easily have brought this girl out with me to civilization. But what good would that have been? I would have to keep her in a cage for public display. I don't think we have any right to force our way of life on other people, and I don't think they would like it. (The noise and racket in a modern city they would not like any more than I do.)

The young fellow might have been between 11-18 years old and about seven feet tall and might weight about 300 lbs. His chest would be 50-55 inches, his waist about 36-38 inches. He had wide jaws, narrow forehead, that slanted upward round at the back about four or five inches higher than the forehead. The hair on their heads was about six inches long. The hair on the rest of their body was short and thick in places. The women's hair on the forehead had an upward turn like some women have—they call it bangs, among women's hair-do's. Nowadays the old lady could have been anything between 40-70 years old. She was over seven feet tall. She would be about 500-600 pounds.

She had very wide hips, and a goose-like walk. She was not built for beauty or speed. Some of those lovable brassieres and uplifts would have been a great improvement on her looks and her figure. The man's eyeteeth were longer than the rest of the teeth, but not long enough to be called tusks. The old man must have been near eight feet tall. Big barrel chest and big hump on his back—powerful shoulders, his biceps on upper arm were enormous and tapered down to his elbows. His forearms were longer than common people have, but well proportioned. His hands were wide, the palm was long and broad, and hollow like a scoop. His fingers were short in proportion to the rest of his hand. His fingernails were like chisels. The only place they had no hair was inside their hands and the soles of their feet and upper part of the nose and eyelids. I never did see their ears, they were covered with hair hanging over them.

If the old man were to wear a collar it would have to be at least 30 inches. I have no idea what size shoes they would need. I was watching the young fellow's foot one day when he was sitting down. The soles of his feet seemed to be padded like a dog's foot, and the big toe was longer than the rest and very strong. In mountain climbing all he needed was footing for his big toe. They were very agile. To sit down they turned their knees out and came straight down. To rise they came straight up without help of hands or arms. I don't think this valley was their permanent home. I think they move from place to place, as food is available in different localities. They might eat meat, but I never saw them eat meat, or do any cooking.

I think this was probably a stopover place and the plants with sweet roots on the mountain side might have been in season this time of the year. They seem to be most interested in them. The roots have a very sweet and satisfying taste. They always seem to do everything for a reason, wasted no time on anything they did not need. When they were not looking for food, the old man and the old lady were resting, but the boy and the girl were always climbing something or some other exercise. A favorite position was to take hold of his feet with his hands and balance on his rump, then bounce forward. The idea seems to be to see how far he could go without his feet or hands touching the ground. Sometimes he made 20 feet.

But what do they want with me? They must understand I cannot stay here indefinitely. I will soon have to make a break for freedom. Not that I was mistreated in any way. One consolation was that the old man was coming closer each day, and was very interested in my snuff. Watching me when I take a pinch of snuff. He seems to think it useless to only put it inside my lips. One morning after I had my breakfast both the old man and the boy came and sat down only ten feet away from me. This morning I made coffee. I had saved up all dry branches I found and I had some dry moss and I used all the labels from cans to start a fire.

I got my coffee pot boiling and it was strong coffee too, and the aroma from boiling coffee was what brought them over. I was sitting eating hard tack with plenty of butter on, and sipping coffee. And it sure tasted good. I was smacking my lips pretending it was better than it really was. I set the can down that was about half full. I intended to warm it up later. I pulled out a full box of snuff, took a big chew. Before I had time to close the box the old man reached for it. I was afraid he would waste it, and only had two more boxes. So I held on to the box intending him to take a pinch like I had just done. Instead he grabbed the box and emptied it in his mouth. Swallowed it in one gulp. Then he licked the box inside with his tongue.

After a few minutes his eyes began to roll over in his head, he was looking straight up. I could see he was sick. Then he grabbed my coffee can that was quite cold by this time, he emptied that in his mouth, grounds and all. That did no good. He stuck his head between his legs and rolled forwards a few times away from me. Then he began to squeal like a stuck pig. I grabbed my rifle. I said to myself, "This is it. If he comes for me I will shoot him plumb between his eyes." But he started for the spring, he wanted water. I packed my sleeping bag in my pack sack with the few cans I had left. The young fellow ran over to his mother. Then she began to squeal. I started for the opening in the wall—and I just made it. The old lady was right behind me. I fired one shot at the rock over her head.

I guess she had never seen a rifle fired before. She turned and ran inside the wall. I injected another shell in the barrel of my rifle and started downhill, looking back over my shoulder every so often to see if they were coming. I was in a canyon, and good travelling and I made fast time. Must have made three miles in some world record time. I came to a turn in the canyon and I had the sun on my left, that meant I was going south, and the canyon turned west. I decided to climb the ridge ahead of me. I knew that I must have two mountain ridges between me and salt water and by climbing this ridge I would have a good view of this canyon, so I could see if the Sasquatch were coming after me. I had a light pack and was making good time up this hill. I stopped soon after to look back to where I came from, but nobody followed me. As I came over the ridge I could see Mt. Baker, then I knew I was going in the right direction.

I was hungry and tired. I opened my packsack to see what I had to eat. I decided to rest here for a while. I had a good view of the mountain side, and if the old man was coming I had the advantage because I was up above him. To get me he would have to come up a steep hill. And that might not be so easy after stopping a few 30-30 bullets. I had made up my mind this was my last chance, and this would be a fight to the finish . . . I rested here for two hours. It was 3:00 p.m. when I started down the mountain side. It was nice going, not too steep and not too much underbrush.

When I got near the bottom, I shot a big blue grouse. She was sitting on a windfall, looking right at me, only a hundred feet away. I shot her neck right off.

I made it down the creek at the bottom of this canyon. I felt I was safe now. I made a fire between two big boulders, roasted the grouse. Next morning when I woke up, I was feeling terrible. My feet were sore from dirty socks. My legs were sore, my stomach was upset from that grouse that I ate the night before. I was not too sure I was going to make it up that mountain. I finally made the top, but it took me six hours to get there. It was cloudy, visibility about a mile.

I knew I had to go down hill. After about two hours I got down to the heavy timber and sat down to rest. I could hear a motor running hard at times, then stop. I listened to this for a while and decided the sound was from a gas donkey. Someone was logging in the neighborhood.

I told them I was a prospector and was lost . . . I did not like to tell them I had been kidnapped by a Sasquatch, as if I had told them, they would probably have said, he is crazy too.

The following day I went down from this camp on Salmon Arm Branch of Sechelt Inlet. From there I got the Union Boat back to Vancouver. That was my last prospecting trip, and my only experience with what is known as Sasquatches. I know that in 1924 there were four Sasquatches living, it might be only two now. The old man and the old lady might be dead by this time.

Albert Ostman is dead now, but I enjoyed his friendship for more than a dozen years, and he gave me no reason to consider him a liar. I have had him cross-examined by a magistrate, a zoologist, a physical anthropologist and a veterinarian, the latter two being specialists in primates. In addition to that all sorts of sceptical newsman have grilled him. Those people didn't necessarily end up believing him, but none was able to trap him or discredit his story as a result of their questioning, although the magistrate in particular tried very hard to give him a rough time.

The story does have at least two things very much wrong with it. Most obvious, to anyone who has seen the country involved, is the

110

difficulty of going from where he says he started to where he says he came out, if his basic direction of travel was northeast. The place where he came out is over 60 miles south southeast of the head of Toba Inlet. Unless the sasquatch carried him at least 50 miles over several mountain ridges in a few hours the route can't be fitted together. But there would be no reason for that problem in a manufactured tale. Unlike some of the people who have told of fabulous travels in sasquatch country, he was undoubtedly an experienced woodsman. Had he been making up a route there would be no problem about having it make sense. It is also doubtful, probably impossible, that he could have seen Mount Baker. However his story would be truthful if he just thought he had seen it when actually looking at another mountain.

The other problem is that the picture he gives of the activities of the sasquatches does not fit with the concensus of other information. Generally speaking he makes them far more human than anyone else whose story has any apparent claim to serious consideration. In fact, if anyone came along with such a story today, I wouldn't pay any attention to him. It isn't just the episode with the snuff, or even the carrying off of a man in a sleeping bag that creates the problem. I wouldn't really care if those things were made up. Any circumstance under which a man was able to make such detailed observations of four such animals would have to be rather remarkable. What doesn't fit is the way they lived—as a family group, with the old lady gathering food apparently for all of them, and with a permanent sleeping place equipped with what appeared to be crudely-woven blankets.

There is no way that those things can be resolved. They remain a problem. What counterbalances them is the fact that a man with no pattern to follow was able to describe not just the size and shape and hairiness of what people have been reporting ever since, but facial features, teeth, fingernails, and a lot of other details that aren't in the written story, with nothing that jarred on the experts in ape anatomy or that conflicted with the concensus of later detailed descriptions. Yet when he wrote his descriptions down the common idea of sasquatches was that they were giant Indians.

When I first saw the Patterson movie it looked to me just like what Albert Ostman had described. Oddly enough, he didn't think so. He always insisted that the creature in the movie was not one of the ones he saw. Those who depend on the Ostman story to support their opinion that sasquatches are some kind of humans would be disappointed at the one drawing of such a creature that did appeal to him. He was not satisfied with any of several attempts to draw a sasquatch under his own direction, but greeted me one day with the

111

information that he had found a picture that looked just like the "old man". It was Roger Patterson's sketch from a description by a Yakima youth (now a deputy sheriff) who had a close look at a white sasquatch standing in the road at night, and it was probably the least human-looking sasquatch that Roger ever drew.

Long before he first told his story publicly, Albert Ostman had become a semi-invalid, unable to return to the mountains. Since he was the only witness and did not know the identity of any of the people at Lund or Salmon Inlet 33 years ago, there was no way to check anything at all. He recalled having told the story a few times shortly after it happened but had stopped doing so long ago because no one would believe him, and he could not remember anyone who would have heard it. In 1969, however, an old friend of mine told me that he had first heard of the sasquatch in the early 1930's from a trapper at Toba Inlet who said he knew a young Swede who claimed to have been carried off by one.

-6-

Roger Patterson's Movie

The one most outstanding exhibit in the evidence for the sasquatch is Roger Patterson's movie. It has been shown in halls and theatres all around the continent more than once, and has also been shown on network television and used in television commercials. Most people must have had an opportunity to see it and many probably feel that they know all about it. Indeed it would not be very difficult for anyone to become conversant with most of the available information. Nevertheless, I have the impression that most people do not know much about it, and that what they do know is mostly wrong.

There is a widespread impression, for instance, that the movie has been studied by experts who have decreed it to show a man in a suit. Yet it has actually never been subjected to any multi-faceted study by experts anywhere, and to the best of my knowledge no individual expert who has brought to bear on it the techniques of his own particular field has ever taken the position that it is faked in any way.

I do not mean to say that no zoologist or physical anthropologist has ever watched the film run through a couple of times and then said it looks like a man in a suit to him. That has certainly happened. But I have heard of no one, in any field, who has obtained a copy of the film for thorough study and then publicly taken and defended the position that there is anything the matter with it.

Another widespread misconception is that the film is too blurred to be of much use. It is certainly not as sharp as one could wish, nor as long as one could wish, but it is not at all bad for a film of a wild animal that did not hang around to have its picture taken. A great deal could be done with it if anyone with the necessary skills and equipment or the money to pay for them would take an interest. In

113

fact, it may well prove eventually that the evidence to establish the existence of the sasquatch has been available on that piece of film since 1967.

That statement may seem contrary to assertions I have often made that photographs are no use, that it is necessary to have physical remains. I do take that position, but not on the grounds that it is impossible to settle the matter by the study of photographs. I take that stand on the grounds that experience with the Patterson movie has shown that photographs will not be studied.

Since there is no such animal, any photographs showing such an animal must be faked, and reputable scientists will not have anything to do with them. It's a very effective argument, completely circular without a weak spot in it anywhere.

I first met Roger Patterson in 1965, when he was working on a book about Bigfoot. He had become an enthusiast after reading Ivan Sanderson's articles in *True* magazine in 1959 and 1960, but the first thing he actually did about it was to make a trip to Willow Creek and then to Bluff Creek in 1964. Some 20 miles up Bluff Creek he was fortunate enough to meet Pat Graves, who had seen tracks many times and knew where there were some fresh tracks just a little way down the road. Here is part of what Roger had to say about the incident in his book *Do Abominable Snowmen of America Really Exist?*

> Throughout our conversation we learned that he definitely was a firm believer in the existence of "Bigfoot". He had seen hundreds of tracks year after year, some of them on creek beds thirty or forty miles back from any logging road!
>
> Pat wasn't always a believer that such things exist. During his first six months on the job, stories from other forest service men and loggers came to him about huge human-looking tracks being found in different places. Before he had seen them he thought they might be those of a deformed bear or possibly someone trying to fix up a hoax. But after seeing the tracks for himself, Pat states flatly, "The thing that makes those tracks weighs around 1,000 pounds, walks upright, takes strides from five to ten feet long (10-foot strides probably running), goes up and down rugged terrain where a man just can't go, and has a foot very humanlike."
>
> As we talked more on the subject, Pat said he had seen these giant tracks on Leard Meadow Road by another old logging landing only the day before. We said a hurried goodbye and hurried over there. What we found was an amazing sight.
>
> The creature had come down the mountain, crossed a

road, gone down around an old logging landing, then over the bank into the brush, taking an average 52-inch stride. The prints were of enormous size — 17 inches long and five inches across the heel. I was so astonished I could only stare and try to picture the creature that had made those tracks only the day before. I believe that anyone who sees tracks like Rod and I saw will have to admit there would be no faking them. The imprint of each foot pressed into the ground an inch and a half while our own tracks were barely visible. It was plain to see the foot was flexible as it stepped on small rocks as it travelled down the road. If a rock happened to be where the ball of the foot stepped where the most weight was it was smashed down into the hard road. Where the rocks were up by the toes the foot curled over them like a bare foot would do.

An experience like that can have a powerful effect on a person, and it certainly did on Roger. After that he began spending a great deal of time on trips to places like Mount St. Helens, which is not far from his home in Yakima. He had never heard of the Ape Canyon incident until he went to California.

In July, 1966, he saw 18-inch tracks on Charlie Erion's ranch near Woodland, Washington. The tracks were much like those of "Bigfoot", but a little narrower and with a more elongated heel less than four inches wide. Later that year he published his book, which was composed largely of reprints of stories from newspapers and magazines. It did not make enough money to finance a full-time search as he had hoped it would, but in the course of promoting it he got a lot of public exposure and obtained a good deal of information.

He had come to see me about using some of my newspaper articles in the book, including the Ostman and Roe stories, and I gave him permission with some silent reflections that I ought to have used them in a book myself. When Rene and I made a trip south the following February we visited with Roger, both going and returning. He obviously had no money, and was plainly a man given to obsessions, not only about the sasquatch but about horses, health foods and a couple of inventions he was promoting. He was certain that sasquatches were human and must not be shot, and was deaf to any argument to the contrary. I have seen him described as a rancher, but that was not the case. He had been a rodeo rider, however, and was never to be found on foot if there were any way he could be on a horse. He was a small man, though very muscular, and his horses were small too. He actually had a Volkswagen van rigged up so he could transport two horses in it. His obsession with horses may have had something to do with his success in getting the movie, while his insistence on not shooting was certainly the reason why his success was not conclusive. At that time he had already had one brush with death from Hodgkin's disease, which was then considered incurable, and from which he died about five years later, but no one who met him then would have suspected it.

It was in the fall of that year, about six weeks after Rene and I had seen the tracks on Blue Creek Mountain and a Bluff Creek sandbar, that Roger and his friend Bob Gimlin went to Bluff Creek in Bob's truck with three horses, to look for Bigfoot. Roger had already taken a great deal of 16 mm. movie footage of outdoor scenes and other material to use in a show he planned about his hunt for the hairy giants, and he had a reasonable chance of getting some footage of Bigfoot tracks if he could stay with the search for even a few weeks. As it was they had been there only a few days, riding the roads and creek beds looking for tracks, when they stumbled on Bigfoot herself.

The encounter took place early in the afternoon of October 20, as the two men rode around a bend in the creek. Perhaps the noise of running water had been enough to muffle the sound of their approach. Perhaps the sasquatch did not relate the sound of hooves to the approach of humans. In any event it was still squatting beside the creek when they first saw it, and it took the confrontation calmly. The horses did not. They reared and Roger's horse fell on its side. He had to scramble clear and then get around to the saddle bag on the opposite side to get the movie camera. Bob, in the meantime, was still on his horse and had his 30.06 ready. The animal did not show any concern about the gun, and remained watching them briefly before walking away across a sand bar at a deliberate pace. By the time Roger got hold of the camera he had to run after the sasquatch to get close enough for a good picture, and that is what he did. He tried to film as he ran, getting only a series of blurs and losing his footing as he crossed the creek, but finally he got within about 80 feet and stopped. The animal had ignored him while he was running, but almost as soon as he stopped it turned its upper body sideways for a single stride and looked straight at him. Roger stayed right where he was after that.

"You know how it is when the umpire tells you 'One more word and you're out of the game!' " he told me once. "That's the way it felt."

Roger got clear pictures of only about nine paces before the creature started to pass behind some trees, and for much of that distance the feet and parts of the legs were hidden behind some drift logs on the bar. The exposure is approximately correct for the general scene, but the black animal is underexposed, and even the sharpest frames are slightly blurred by camera motion. The original film looks

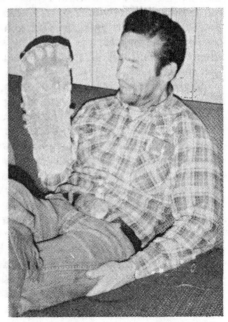

Roger Patterson, with one of the casts he made on a Bluff Creek sandbar after taking the photograph above.

fairly sharp, but the image of the creature is only about one sixth of the height of the frame. It isn't even as big as one of the sprocket holes in the film. When the picture is enlarged the blurring becomes more obvious, and no more details are apparent.

Roger got some more footage when the animal came into view beyond the trees, but it was considerably farther away and he was not holding the camera very steady. There was also some film taken later when they were making casts of the tracks. It seems to have been lost somehow—I have seen it only once—but it clearly showed that when the men walked beside the tracks their feet did not sink appreciably into the packed sand. The prints of the creature, on the other hand, sank about an inch deep, indicating tremendous weight. Its feet measured 14 inches in length, 5½ inches in width at the ball and four inches at the heel. The prints were flat, and there were five toes of fairly human pattern, except that there was less difference in size from largest to smallest. The men made beautiful casts of both left and right feet.

After a rush trip out to civilization to send the film to Roger's brother-in-law for processing, and phoning the B.C. Museum to try to get the scientists and a tracking dog sent down, the two men returned to Bluff Creek, but their good luck was over. A torrential rain started, and they were lucky to get out again before mud slides blocked the road. Meantime I was contacted from the museum, but no one from there would go to California and I spent about $100 on long-distance calls trying to find some scientist who would, without success. The answer was that they would wait to see the film and then they could tell all about it. I was left without time or money to get there myself.

Roger's brother-in-law, Al de Atley, was the first person to see the film. I understand that he had put up the money for Bob and Roger to make the trip and had a one-third interest in the results. Roger, Rene, Jim McClarin and I all saw it for the first time on the same day, at Al's house at Yakima. None of us knew much about movies. No copies had yet been made, and by the end of the day the original was thoroughly scratched up. It was certainly a very exciting day, but for me the film itself was somewhat anti-climatic. Just a dark hairy creature like a gorilla but with longer, heavy legs, striding away. I had heard it described so often it seemed quite familiar.

Roger was eager to be off to New York and Los Angeles with his prize. We argued that he would do better to make people come to him, that he would lose credibility away from his own setting, but the best we could do was to persuade him to show the film at the University of B.C., where we at least had some top people softened up enough to take an interest in the subject. The Vancouver showings, both for the university and museum people and for the press, were well received. Nobody was prepared to suggest to their faces that Roger and Bob had faked the film, and the depth of the footprints, which they said had been made as they watched, established that the thing was far too heavy to be a man in a suit. If the film was a fake, those footprints had to be fakes too and Roger and Bob had to be involved.

Bob Titmus had come down from Kitimat for the showing and among the lot of us the sasquatch hunters had quite a few footprint casts on hand. To our surprise they were rejected as being too good, not showing any evidence of being made by a moving foot. We had all seen lots of prints that showed ample evidence of motion, but no one ever cast any of those. We were looking for prints that would show the exact shape of the foot.

The film was the talk of the town for a day or two, but interest died down and it quickly became obvious that no one was going to do anything about it. The results in Los Angeles and New York were even worse. In the big cities no one was about to be impressed by three young men from Yakima who claimed to have a movie of a monster. On the basis of reports from their Seattle representative, *Life* magazine editors showed some interest, until they had the film run through for some people at the American Museum of Natural History—who took one quick look and wrote it off as a fake. Only Ivan Sanderson, and to some extent Dr. John Napier at the Smithsonian Institute, were prepared to take it seriously. When the first stills from the movie were finally presented to the public it was in a men's magazine, *Argosy*, with a story by Ivan.

The subsequent history of that strip of film would make a book by itself, with deals and double deals involving many individuals and some fair-sized corporations, and with several legal actions, at least one of them in the hundred thousand dollar range, and not all settled yet. The only thing of importance for the purposes of this volume is that at no time in any of the disputes has anyone ever questioned the authenticity of the film. In a way that is unfortunate, as there would be great value in having that question dealt with in court, but it has never been an issue. As a result of one disagreement Bob Gimlin was cut out for several years from any participation in the profits, and it is worth reflecting on whether Roger would have dared to let that situation develop if the film were a fake.

No systematic evaluation of the film itself was undertaken at that time, and the only person who did anything about it was Bob Titmus, who went to California and spent several days studying the evidence there. Here, from a letter to me, are his comments regarding

119

the showing of the film at the university in Vancouver and what he found at Bluff Creek.

By the end of the day it became apparent that a few of the viewers felt that there was a possibility the whole thing was a very elaborate and expensive hoax. I felt that this possibility was so extremely remote as to be almost non-existent. (None of these individuals witnessed more than one showing, I believe). However, I did have to take into consideration the fact that I believe that I viewed the film through somewhat different eyes than most of the persons present.

Firstly I think that a taxidermist will see and retain far more detail, while watching an animal, and is probably far more qualified to recognize anything unnatural, than the average person.

Secondly, evidence I witnessed in the mountains of Northern California about ten years ago changed me from a non-believer to a believer and since that time I have spent a major portion of those years, as you know, interviewing witnesses, investigating reports, collecting evidence, casting many, many different tracks, setting up camera and live traps, tracking the creatures dozens of times, etc., all of this was in an effort to capture one of the creatures. All of this experience only strengthened the case of the existence of the creature Bigfoot/Sasquatch.

Thirdly, many years ago I saw one of these creatures at fairly close range and watched it for about ninety seconds before it walked off into the timber.

Almost none of the persons present at the showing of the film had a background of experience like this so it is not surprising that there was some variance in the conclusions arrived at.

Since I know more about tracks than film and generally feel that they will tell me a more accurate story than film, I had a very strong urge to see the tracks that were being made during the time that Roger was shooting his film. I felt that the tracks could very well prove or disprove the authenticity of the pictures. No one else present seemed inclined or able so the following day I went on to California to have a look at the tracks.

My first full day up near the end of Bluff Creek, I missed the tracks completely. I walked some 14 to 16 miles on Bluff Creek and the many feeder creeks coming into it and found nothing of any particular interest other than the fact that Roger and Bob's horse tracks were everywhere I went. I found the place where the pictures had been taken and the tracks of Bigfoot the following morning. The tracks traversed a little more than 300 feet of a rather high sand, silt and gravel bar which had a light scattering of trees growing on it, no underbrush whatever but a considerable amount of drift debris here and there. The tracks then crossed Bluff Creek and an old logging road and continued up a steep mountainside . . .

This is heavily timbered with some underbrush and a deep carpet of ferns. About 80 or 90 feet above the creek and logging road there was very plain evidence where Bigfoot had sat down for some time among the ferns. He was apparently watching the two men below and across the creek from him. The distance would have been approximately 125-150 yards. His position was shadowed and well screened from observation from below. His tracks continued on up the mountain but I did not follow them far. I also spent little time in trying to backtrack Bigfoot from where his tracks appeared on the sandbar since it was soon obvious that he did not come up the creek but most probably came down the mountain, up the hard road a ways and then crossed the creek onto the sandbar. It was not difficult to find the exact spot where Roger was standing when he was taking his pictures and he was in an excellent position.

I spent hours that day examining the tracks, which, for the most part, were still in very good condition considering that they were 9 or 10 days old. Roger and Bob had covered a few of them with slabs of bark etc., and these were in excellent condition. The tracks appeared perfectly natural and normal. The same as the many others that we have tracked and become so familiar with over the years, but of a slightly different size. Most of the tracks showed a great deal of foot movement, some showed a little and a few indicated almost no movement whatever. I took plaster casts of ten consecutive imprints and the casts show a vast difference in each imprint, such as toe placement, toe gripping force, pressure ridges and breaks, weight shifts, weight distribution, depth, etc. Nothing whatever here indicated that these tracks could have been faked in some manner. In fact all of the evidence pointed in the opposite direction. And no amount of thinking and imagining on my part could conceive of a method by which these tracks could have been made fictitiously.

While passing through Weaverville I had phoned my sister and brother-in-law in San Diego and invited them up to Bluff

121

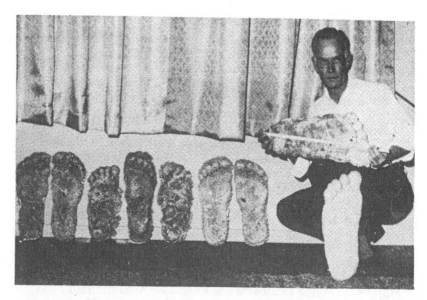

Bob Titmus with some of the casts he has made of tracks found in northern California. The fourth from the left is from the footprint shown in the picture below, which was taken by Lyle Laverty, of Skykomish, Washington, at the site of the Patterson movie sometime before Bob got there.

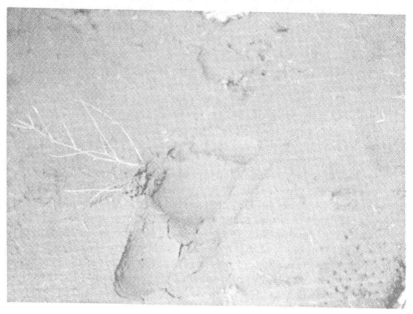

Creek for a visit after my several years away and also to see the tracks. They arrived at my camp this particular evening shortly before I was preparing to leave. We stayed over another day. Allene was a skeptic and Harry a hard-headed non-believer. Both of them left there believing in the existence of this creature. I didn't try to convince them of anything. I simply took them to where the tracks were and let them examine them to their own satisfaction and draw their own conclusions. Harry has hunted big game all of his life. He has been all over Africa, Alaska, Yukon Territory, Canada, Mexico and the U.S. and stated that this impressed him more than anything he had ever seen in the bush in all of his travels. Harry made several tests and observations, one of which was walking briskly beside the tracks to try to match their depth of up to an inch and a quarter and more in places. Harry is a 200 pounder and the best he could do was an imprint of about ¼ of an inch on the rear portion of his shoe heel and one-eighth of an inch and less on the rest of his shoe imprint. We both agreed, considering the depth of the two imprints and the difference in the amount of bearing surface, that the creature that made these tracks would have to weigh at least 600 to 700 pounds.

Jim McClarin, an active sasquatch hunter at that time and for several years afterwards, was a student at Humboldt State University at Arcata, California, about 80 miles from the film site, so he also was able to look it over. I wasn't able to get there until the following June. By that time Rene and I had made a deal for Canadian lecture rights to the film, so I had slides of several frames from it with me. It was my intention to make a similar film with a man walking where the sasquatch had been. Jim, who is six foot five, and who had seen where the tracks went, was to be the sasquatch. I also got Roger's permission to show the film at Willow Creek, as the people there had never seen it, and on that occasion I met for the first time Ken Coon, who is still one of the most active investigators, and George Haas, who later published the Bigfoot Bulletin for several years.

The location proved to be just a few hundred yards upstream from the sandbar where Rene and I had seen tracks the preceding September. To our surprise, a few of the depressions left when the tracks had been cast could still be seen, although none were evident in the critical area shown in the few clear frames in the movie. I found that making things line up in the camera viewfinder the way they appeared in the slides was a remarkably precise way of locating a position. Moving the camera less than a foot in any direction would change the relationship of objects in the background and foreground.

I presume that there might be inaccuracies caused by differences between the lens of Roger's camera and the viewfinder of mine, but I think I must have been very close to the right position. It turned out that Roger had been down near the ground when he took the good footage, and that he had stayed in the same place when he took the later shots of the creature from behind, although by then he was standing up. He had never moved his feet after the sasquatch shot a look at him. The film jiggles violently during the later sequence and we had presumed that Roger was again pursuing the creature, but apparently he was just trying to film and wind the camera at the same time. I don't remember the camera's height for the critical sequence, but it was certainly less than four feet. I assumed that Roger was kneeling, but Bob Titmus tells me there was no indication of that on the ground, so I guess he must have squatted down.

In any event the camera was held so low that slight undulations in the sandbar hide the creature's feet when they are on the ground, so the slides were of limited help in trying to determine the exact position of the feet. That being so, there is little liklihood that Jim walked exactly where the sasquatch walked, so estimates of the size of the creature made by comparing the two movies can not be considered exact. They should be very close, but unfortunately estimates made by different people are not the same. I expect that I have spent more time on that problem than anyone else, since I have gone through it several times on my own and also with people who came to my house to see the film, including Dr. Krantz, and I find that one does not get exactly the same measurements twice running. One problem is that my movie was taken in early summer, Roger's in mid autumn, and the colors in the foliage are a great deal different. The lighting is different too, with the shadows in different places. Had someone made a second movie right after Roger's was made, and at the same time of day with all the tracks still there, it would have been infinitely more useful. As it is, lining up the two very different films for comparison is a fussy business. At one time I was certain that the creature was several inches taller than Jim (at least 6 ft. 5½ ins. in his boots, 180 lbs.), particularly since it is more stooped. Now I am aware that it is possible to line up some frames so that Jim appears as tall or even taller. The question of whether one or the other is a little closer to the camera at a particular spot cannot be answered, so there is little value in debating the point. Dr. D. W. Grieve, an English anatomist specializing in the human gait, studied the two films after Rene took them to England late in 1971, and concluded that the creature was not over 6 ft. 5 ins. However he also found the foot length as shown in the film to be 13.3 ins. Referring to my published estimates, he included in his report, in John Napier's book *Bigfoot*,

124

The Yeti and Sasquatch in Myth and Reality, the following statement:

> Earlier comments that this specimen was just under 7 ft. in height and extremely heavy seem rather extravagent. The present analysis suggests that Sasquatch was 6 ft. 5 in. in height, its weight about 280 lb. (127 kg.) and a foot length (mean of four observations) of about 13.3 in. (34 cm.).

Dr. Grieve's analysis, while undoubtedly more skilful than my own, suffered from the fact that he was working only with the films and apparently had not been supplied with the actual measurements of the footprints and the creature's stride. Having to make assumptions as to such things as the angle between the creature's direction of travel and the surface of the film, he estimated the stride length at 262 centimeters when it was actually 207, and the foot length at 13.3 inches when it was actually 14. He recognizes these points in a note in a recent paperback edition of *Bigfoot,* but still maintains that the height is not over 6 ft. 5 in.

According to my pocket calculator, 14 is more than 1.052 times 13.3, which is a fairly substantial difference. If the creature was estimated at 6 ft. 5 ins., and its foot length in comparison seemed to 13.3 ins., then in relation to the actual foot length the height is 1.052 times 77 inches, or almost exactly 6 ft. 9 in. That happens to be precisely the same as my "extravagant" estimate, and neither of us allowed anything for the fact that the creature's feet were sinking an inch deeper in the ground than Jim's. The same adjustment applied to Dr. Grieve's weight estimate would increase it to 326 pounds, as his figure of 280 would have to be multiplied by 1.052 three times—for the error in height, width and breadth. However Dr. Grieve's figure is based on the assumption that the sasquatch is only one quarter larger than Jim in both width and breadth. I won't contest the estimate as to width, although I think it is low, because there is no way to measure that accurately. By the time it is seen from the back the creature is a long way off and partly hidden by a tree. There are good clear frames showing the size of the animal from front to back, and it is easily one and a half times as large as Jim. That changes the weight calculation to 391 pounds.

Estimating the weight by calculating the volume of each part of the body, Dr. Krantz arrived at a figure of 640 pounds. I have done that calculation several times, and since accurate measurements are impossible I never get the same answer twice, but my lowest estimate would be as much as Grover's. Moreover I cannot get out of my mind the recollection of a female gorilla in a zoo that looked very short, if rather fat, but weighed about 400 pounds. By comparison, the creature in the film should weigh at least 800.

Dr. Grieve notes that the height of the fingertips above ground and of the hip joint are a similar proportion of the total height to what would be expected in a human, but that the shoulders are about six per cent higher and 30 per cent wider. That means that the arms are only slightly longer than a human's, but because of the wide shoulders the arm span would be considerably greater, as it obviously is in the picture. Arms outstretched it would span at least eight feet.

In his analysis of the stride, Dr. Grieve concluded that a lot depended on the speed at which the camera was running, which is not known. He found the similarities between the sasquatch gait and a human gait to be very striking in terms of stride length, speed and time of swing, if the film was taken at 24 frames a second. However the amount of bending in the knee at one point in the stride and the angle of the thigh at another are not typically human. At 16 or 18 frames a second the resemblances to a human gait would be less. Roger normally had his camera set at 24 f.p.s., but he told me that after this film had been taken he found the setting to be 18 f.p.s. and he did not know when it was moved. The most likely time would be during the scramble to get the camera out of the saddle bag, which would mean that the creature was photographed at 18 f.p.s., but that cannot be considered certain.

Igor Bourtsev, in Moscow, did find an ingenious way of determining the film speed. He reasoned that the bouncing up and down of the camera might show the pattern of Roger's movement as he was walking or running, and studying the film frame by frame he found that two such patterns did exist, one indicating a step every seven frames and one indicating a step every four frames. Presumably when Roger made a pace every four frames he was running, and when he made a pace every seven frames he was walking. If the camera was exposing 16 frames a second, then Roger was running four paces a second. If the camera was exposing 24 frames a second, then Roger was running six paces a second. Igor then enquired about the gait of a running human and found that a sprinter makes only five paces a second. Since it is not reasonable to suppose that Roger, while taking pictures, was making more paces in a second than a sprinter, the film speed of 24 frames a second is ruled out. I wouldn't assume that the matter can now be considered settled, but that certainly seems a sound approach to the problem, and it produced a definite answer.

Anyone who studies the film for any reason can expect to find himself forming what might be considered subconscious conclusions about it in addition to the results of his measurements. In this regard, Dr. Grieve made an interesting observation:

My subjective impressions have oscillated between total

126

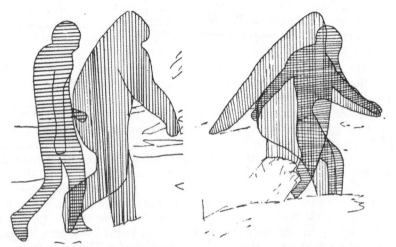

Jim McClarin (6'5", 180 lb.) and the sasquatch traced from movie frames taken from the same spot with background and foreground objects all lined up.

Dmitri Bayanov, Dr. Dmitri Donskoy and Igor Bourtsev with scale models of the creature shown in the Patterson film.

acceptance of the Sasquatch on the grounds that the film would be difficult to fake, to one of irrational rejection based on an emotional response to the possibility that the Sasquatch actually exists.

What intrigues me is that this statement recognizes that it is the rejection that is irrational and emotional, while the evidence suggests that the creature does exist.

Another thing about Dr. Grieve's study leaves me very puzzled, and that is how he can consider the walk at any speed to be similar to a human gait when his own measurements show that when one leg is in the air passing the other, the leg bearing the weight is bent 30 degrees at the knee. At that point in a typical human stride the leg with the weight on it is virtually straight. That difference is tremendously noticeable when you have been watching the sasquatch film for a while and then run the Jim McClarin sequence. Jim seems to move very jerkily, bobbing up and down as his body goes up and over the straight leg in the middle of each stride. The sasquatch, on the other hand, moves smoothly, with the knee dipping as the body passes above it.

Rene also succeeded in having a biomechanical study of the film done in the U.S.S.R. (None has ever been done in North America.) Dr. Dmitri D. Donskoy, Chief of the Chair of Biomechanics at the U.S.S.R. Central Institute of Physical Culture in Moscow, examined both the movie and stills taken from it, and concluded that it showed a highly efficient pattern of locomotion which differed from that used by humans. The full report is in the book *Sasquatch* which Rene co-authored with Don Hunter. Dr. Donskoy states that the arm movements indicate that the arms are massive and the muscles strong. The leg movements he finds to be typical of massive limbs with relaxed muscles, while the amount of knee flexion far exceeds that of a normal human walk, resembling instead the way humans move while cross-country skiing.

He finds the walk of the creature to be a natural movement without any sign that it is the result of an attempt to move in an unusual way. All its movements indicate that the creature is much heavier than a man and its muscles much stronger. In fact, he concludes, despite all the diversity in the ways various humans walk, the way this creature walks is absolutely different from any human gait.

That is the voting so far by the biomechanics—the specialists in the functioning of living machines. One is non-commital, the other certain that the creature in the film is not human. Strange that so many people without such qualifications have been able to see immediately that the creature is a human in a fur suit.

While there has been no systematic investigation of the film and the circumstances in which it came to be made, a lot of people over the years have done a lot of inquiring, and no one has found any indication of how the film could have been faked. When I bought lecture rights to it I had the original in my possession for some time, and was told by the technical people with Canawest Films in Vancouver, that it was indeed an original, not the product of any fakery in a lab, but a strip of film exposed in a camera and showing something that had walked in front of the lens. Patterson and Gimlin are reported to have taken it to Universal Studios and to have been told that matching it artificially would be all but impossible. I have no proof of that, but I do know that someone took it to the Disney Studios, because I went there myself in 1969. Ken Peterson, a senior executive, told me that their people had already studied the film, and that if they wanted something like it in one of their own movies they would not attempt to film it, they would draw it. None of their mechanical creations for Disneyland was sophisticated enough to walk free, they were all attached to a base at some point. The only way to imitate the Patterson creature would be with a man in a suit. Top man in that field, they told me, was Janos Prohaska, a stunt man who had made several ape suits for himself. I went to his home and showed him the Patterson movie in slow motion. He said that it could not be a padded suit like his gorilla suits, it would have to be skin tight to show muscle movement as it did. Since that time Wolper Productions have shown the public what human imitations of a sasquatch look like, in the movie *Mysterious Monsters*. The Patterson footage was shown in the same movie, and the conclusion was obvious that if a couple of small-time sharpies had actually created an imaginary monster they were somehow able to do a far better job of it than the professionals were able to do.

There have been occasional rumours that one person or another has admitted to having helped to make the suit for Patterson. One man even claimed to have been the person in it. On investigation these stories have come to nothing. It should be noted also that although Roger Patterson eventually received some substantial sums of money for the film, he spent them again in a continued quest for another encounter with the creature, and he continued to do so even when he knew he did not have long to live. His action immediately after taking the movie was also inconsistent with fakery. When he phoned to try to get sceptical scientists and a tracking dog down to California from British Columbia he had no way of knowing that there was going to be a heavy rain before anyone could get there. Had the whole thing been a hoax there is every chance that the dog would have made that obvious. People who have actually faked sasquatch films have always concealed the location where they were made, and the one time when such a location was discovered the whole case immediately fell apart, even though the man involved was far more experienced and better equipped than Roger Patterson, both as a wildlife photographer and as a student of animal behavior.

I have given a fair sample of the kind of evidence that has accumulated favoring the authenticity of the film, but what about the other side? Who speaks for the scientists who have seen the film and rejected it? Well, how about Dr. William Montagna, director of the federal primate center at Beaverton, Oregon, writing in *Primate News*, September, 1976:

> Along with some colleagues, I had the dubious distinction of being among the first to view this few-second-long bit of foolishness. As I sat watching the hazy outlines of a big, black, hairy man-ape taking long, deliberate human strides, I blushed for those scientists who spent unconscionable amounts of time analyzing the dynamics, and angulation of the gait and the shape of the animal, only to conclude (cautiously, mind you!) that they could not decide what it was! For weal or woe, I am neither modest about my scientific adroitness nor cautious about my convictions. Stated simply, Patterson and friends perpetrated a hoax. As the gait, erect body, and swing of the arms attest, their Sasquatch was a large man in a poorly made monkey suit. Even a schoolchild would not be taken in. The crowning irony was Patterson's touch of glamor: making his monster into a female with large pendulous breasts. If Patterson had done his homework, he would have known that regardless of how hirsute an animal is, its mammary glands are always covered with such short hairs as to appear naked.

The voice of science has spoken? Well, not really. No one who has spent any time studying the film in slow motion has ever, to my knowledge, suggested that it is not at least a very well-made suit. The objection regarding the hairy breasts has been raised on other occasions, but people with equal qualifications consider that hair on the breasts is a logical adaptation to cold winters for a creature with its breasts exposed on the front of its body.

It is quite normal to find that the objection that leads one scientist to conclude that the creature is a fake will be contradicted by the next one, although he in turn has some reason of his own for considering the film to be fraudulent. The apparent peak on the top of the head has drawn criticism. In other apes such a peak is caused by a ridge of bone anchoring huge jaw muscles. Females are not supposed to have it, and it is supposed to be accompanied by a pot belly to accommodate all the roughage that the jaws have to grind up. That is how some students of primate anatomy have reacted. Others see no reason why a female could not have such a crest, and find the abdominal dimensions quite acceptable. They in turn may contend that the creature walks like a male. Of course they mean a human male, since humans are the only primates known to walk like that at all. Others point out that in humans the female walks differently because she has a wider pelvis, and she has a wider pelvis because infant humans have inordinately large heads. To them, it would spell fraud if an upright female ape did not walk like a male. The layman is left to conclude that the anatomists will be able to adjust their thinking to accommodate the peculiarities of the sasquatch, whatever they may be, once one is collected. They are not likely to include anything as outlandish as the elephant's trunk or the narwhal's tusk.

But let's hear a little more from Dr. Montagna:

To believers who complain that we scientists are too opinionated to look at the evidence, I reply: Is a scientist to listen to every zealot who regales him with tales of a putrid stench, who shows him fake footprints, or makes films of a man wearing a badly tailored monkey suit? The scientist who is reviled because he won't listen to fantasy goes securely on his way, knowing that life is so full of real wonderment and mystery that he does not have to fantasize.

But perhaps I ought to add that man's need to fantasize is a vestigial remnant of his past. It created mythological characters, good and evil; visions of miraculous events, heaven, purgatory and hell. It created the oracles, the art of palmistry, phrenology, astrology and all sorts of other occult sciences. And finally, it peopled man's world with monsters.

It must be comforting indeed to have such faith in one's beliefs, Dr. Montagna.

-7-

Furry Fellows Around the World

Reports of hairy bipeds are a world-wide phenomenon.

The yeti or "abominable snowman" of the Himalayas is well known and can be passed over with a couple of comments. First is that the yeti and the sasquatch do not appear to be the same thing. The yeti as generally described is a good deal smaller and less erect than what is reported in North America. It also leaves a footprint that does not resemble either the standard five-toed "Bigfoot" or any of the variety of other tracks that have been reported on this continent recently. The other thing is that the evidence for the yeti's existence, while sufficient to have half-convinced the world, is insignificant compared with what is available in North America. Besides the recording of probably 20 or 30 times as many reports, the North American research has produced one good movie and several poor photos as well as an almost unlimited supply of casts and photographs of footprints in sand and dirt. From the Himalayas there was for 20 years just a single footprint photo, and that was taken in snow. Then in December, 1972, in the Arun Valley of eastern Nepal, photos and casts were made of nine-inch prints in snow showing an opposable big toe. North American researchers are inclined to look on prints in snow as second-rate evidence because making them does not require a superhuman amount of weight. Sometimes the prints do indicate that they were made by a creature able to do things that would be impossible for a human. According to an article by Edward W. Cronin Jr. in the *Atlantic Monthly*, November 1975, that was the case with the Arun Valley prints.

132

In his book *Abominable Snowmen, Legend Come to Life,* Ivan Sanderson recounted, in detail, reports of hairy bipeds, large and small, from many different parts of the world. From Central America he introduced to the reading public the giant Sisimite of Guatemala, and the little Dwendi of British Honduras. Richard Oglesby Marsh in *White Indians of Darien,* told of a man who claimed to have killed a manlike creature in Central America in 1920. Edward Jonathan Hoyt in *Buckskin Joe* told of killing a five-foot creature of the ape family, resembling a man more than anything else, that climbed over the end of his bunk in Honduras in 1898. A. Haworth, a New Jersey man, wrote to Ivan Sanderson that a friend from El Salvador told him of a hairy human-shaped creature that killed cattle on his uncle's farm near the Honduras border.

In South America, *Abominable Snowmen* lists the Shiru, a four to five-foot hairy biped with short dark brown fur, in the mountains of Ecuador; the giant Mapinguary of the Matto Grosso that leave 20-inch tracks and kill cattle by tearing out their tongues, and the man-sized Didi of the Guianas. I can add two specific reports. In 1969 a Seattle man told me that he had seen four-toed tracks 15 or 16 inches long, high in the mountains in Ecuador in 1963. He said that scientists from a university at Quito were studying them while he was there, and that local residents had reported seeing a sasquatch-like creature a few days before. Also, *Flying Saucer Review,* October-November, 1966, carried a story of a truck driver named Alberte Kalbermatter seeing a huge humanlike form nine feet high and covered with abundant black hair, somewhere in southern Chile or Argentina in May, 1964.

From Africa, *Abominable Snowmen* contains accounts of the little russet-haired Agogwe from Tanzania and Mozambique and something similar from the Ivory Coast. In the giant category is the Muhalu of the Congo, known only from stories told by pygmies to explorer Attilio Gatti, and from a 15½-inch footprint that Gatti saw. In *Year of the Gorilla,* George Schaller tells of a famous animal collector, Charles Cordier, hunting for an almost man-sized inhabitant of the Congo called the Kakundakari. It was described as being 5½ feet tall, covered with hair and upright. Cordier told Schaller that he had seen a footprint of one, and that one had been caught briefly in one of his bird snares.

> It fell on its face, turned over, sat up, took the noose off its feet, and walked away before the nearby African could do anything.

George Schaller was very impressed with Charles Cordier and took the Kakundakari story seriously, but he was nevertheless dealing with something second-hand. Another student of ape behavior had a personal experience involving such a creature. John McKinnon, while observing orangutans in Borneo in 1969, came upon some unexpected footprints on a muddy path in the forest. In his book *In Search of the Red Ape* he describes the incident as follows:

The rhino may be rare but at least it is a well-known and scientifically documented animal, which is more than can be said of Batutut. I was travelling alone along a hill ridge on the far side of the river where I had never ventured before. The path was good, though rather muddy, and I hadn't a care in the world. Suddenly I stopped dead, amazed at what I saw. I knelt down to examine the disturbing footprint in the earth, a print so like a man's yet so definitely not a man's that my skin crept and I felt a strong desire to head home. The print was roughly triangular in shape, about six inches long by four across. The toes looked quite human, as did the shapely heel, but the sole was both too short and too broad to be that of a man and the big toe was on the opposite side to what seemed to be the arch of the foot.

Further ahead I saw more tracks and went to examine them. There were imprints of both left and right feet, though which was which I could not tell from their curious distribution. Many of the prints had been obliterated by recent pigs but a few were quite clear and I made drawings of some of these and notes of their relative positions. I found two dozen footprints in all, scattered along some fifty yards of path.

Back at camp he showed his sketches to his Malay boatman and asked him what animal could make such tracks:

Without a moment's hesitation he replied "Batutut" but when I asked him to describe the beast he said it was not an animal but a type of ghost. Bahat gave an imitation of its plaintive call, a drawn-out tootootootootoo, from which it derives its name, and told me many stories about this shy, nocturnal creature, who lives deep in the jungle feeding on river snails, which it breaks open with stones. Batutut, he told me, is about four feet tall, walks upright like a man and has a long black mane. It is said to be fond of children, whom it lures away from their villages but does them no harm. To adults, however, it never shows itself, but occasionally men had been found that Batutut had killed and ripped open to feast on their liver (to Malays the seat of all emotions, analogous to the European heart).

Like the other spirits of the forest the creature is very shy of light and fire. Bahat said that, as a young boy, he, too had seen the footprints of Batutut and other villagers had also seen them from time to time . . .

When I made further inquiries in the kampong I found that Batutut was quite well known and other stories confirmed what Bahat had told me. I secured photographs of the feet of sun-bears and indeed they were too small and differently shaped to be responsible for the tracks I had seen. Later I saw plaster casts of even larger footprints from Malaya that had definitely been made by the same animal, there known as orangpendek, or 'short fellow'. Again, natives spoke of a small creature with long hair, who walks upright like a man. Drawings and even photographs of similar footprints found in Sumatra are attributed to the Sedapa or Umang, a small, shy, long-haired, bipedal being living deep in the forest.

Abominable Snowmen also mentions the Sedapa and the Orang Pendek, as names for the Indonesian creature. Of course the Dwendi, the Agogwe, the Shiru and the Sedapa are too small to be the same as the sasquatch, but the little fellows could be closely related to each other, and something similar could account for the few reports of small creatures in North America. The Sisimite, Mapinguary and Muhalu could be the same thing as the sasquatch, and giant forms are also reported from several places in Eurasia.

There are some Australian reports, but so far I have seen none of sufficient substance to make a case for such creatures having reached the island continent. There are even a few newspaper stories of a seven-foot apelike "Moehau monster" in New Zealand but how a giant ape could get there seems to me to be beyond explanation.

From Malaya two Americans produced color photos of 18-inch footprints they claimed to have found near the Endau River. One of them, Harold Stephens, wrote an article for the August, 1971, issue of *Argosy* magazine which also covered a number of earlier reports from that area of an upright ape, some of which I also have from earlier sources. Two giant "Monkey Men" were reported near the Burma-Laos border close to the Mekong River in June, 1969, according to a Reuters report. They were 10 feet tall and covered with "khaki-colored hair". From the opposite side of Burma, Ivan Sanderson received a letter from a veterinary surgeon in 1969, telling of a villager killing a sasquatch (that is the name used in the letter) and keeping some of its teeth. The letter writer referred to other evidence and stated flatly, "We have the sasquatch in the Chin hills." *Abominable Snowmen* also mentions giants known as Kung-Lu and Tok in Burma.

In Western Burma we are just about back to the Himalayas. I have said that the abominable snowman, or yeti, is not the sasquatch, but Ivan Sanderson, and also Tom Slick, who sponsored expeditions there, were of the opinion that there was more than one type of creature in the Himalayas, and one kind was a giant. A remarkable account of such a creature is contained in a book, *The Long Walk*, written by Slavomir Rawicz, who claims to have been one of a group that escaped from Siberia by walking south to India in the early 1940's. He tells of seeing two animals far away on a snowfield in the Himalayas and heading towards them in hopes that they could be caught for food. They reached a position from which the animals were again in view, about 100 yards away:

Two points struck me immediately. They were enormous and they walked on their hind legs. The picture is clear in my mind, fixed there indelibly by a solid two hours of observation. We just could not believe what we saw at first, so we stayed to watch. Somebody talked about dropping down to their level to get a close-up view.

Zaro said, "They look strong enough to eat us." We stayed where we were. We weren't too sure of unknown creatures which refused to run away at the approach of men.

I set myself to estimating their height on the basis of my military training for artillery observation. They could not have been much less than eight feet tall. One was a few inches taller than the other, in the relation of the average man to the average woman. They were shuffling quietly round on a flattish shelf which formed part of the obvious route for us to continue our descent. We thought that if we waited long enough they would go away and leave the way clear for us. It was obvious they had seen us, and it was equally apparent that they had no fear of us.

The American said that eventually he was sure we should see them drop on all fours like bears. But they never did.

Their faces I could not see in detail, but the heads were squarish and the ears must lie close to the skull because there was no projection from the silhouette against the snow. The shoulders sloped sharply down to a powerful chest. The arms were long and the wrists reached to the level of the knees. Seen in profile the back of the head was a straight line from the crown into the shoulders — "like a damned Prussian", as Paluchowicz put it.

We decided unanimously that we were examining a type of creature of which we had no previous experience in the wild, in zoos or in literature . . .

They appeared to be covered by two distinct kinds of hair

136

— the reddish hair which gave them their characteristic colour forming a tight, close fur against the body, mingling with which were long, loose, straight hairs, hanging downwards, which had a slight greyish tinge as the light caught them.

In the U.S.S.R. there is a situation comparable to that in North America, with far more information available than from the Himalayas. It has not been widely publicized in English however, except in Odette Tchernine's books. I have corresponded for years with Professor B. F. Porshnev, who is now dead, and with Igor Bourtsev and Dmitri Bayanov of the Hominid Problem Seminar at Moscow. There is an impression abroad in North America that the Russian government is involved in sending out expeditions to look for their versions of the snowman or sasquatch, but that is not the case. The Russians seem to have much the same difficulties with their scientific establishment that we have in North America, and progress there has depended entirely on the efforts of interested individuals, just as it has here. Professor Porshnev did have the advantage of high standing in the academic community which enabled him to get papers published in scientific journals. One such paper appeared after his death in the December, 1974 edition of *Current Anthropology*, starting the only discussion I am aware of in a major anthropological publication.

In a much longer paper published in *Soviet Ethnography* in 1969, Professor Porshnev gave many details of the research done in Russia and the conclusions arrived at. I have had that paper translated, but find the translation still rather difficult to follow. Because it contains so much interesting information that is not otherwise available in English, I will risk some distortions in meaning by attempting a paraphrase of the sections of the paper dealing with the history of the investigation in Russia and the description of the creatures studied.

There is an immediate problem in that none of the names Professor Porshnev used for the creatures is really acceptable to me, since they all carry connotations of primitive man, which I feel prejudges the situation. However it would be an even greater distortion for me to use the word sasquatch for creatures in Russia that may be something entirely different. One name that Professor Porshnev favored was Homo troglodytes. That was the name Karl Linnaeus used more than two centuries ago in his *Systema Naturae* for creatures of which he had read and heard that resembled man physically but were hairy and lacked man's power of speech. "Troglodyte" is a word generally taken to mean "cave man", but it can also be applied to the anthropoid apes. Where a name is needed I will use that one.

Professor Porshnev states that Linnaeus expressed astonishment that everyone could get so excited about monkeys yet natural scientists were able to ignore the troglodytes as if they didn't exist. His paper continues:

During the latter part of the 19th century, Linnaeus' Troglodyte again came to light, but not through scientific enquiry. N. M. Przhevalsky, on trips into Central Asia, received information about a "man beast" but did not investigate it further. Nothing was done in that area until the 1920's, when T. Jamtsarano, a Mongolian scientist, began compiling information about the manlike creatures and over a 10-year period mapped their distribution in Mongolia. His work was not influenced by any knowledge of Linnaeus' belief in such creatures.

Just as Jamtsarano had not known of Linnaeus' work, zoologist V. A. Khakhlov did not know of Jamtsarano's. He worked between 1907 and 1914, gathering information on the manlike creatures in the Tien Shan region. The people of Khazakstan called them "Kshee-guiek". Writing to the Russian Academy of Science on June 1, 1914, he named this creature Primihomo Asiaticus. It should be stressed that Jamtsarano and Khakhlov arrived at their conclusions independently, which establishes that the natives of their respective regions both believed in the existence of similar creatures. Furthermore these independent descriptions agreed as to the habitat of the creatures, what they looked like and how they behaved. That can hardly be considered just a case of similar folklore in two different areas of Central Asia, since the descriptions correspond with fossils which were discovered later, which neither the uneducated people of those regions nor Jamtsarano and Khakhlov knew anything about. Later the Mongolian professor G. P. Dementiev and others carried out extensive field research and obtained more precise anatomical, morphological and biological data.

Evidence gathered by Khakhlov and also by some travellers throughout Central Asia inspired further academic research by P. P. Sushkin, who concluded that the transformation from ape to man took place in the high mountains of Central Asia, passing through the intermediate form of an upright-walking animal. The fact that Sushkin restricted his research to information from a limited region of Central Asia caused him to arrive at this mistaken theory. Nonetheless, Linnaeus, Jamtsarano, Khakhlov and Sushkin all related the problem to man more than ape. On the other hand, Western European scientists, while pursuing research in Nepal and the

Himalayas, related the problem first to bear, then to ape, but not to man . . .

Nevertheless, scientists at different times, in different countries, dealing with different local material completely independently, obtained similar information. It is not often that zoologists, anthropologists and ethnographers reach the same conclusions through their different lines of research, but it has happened in this case.

The leading student of vertebrates in the Caucasus, K. A. Satunin wrote at the end of the 19th century of an accidental encounter in the Talysh Mountains with a female "Biaban-Guli", which is a name given in the southern Caucasus to a hairy, manlike animal without speech. Later, at a nearby settlement, he obtained further information about such creatures. In 1959, Y. I. Merezkinski, a senior lecturer in anthropology and ethnography at Kiev University, was taken to a place in Azerbaijan where a "Kaptar" was known to come to drink. He had promised that he would only photograph the creature, but instead he tried to shoot it. Besides missing the creature, he alienated his local guide. Professor Merezhinski observed that the creature was skinny and covered by white hair from head to foot. What Satunin and Marezhinski saw confirmed the descriptions local people had given of the creatures.

These sightings were both in the Caucasus, but there have been many more elsewhere, for instance in Manchuria, and they agree with the information obtained in the Caucasus. Around 1914 an eminent zoologist, N. A. Baikov, met a Manchurian hunter who had with him as a helper a "half man", hairy, stooped and unable to talk. Despite his amazement at seeing such a creature, Baikov did not describe it in detail, but the basic description he gives is enough to identify it.

From various places where there are traditions of the troglodytes, supporting information is available from two other sources; from educated visitors who were not aware of local stories and from people who live in the area but are not the type to put any stock in folktales. Those in the first group are particularly significant. How could they be influenced by local beliefs if they had never heard of them?

Here are some examples, beginning with observations by physicians. In the 17th century one of the founders of contemporary medicine, a Dutchman, N. Tulp, examined such a creature that had been caught in the mountains of

Ireland. He has left a description of its anatomical peculiarities that has led us to conclude it was a troglodyte. The same goes for several other observations by other learned men of similar "bear men", including a hairy speechless "fellow" caught in 1661 in a Lithuanian forest which was tamed and lived for many years at the Polish court. Later there is a more definite story of a similar creature examined by a Hungarian physician. This one had been caught in the Transylvanian forests and was also domesticated. Another physician, V. S. Karapetian examined a similar creature, a male, in 1941 in Daghestan.

An army general, M. S. Topilski, described in detail his examination, with physicians assisting, of a similar type of creature that was accidentally killed high in the Pamirs in 1925. Marshal P. S. Rybalko, in 1937, while commanding an army unit in Sinkiang, describes how they caught a creature he describes as a "wild man" in the marshes. A soldier, G. N. Kolpashnikov, during fighting in Mongolia in 1937, saw and described two "wild men" that had been accidentally killed by a sentry. In 1906 an explorer from St. Petersburg, B. Baradin, encountered such a creature at close range during an expedition in Central Asia. In 1905 while returning from Tibet the British adventurer Knight had a similar encounter, as did the British botanist H. J. Elwes in 1906. In 1925 one was seen in the mountains of Sikkim by the Italian topographer, Tombazi.

In 1957 another observation was reported by a hydrologist, A. G. Pronin, in the Pamirs. Fifteen years before that the artist, M. M. Bespalko, had seen something similar in the Pamirs. Geologist B. M. Zdorik, in 1934, saw one in the mountains of Tajikistan that was sound asleep. In 1948, geologist M. A. Stronin saw one of the creatures in the Tien Shans that became frightened and ran across a nearby slope. In the same year a geological engineer, A. P. Agafonov, while in the Tien Shans, saw in a Khirgiz home a family relic, a cut-off and dried-out hand of a "wild man" that was covered on the back with brown hair. In 1954 the Chinese historian, Professor Hoy Vai-Loo, while in a mountain village in Shansi province, caught and used for simple labor a kind of "wild man". Again we turn back to the Caucasus, where a Russian livestock specialist, N. Y. Serikova, in 1950 had hardly begun her new job in Kabarda, and had not yet heard from the local inhabitants about the "Almasti" when she saw one of

140

the creatures at a short distance. She had no idea what it was, but was able to give a detailed description.

This series of observations is not presented for any biological analysis. Considered one by one, each story may be questioned. What cannot be questioned is that such a series of reports exists, that they are independent of one another, and that none of them could have any connection with local folklore. They do not contain any contradictions, rather they support each other.

We can turn now to the information gathered from the local inhabitants in different ethnographic regions. This information is plentiful. Each item by itself is not proof of anything, and if we had only folklore to rely on we would probably reject all of it, since folktales are commonly embellished and transformed by fantasies. However researchers who deal with these traditional beliefs have to face three difficult questions:

Why is it that in each interview within a region, although the topic under discussion, the situation, the whole circumstances may be different, the references to anatomical features of such creatures are consistent and are biologically sound?

Why are there no biological contradictions in information that has been gathered from widely-separated peoples with varying historical, linguistic and religious backgrounds? Variations in names for the creatures are endless, yet sometimes the names are similar among groups that are very far apart, like Amasti, Almas, Albasti, and Goolbiyavan, Biabangooli, Yavan, Gool. Whatever the diversity in names, biologists and anthropologists find only similarities.

Why is it that the enormous amount of folklore is in basic agreement with the observations of strangers who have never heard of the local traditions?

To my knowledge, none of the doubters has ever been able to provide a satisfactory answer to any of those questions. Common sense provides an obvious answer, that the subject of the stories exists, or once did exist, in each of the regions.

As a result of analysis of the abundant information available it is possible to give a preliminary description of the creatures. Their average height is from five to six feet, but there are great variations in size. This is the case also with humans, but to a lesser extent. The entire body is covered with hair from three quarters to three inches in length, but uneven, for instance the cheekbones are covered very slightly.

There is no underlayer of hair, so that sometimes the skin can be seen. The hair grows longer in cold weather. Infants are born without hair. There is not much hair on the hands and feet, and none on the palms or soles.

Color varies with age and with locality, but can be black, brown, reddish, pale yellow or grey. A few are white. The color is not always the same on all parts of the body, and hair does not turn grey evenly all over the body. There is no hair on the face, usually no beard or moustache, but the eyebrows are unusually thick. Sometimes there is sparse hair around the mouth and on the cheeks. The skin of the face is dark, or grey or reddish brown. Hair on the head is usually of a different color from that on the body, and noticeably longer. It is sometimes matted and sometimes falls onto the shoulders or even down to the shoulder blades.

The head leans forward more than a man's, and is supported by strong muscles which make the neck appear to be short and wide with the head right on top of the trunk. The back of the head rises high to a cone-shaped peak. The forehead is low and receding, with prominently protruding eyebrows, and eyes deeply buried in the skull. The bridge of the nose is usually flat, with the nostrils turned outward, however the shape of the nose tends to be variable. Cheekbones are wide and protruding. Jaws are heavy, strong and greatly protruding. The mouth is wide, but almost without lips. Teeth are like a man's but larger, with the canines more widely separated. The ears differ little from a man's ears but have longer lobes and are occasionally somewhat pointed at the top. The eyes appear slanted, and the face seems sometimes Mongoloid, sometimes Negroid.

The creatures are generally upright but shoulders are rounded and bodies somewhat stooped, or forward leaning, with arms hanging in front of the body, especially when walking. This makes the arms appear longer than a man's. The females have large and long breasts which they flop over their shoulders when they run or walk. Probably this enables them to feed their offspring while walking, because they generally carry them on their backs.

The legs are usually slightly bent at the knees, as are the arms at the elbows. Their stride is clumsy, when they walk they sway from side to side as well as front to back with the swing of their arms. Difference in size of the hands and feet is less than in men, and the thumb is less opposed than man's, so that they often grasp objects between fingers and palm, without the thumb. The big toe has independent muscular action, including the ability to turn in opposition to the rest of the foot, and the other toes also are flexible and capable of independent movement. As a result they can walk on tiptoe, using only half of each foot, or put pressure on the forward edges of the foot whether pushing together or pulling apart. Often there is no lengthwise arch. The toes and fingers have nails, not claws.

The creatures are capable of running as fast as horses, of climbing cliffs and trees and of swimming in swift currents. However they cannot climb something smooth, like a pole, where gripping with an opposed thumb is necessary. They move easily in varying terrain, and in snow, although snow is not typical of their habitat. They can squat, and sit on their buttocks. They have been seen sleeping in the daytime face downward with knees and elbows drawn in under their stomachs and their hands behind their heads. For their size they are significantly stronger than men.

Breeding pairs remain together, but the males are inclined to range over a wider territory, whereas females tend to remain in one area while their offspring are young. They give birth to single and twin offspring which resemble human babies. They do not have permanent homes but often find shelter in holes in the ground. The young eventually leave their mothers to find territories of their own. Large old individuals, past the age of breeding, live alone in the most mountainous and heavily forested regions.

Witnesses often report a very distasteful smell. The creatures are capable of many very different sounds and call to each other over long distances, particularly at twilight and dawn. Mooing, mumbling, whining and whistling are reported, but nothing resembling speech. They do not make any tools, but can throw stones and even carry them and use them to build windbreaks. They may break stones one against the other. They do not make fire, but are glad to warm themselves at the embers of a fire that has been abandoned. They may use sticks and clubs.

They eat both meat and vegetable foods. They may dig roots, they eat different sprouts and leaves, all kinds of berries, fruits and nuts. Given the opportunity they will take cultivated vegetables and fruit, such as corn, sunflower seeds, hemp. They catch frogs, crabs, turtles and snails, and will eat frogs' eggs, insects, worms, nestlings and eggs. They may catch small animals such as gophers, moles, rabbits and other rodents, and scavange from the carcasses of larger animals, but they do not hunt large animals themselves. On the few occasions when they kill larger animals it is by breaking the back. They steal lambs, and eat the entire body. It may be that they eat their own dead, since none are ever found.

They are active mainly in twilight and at night. They will use caves or dig their own dens. In northern regions they sleep through the winter, after putting on fat in the fall. However they may wake up and move around occasionally.

Their senses of sight, hearing and touch are acute, and they can move without making a sound and are expert at concealing themselves and otherwise taking advantage of their surroundings. They avoid leaving tracks by walking on hard ground. They are expert mimics, and have been seen to imitate men after an accidental encounter, as if they had a compulsion to do so. They imitate the sounds of animals and birds with great skill, and sometimes seem to have fits of laughter.

Towards man they are usually not aggressive, although in ancient times the Persians and others are supposed to have used them as fighting animals which showed great ferocity on the battlefield. Those domesticated in recent years have been quiet and complacent, but hostile to domestic animals, particularly to dogs. Horses and cows fear them. In the wild their relations with other animals appear to be peaceful.

Professor Porshnev noted that the great obstacle to recognition of the existence of creatures like this seemed to be the fact that church authorities, scientists and other learned persons had been insisting for two or three centuries that such things could not be. People who see one are expected not to believe their eyes. He noted also that the situation is different with Moslems. They accept that the creatures exist, but consider them beings punished by Allah by being reduced below the level of humans to animal men. Where the Christian rejects them as mere superstition, the Moslem considers it shameful to be involved with them, and either way someone encountering such a creature has good reason to say nothing about it. That has been a source of continuing difficulty to researchers.

Another curious problem Professor Porshnev noted, is the insistence of students of folklore that things which exist in old tales can not also exist in reality. They have to be one or the other, so these creatures, which exist in folklore, can not be real. Professor Porshnev could not see that folklore specialists had any particular qualifications to decide whether living beings exist or not.

He refers to organized efforts to find the manlike creatures in various parts of the world and noted that after 1961 the government of Nepal banned expeditions for that purpose. In Russia the Academy of Science had sent one expedition to the Pamirs in 1958, of which he was second-in-command, but it was concerned with much more than looking for the snowman and did not even include a mammologist or anthropologist. From 1959 on, there had been annual expeditions to the Caucasus mountains, all but the first one led by Marie-Jeanne Koffman, a medical doctor, and all conducted without financial support.

Odette Tchernine, in her book *In Pursuit of the Abominable Snowman*, dealt at length with the Russian material, and she has a second volume being published. One of the reports she detailed is that of a female caught and tamed in the Caucasus in the 1800's. Called Zana, the creature was tall and massive, with dark skin and a complete coat of reddish-black hair. She never learned to speak, and was described as having a terrifying face with a purely animal expression. She could outrun a horse and swim a swift river. However she obeyed her master and could perform simple tasks like carrying firewood.

This Zana was reputed to have borne several half-human children, four of whom grew to adulthood and were fairly normal humans, though powerful and dark. Her younger son lived until 1954 and many of her grandchildren are still alive. Zana died in the 1880's or 90's, but there are many extremely long-lived people in the Caucasus region, and Professor Porshnev talked to several who could remember her. Miss Tchernine quotes Dr. Porshnev as saying that one of Zana's grandsons had such powerful jaws he could pick up with his teeth a chair with a man sitting on it.

One of the most intriguing reports from Russia concerns the examining of a hairy biped by a Soviet army doctor, Lieut. Col. V. S. Karapetian in 1941. Text of his report on the creature, as supplied by the Russian Information Service to Ivan Sanderson, and quoted in *Abominable Snowmen* is as follows:

> From October to December of 1941 our infantry battalion was stationed some thirty kilometers from the town of Buinaksk (in the Daghestan A.S.S.R.). One day the representatives of the local authorities asked me to examine a man caught in the surrounding mountains and brought to the district centre. My medical advice was needed to establish whether or not this curious creature was a disguised spy.
>
> I entered a shed with two members of the local authorities. When I asked why I had to examine the man in a cold shed and not in a warm room, I was told that the prisoner could

not be kept in a warm room. He had sweated in the house so profusely that they had to keep him in the shed.

I can still see the creature as it stood before me, a male, naked and bare-footed. And it was undoubtedly a man, because its entire shape was human. The chest, back and shoulders, however, were covered with shaggy hair of a dark brown color (it is noteworthy that all the local inhabitants had black hair). This fur of his was much like that of a bear, and 2 to 3 centimeters long. The fur was thinner and softer below the chest. His wrists were crude and sparsely covered with hair. The palms of his hands and soles of his feet were free of hair. But the hair on his head reached to his shoulders partly covering his forehead. The hair on his head, moreover, felt very rough to the hand. He had no beard or moustache, though his face was completely covered with a light growth of hair. The hair around his mouth was also short and sparse.

The man stood absolutely straight with his arms hanging, and his height was above the average, about 180 cm. He stood before me like a giant, his mighty chest thrust forward. His fingers were thick, strong, and exceptionally large. On the whole, he was considerably bigger than any of the local inhabitants.

His eyes told me nothing. They were dull and empty— the eyes of an animal. And he seemed to me like an animal and nothing more.

As I learned, he had accepted no food or drink since he was caught. He had asked for nothing and said nothing. When kept in a warm room he sweated profusely. While I was there some water and then some food (bread) was brought up to his mouth; and someone offered him a hand, but there was no reaction. I gave the verbal conclusion that this was no disguised person but a wild man of some kind. Then I returned to my unit and never heard of him again.

Just south of the Caucasus is the border of Iran, from which the Society for the Investigation of the Unexplained has one second-hand report. It is a letter from a New Jersey man to Dr. Bernard Heuvelmans, a pioneer in the search for unknown animals. The writer quotes an army buddy as telling him that while working in Iran before the Second World War he had seen what he called a "gorilla" which some local men had killed in the mountains. The friend knew that there was not supposed to be any apes in that part of the world, but the thing was as big as a gorilla and looked like one. The friend's name was Daniel Dotson, but the letter writer had lost track of him.

Other than from the Caucasus, which are considered to be the southern border between Europe and Asia, I have only one European report on file. It was published in the newspaper *Arriba* in Madrid on February 27, 1968, and refers to a strange creature reported seen at night near Gerona at the eastern end of the Pyrenees. There had been several reports from children and then from adults regarding it, but they were ignored until it was finally seen in the daytime.

The animal was drinking from a pond near the house of the witnesses. It fled, leaving in the clay soil a number of great footprints 40 centimeters long (nearly 16 inches) and resembling those of a plantigrade being (one that walks on the entire sole of the foot). These footprints agree with the description given by the motorist Ruperto Juher, who said that he had seen, near Hostalrich a few days ago, an animal with a large hairy body and long arms, that crossed the highway in front of him, walking with a weary sort of gait.

One would think that there might be reports from the forests of Scandanavia, and Norwegian friends of mine have told me that they have heard of giant footprints in the snow in the mountains there. There is also a similarity between the sasquatch and some descriptions of trolls. However I have nothing specific from that area.

Plainly the Russian information is significantly different from that from other parts of the world. The question is whether it refers to a different creature. I think that it probably does. There is nothing in the reports from the Himalayas or from North America to indicate the slightest possibility that the creatures in those places ever were or could be domesticated and taught to participate in human activities. Professor Porshnev eventually concluded that the material he had accumulated referred to Neanderthal Man, and if the story of Zana is true she must indeed have been extremely close to Homo sapiens genetically, able to produce cross-bred offspring that were fertile themselves. The matter is by no means certain, however, since the way of life of the Russian creatures in the wild is apparently very similar to that of the sasquatch and some other animals, while it does not resemble at all that generally ascribed to Neanderthal Man. There is no proof that the Neanderthals were not hairy, but they did use fire, and a good variety of stone tools, and they had been doing so for a hundred thousand years.

The Russian creatures are apparently little more than man-sized. In North America we are dealing with a giant — estimates of height averaging more than 7½ feet. A difference of about 20 inches in height may not seem like a great deal, but it is. If the height of the North American creature is one quarter greater than that in the U.S.S.R., and if the other two dimensions are in the same proportion,

147

the weight must be greater by the cube of 1.25, which is 1.953, or approximately double. That calculation assumes that both have the same proportions, but they do not. Although he is constantly making comparisons with man, Professor Porshnev does not even mention body build or weight, whereas North American descriptions frequently stress sheer bulk. The female in the Patterson movie, less than seven feet high, scales out at no less than 500 pounds, probably a good deal more. No doctor examining such a creature would settle for "considerably bigger than the local inhabitants." On the Pacific Coast, which is the area primarily under study, average height is not 7½ feet but eight feet. Using that height estimate involves cubing one and one third, and if in addition the creatures were built one third heavier for their height, which seems to me a conservative estimate, then their average weight is triple that of the Russian creatures. In my opinion it is probably more.

That does not by any means establish that they do not belong to the same genus, or even the same species, but it puts them in a different relationship to their environment. For instance the larger variety would require a great deal more food, so presumably would need a greater range and have a smaller population. But it would be capable of killing larger animals and would have fewer predators to fear. As a matter of fact, a full-grown sasquatch in North America would have no need to fear any predators at all, unless it met a killer whale while swimming. It might give a wide berth to a grizzly or a polar bear, but it would presumably be a relationship of mutual respect. In Mongolia, on the other hand, the smaller 'Almas" shares its range with the largest tigers in the world. Also there are numerous references to the creatures being killed or captured by humans in various parts of Russia.

There are also a number of lesser differences that should be noted, although remarkably few. The North American creatures do have hair on the face and the color and length of hair are normally not noticeably different on the head and body. Their shoulders are not rounded, their stride is smooth-flowing, not clumsy, and there are no descriptions of breasts that could be flopped over the shoulders. Sasquatches have not been reported to build windbreaks or to carry stones for any reason, and if they use sticks it is not to any significant extent. They are not known to dig dens or use caves. Attributing sounds to creatures not seen in the act of making them is a very chancy business, but the impression I have of sasquatch vocalizations is a lot noisier than what the Russian list suggests. Whistling is on both lists, but mine stresses screaming, bellowing, growling and grunting, and does not include mooing or mumbling.

A lot of things that Professor Porshnev mentions, such as hairless

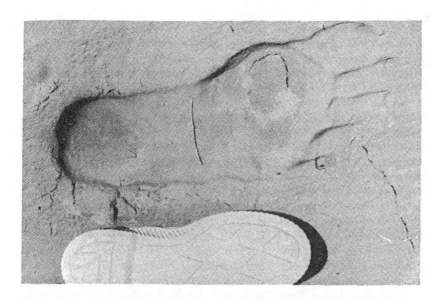

Photo of a 15" footprint of a Kaptar in the Tien Shans.

offspring and the occurence of twins, touch areas about which there is no information in my files. It should be noted, however, that he is using folklore as a source and I am nòt. Russian researchers have recorded hundreds of observations in modern times, but whether all of the behavioral information and physical descriptions Professor Porshnev gives are covered by those observations or whether some are based only on folk beliefs, I don't know.

There are two points on which I do have information that particularly intrigue me. One is regarding sleeping position. The only report I have of a sasquatch seen sleeping, in fact two of them, describes them as having their arms and legs drawn in underneath them as they lay with their backs to the sky. The position was identical to that described in Russia except that their hands were under their heads, not on top of them. The other concerns the grasping of objects without using the thumb. Observations of sasquatches grasping things are rare, but I can recall none that specifically mention the use of thumb and fingers in opposition, whereas one man who had watched two of them from a distance using their hands for picking up rocks and stripping leaves off bushes wondered whether they had any thumbs.

I don't know much about footprints in Russia except for one set from the Tien Shans which Professor Proshnev said some years ago was the best to have been found up to that time. He sent me a photo of one of the prints, which was 15 inches long, 3½ inches wide at the heel and five inches wide at the ball. That is long enough for a fair-sized sasquatch print, and it is suitably flat, but in width and in the shape of the ball and the toes it is far more like a human print than any I have seen in North America.

To sum up, there are reports of hairy bipeds, both large and small, from all over the world, so that it appears entirely possible that there could be more than one species and that one or more species could have world-wide distribution. Reports of their appearance and behavior indicate more similarities than differences, again leaving open the possibility that they could be either one type of creature occupying a variety of habitats or several different types with a lot of characteristics in common. In most parts of the world no significant amount of research has been done, the exceptions being several areas in North America and the U.S.S.R., and to a lesser extent in Nepal. In every area there is information suggesting that there may be more than one type of creature, but nevertheless one type predominates in each case. In very approximate terms, the North American variety is a good deal larger than the others, while the Russian one is taller than the Himalayan but perhaps not heavier. The Himalayan creature, on the evidence both of its description and its footprint, is entirely unlike a human. The Russian variety, on the other hand may be very human indeed. Recall the medical officer's statement, "it was undoubtedly a man, because its entire shape was human." The Tien Shan footprint bears that out. In its similarity to mankind, the sasquatch ranks between the other two in all but size.

It is perhaps a mistake, however, to ignore the immense distances within Russia. Mongolia, the Tien Shans and the Pamirs are all much closer to the Himalayas than to the Caucasus, in fact the Pamirs and the Tien Shans are almost a northern extension of the Himalayas.

It may well be that the predominant species in one part of Russia is not the same as in another part. In any event the general opinion of the Russian researchers seems to be that they are not dealing with ordinary animals, and Dmitri Bayonov and I have engaged in a protracted debate by mail in which he contends that no "Homi", as he calls them should ever be killed.

My own opinion, which isn't worth much, is that these world-wide reports indicate the existence of not one, but several bipedal primates other than Homo sapiens. I think that the sasquatch is probably Gigantopithecus, while the Kaptar, Almas and company descend from one of the more manlike fossil types. I know of no probable ancestor for the yeti or for the various little creatures, and there is no reason why all four could not be animals for which no fossil forbears have yet been found.

150

-8-

Expeditions and Institutions

To watch the sasquatch hunt you really should have a program, but there aren't any. Without one there's just no way to follow what is actually going on. Reading the newspaper and magazine articles or watching the interviews on television, a person with a casual interest in the sasquatch must get the impression that the woods of the West are overrun, in season and out, by teams of scientists and expert hunters, equipped, directed and supported by major scientific institutions of infinite resources and international repute, pursuing elusive hairy giants round and round the mountains. Then as if that were not enough, there are the people who are forever being visited in their camps by such creatures, people who have found the caves where they live, people who can take a movie of one almost any time they feel the urge, and people who find footprints so often that they have to keep rubbing them out to hide them from the gun-happy hordes.

The Interplanetary Wildlife Conservation Society has thrown all its resources into the search. So has International Wildlife Research. You can read all about it in the papers. The North American Anthropological Foundation is active. The Monster Information Center and Exhibition has its top men on the job, "evaluating" reports for the guidance of those less knowledgeable than themselves. American Yeti Research was recently in the field, parading the support of a national wildlife organization. (Those names are changed just a little. I hope I didn't name any actual organization by accident.) In earlier years the Northwest Research Association was in

151

the forefront of the search, and back at the beginning there was the Pacific Northwest Expedition. Nor have those mighty organizations been half-hearted in their efforts. They have flung in their leaders and their deputy leaders and their co-deputy leaders. Their directors and their executive directors go bravely into the field, even their presidents. It is just as well, because if they didn't there would have hardly been anybody there. The British Columbia Expedition, besides the leader, deputy leader and co-deputy leader, had a cook.

There's nothing stopping people from calling themselves anything they like. There have been lots of other names, not so all-embracing: Vanguard Research, Project Discovery, South Mountain Research Group, to mention a few. The thing to remember is that the name doesn't tell you anything. There may be an actual organization there, as was the case with the California Bigfoot Organization, or there may be just one man. Behind the name may be a dedicated effort to contribute something to human knowledge, a con job aimed at a fast buck, or something in between. There is no way an outside observer can tell which is which. What he reads in the papers or sees on TV is not likely to be any help, since the self-seekers, not the honest plodders, are most likely to bring themselves to the attention of the media.

The fact of the matter is that no institution with the kind of personnel and resources that could be expected to achieve anything has ever shown the slightest interest in investigating the mass of information already available indicating the existence of hair-covered manlike animals in North America. A few individual scientists have volunteered to attempt to identify hair or droppings, and a few have written something on some aspect of the subject, but those are strictly private efforts. There have also been some people taking part for a month or so in one or other of various "expeditions" who had scientific degrees. That has made it possible to issue truthful press releases referring to "expedition scientists", but I am not aware of any case in which such a person had a scientific training giving him any particular qualification for what he was actually doing on the "expedition." A biologist, for instance, may be concerned with cell structure, or enzymes, or algae. He is much more likely, in this day and age, to be trained in something like that than in animal behavior in the wild. If what he is actually doing is blundering through the bush hoping to catch sight of an apelike biped, his real qualifications are those he acquired through any hunting experience he may have had. If he is not a hunter, then he hasn't much going for him—but to the press he is a scientist on a scientific expedition.

As to the "expeditions" themselves, the use of the word is not improper, since its range of meanings includes "journeys made for some definite purpose". The picture the word paints for the public, however, is probably far different from the reality. If you can have an expedition living in a tent or a motor home in a campsite along with assorted vacationers, then there have been expeditions. If you can have an expedition walking on the same trails or driving the same back roads as people who are going about their daily business logging or surveying, hunting or prospecting, or berry picking, fishing, hiking or just goofing off, then there have been expeditions. Apparently some people do find that sort of activity impressive. The Explorers Club actually gave one of its flags to two men who went sasquatch hunting in British Columbia in a land plane, "exploring" territory perhaps never penetrated before, except, of course, by the people who built the airstrips.

I am not making such statements to make fun of the people involved. I am one of them myself, a retired co-deputy leader. In many cases they may have been doing the very best thing that they personally could do to try to find a sasquatch. A few of them may actually have succeeded in seeing a sasquatch, although only once was any kind of proof obtained. Several have found footprints. What I do want to do is dispel any impression that substantial resources have ever been devoted to the search for the sasquatch. That has never happened, and there is nothing to suggest that it is likely to happen in the forseeable future.

Publicity given to people involved in the sasquatch search is of great value, since it makes it possible for those who have information to contact them, and that frequently happens. But those who seek publicity by exaggerating the significance of their efforts bring about some unfortunate results. Many people must by now be thinking that if there really were any unknown animals to find, one of the "scientific expeditions" would surely have found one. Other people, while still open-minded about the possibility that the creatures actually exist, are increasingly susceptible to the argument that they must have great intelligence in order to elude the legions of experts who are pursuing them. Both impressions are without any actual foundation, since the impressive search efforts are an illusion.

Another misleading term is "scientific advisory board." That or some similar term indicating the participation of large numbers of scientists is frequently used, and sometimes the names of the scientists are published. The people are almost certainly real, but the "board" is not, in any active sense. The scientists have been asked to lend their support, perhaps only asked if they would be willing to examine a sasquatch when one is caught, and they have agreed. They don't have anything to do with the organization's activities: if there are any activities: if there is an organization. The situation could change, of course. At any time some real scientific institution could mount a major expedition in search of a sasquatch, but so far there is no sign of that happening.

Then there are "computer studies". I organized one myself in which a dozen investigators participated, and plugged away at it for four years. I know of other people who have also used printed questionnaires, but have not seen any indication that they went very far with them. By the summer of 1974 I had about 300 completed questionnaires, and two thirds of them had been key punched at the University of British Columbia. Since the professor involved, Dr. Carl Walters, was going away for a year, we spent a few hours seeing what could be done with the information in the computer, and that's all the time it took to show that we weren't getting anywhere.

For simple correlations you don't need a computer. Anyone who claims to have found migration routes with a computer is talking nonsense. If there were a pattern of reports indicating migration anyone could spot it without even a pencil and paper. Unfortunately there is no such pattern. Even such a question as whether there is a correlation between night sightings and the amount of moonlight can be worked out for a thousand reports in a few hours, with no electronic assistance. I have done it myself. (There isn't any.) When dozens of people have been looking for answers for decades you can count on it that they have thought of most things. If you want to test a whole series of things for relationships to each other, such as whether sasquatches seen early in November are more likely to be below 3,000 feet on rainy days than when the sun is shining, then the computer should be the thing to provide an answer. The only trouble is that first the computer has to have the information. When you consider that an average sighting might consist of something hair-covered running upright across the road in front of the observer's car, and that in most footprint reports there is no way of knowing when the prints were made, let alone anything about the creature that made them, it is obvious that a lot of the questionnaires contain no answers to a lot of the questions. With only a few hundred reports you may average less than 100 answers to many of the specific points you are investigating, and if you are trying to correlate three or four things, the number of reports in which all your points are covered may be only a dozen or so.

If the result is that every sasquatch seen early in November on a rainy day was below 3,000 feet and every one seen on a sunny day was higher up, then you may be getting at something even though you have only a dozen answers. If you find it's 60 per cent one way and 40 per cent the other out of 1,000 answers, then you are probably learning something too. But if you get a 60-40 answer out of

20 reports it doesn't mean a thing. The results we got out of the computer were of that sort—of no statistical significance. If we had 3,000 questionnaires done instead of 300, and the ratios stayed the same, some of our results would mean something, but the resources to gather that quantity of information are not available, even though that number of informants undoubtedly is.

Doing the computer survey was valuable in that it resulted in re-interviewing a lot of people some years after the original contact, and contacting for the first time people involved in reports that had been known for a long time but never investigated. The printed questionnaires also resulted in obtaining more specific details than had ever been recorded before. But if anyone claims to have learned where to find a sasquatch as a result of a computer study, take it with a pinch of salt.

Then there are the people who have already found the sasquatch, so they say. Not all of them catch the attention of the media, but some do. It is by no means impossible that there are people somewhere who have a sasquatch picking over their garbage dump or that someone knows where a sasquatch can be seen frequently for some other reason. For years I have been half expecting to hear from such a person. It is also very probable that there have been repeated sightings at the same spot or within a small area for brief periods of time. A number of such episodes have produced some of the most convincing reports, with many witnesses. Sometimes investigators have been on hand before things stopped happening, and one or two claim to have seen something themselves. But in every case the reports have petered out without anything being proven.

That kind of report appears to be legitimate, but there is another kind. There are individuals and groups who claim repeated sightings at some secret place, or a place which they will show to only a few selected people. Sometimes the secret is for sale. So far, people making such claims have never produced anything to substantiate them. Such stories can do a lot of damage, by giving wide publicity to imaginary information about what sasquatches are and do. They also nourish the general tendency to reject the whole subject, by building and then disappointing expectations that the mystery is close to solution.

To avoid being taken in by the more obvious frauds, here are some things to watch for:

Mention of long fangs or claws.

Sasquatches using caves or having homes.

Stories of tracking sasquatches through the woods, or being able to find them at will, or having seen one—particularly the same one—at more than one location.

Sasquatches supposedly captured, or filmed, but for some reason the animal or film can't be produced right now.

Claims of being able to find signs of where the sasquatch have been eating, or sleeping, or marking their territories.

Communicating with sasquatches in any way.

Exchanging things with sasquatches.

Secret methods of attracting sasquatches.

Knowledge of sasquatch migration routes.

With one exception, those things are not out of the question entirely, but they are extremely unlikely, and in the past they have cropped up frequently in stories that turned out to be fraudulent. In no case have they been shown to be true. The exception is any reference to claws. Sasquatches do not have claws. They have nails. They don't have fangs either, but descriptions indicate that their eye teeth are sometimes longer than the other teeth, and an honest observer might overestimate that characteristic. Still, they don't appear to have teeth evolved for fighting; they don't look like a carnivore or a TV monster.

Some of these points are dealt with thoroughly elsewhere, but here is a quick run-through: There seems to be a natural tendency to assume that sasquatches use caves. Newspaper stories about sightings sometimes mention that there are caves in the vicinity, as if that were automatically of some significance.

Legends very often make cave dwellers of the sasquatch, and people with elaborate made-up stories follow the same tendency. They find sasquatches sleeping in caves, or learn the location of the caves in which they raise their young, or where they store the things stolen from humans. In fact, animals do not make much use of caves. They get along fine without them, either living in the open or digging their own dens. There is nothing to indicate that sasquatches have homes of any sort, any more than any of the other great apes do. I suppose a sasquatch might go in a cave to look around or to get out of the rain—if they bother about getting out of the rain—but I don't know of one single substantial report directly linking a sasquatch to a cave. A point worth remembering in that connection is that there are an awful lot of people involved in exploring caves, far more than are looking for sasquatches, yet no sasquatch information comes from that source.

A built-in pitfall for people with fake stories is that they have to make them sound too good. Reports of ordinary sightings are a dime a dozen. They don't attract the kind of attention that would justify bothering to make them up. If a faker wants notoriety, or wants anyone to make a fuss over him, he has to come up with something special, but if he claims to know where the sasquatch is, or how to track one, he can be called upon to prove it. Quite a few have made such claims, but none has yet been able to show proof. One or two have even claimed to have established contact with individual sasquatches. The more elaborate the story, the more likely it is to be fiction. It isn't possible, in most cases, to prove that anyone is lying, but when someone makes such claims and yet produces nothing year after year there is no mistaking the inference.

The same type of conclusion is even more obvious regarding the few people who have claimed that they have already captured a sasquatch, or that they have a film of one. There was even a man who combined the two claims. He had a captured sasquatch and had thousands of feet of film of it. However the world is still waiting to see either the animal or the film.

The media haven't shown much inclination to fall for anything that obvious, but they often do bite, perhaps tongue in cheek on some occasions, on stories told by people who emerge from nowhere claiming to be sasquatch experts, able to do such things as pointing out where they have been feeding and sleeping. One man said he had discovered their survey markers (any broken-off sapling) and their comfort stations (old stumps). He actually got a lot of newspaper space. In another story, carried by the wire services, several members of a family told of seeing sasquatches repeatedly at a certain place and finding their footprints so often that they had to keep obliterating them for fear some hunter would see them. There have also been people who claimed to be able to communicate with sasquatches. In spite of the fact that these things are never demonstrated the people who tell such tall tales are sometimes given considerable publicity.

The things mentioned so far usually show up in stories told by people who emerge from obscurity, make a bid for attention and eventually fade away again. It is different with the last two items on the list: secret methods of attracting sasquatches, and knowledge of sasquatch migration routes. Those claims have sometimes come from people who actually are active in sasquatch investigation. Typically they will be made by someone looking for a backer to finance his efforts and in need of some plausible reason why his chances of success would be good enough to justify the investment.

It goes almost without saying that the only way to prove the effectiveness of any method for attracting sasquatches would be to actually attract some. Anyone who succeeded in doing that should have something to show for it, and would hardly be in need of publicity, or a backer. As to migration routes, almost everyone who has taken an interest in finding the creatures has looked for evidence that sasquatches travel past certain points at definite times of year, and many have gone through periods of thinking that indications of such behavior had been found. With one exception, in every case that I am aware of the gathering of additional information or an attempt to make use of the supposed information has resulted in the apparent pattern breaking down. The exception involved finding the footprints of the same individual in places about 100 miles apart at different times of the year. To that extent it may be that sasquatches migrate, but that is not the sort of thing that would enable a person to take up a station and intercept a migrating sasquatch on the way by. If someone who claims to know of such a migration route has been involved in the hunt for any length of time and is still talking that way, it is a fair inference that he is trying to sell a bill of goods.

What is really going on in the sasquatch hunt is that a few dozen people are spending a few weeks a year in actual hunting, and even that small number is divided up. Some are hunting with guns, some with cameras, and some with no preparation to do anything at all should they actually encounter a sasquatch. There may be a few with tranquilizer guns—they belong in the same category as those with nothing at all. If one of those with a camera got a picture it would certainly be interesting, but it would only be more of the type of evidence that has already proved unconvincing. Only the few with guns really offer any prospect of a possible solution to the mystery.

In addition to the people actually out beating the bushes there are people who will travel to enquire into reports of sightings and footprints. There are probably less of them than there are of the hunters, and of course some individuals are in both groups. Most of them are limited to the time, money and equipment they can put together while making a living doing something else. From time to time there have been a few people who have been able to obtain substantial financing from private backers, but that kind of money seems scarce at time of writing, and will probably remain so, as the odds against any particular expenditure paying off become more obvious every year. Some money comes in from sale of books, film showings, lectures and other sources, but only three or four people have been able to make their efforts in the sasquatch field self-financing. It is obvious that under those conditions neither hunting nor research can proceed on any but the most insignificant scale.

A somewhat larger number of people collect any information about sasquatches that comes to their attention, and a lot of the people who are involved in various ways keep in touch with each other and exchange information, but even that activity is fragmented. To quite a few of the participants the sasquatch hunt is a competitive activity, and information that they think might prove useful they keep to themselves. That attitude may seem strange to the outsider, but it is

158

entirely understandable. Sasquatch hunters don't get salaries or research grants, nor do they accumulate anything of value. Some are content with the satisfactions they receive from participating in an interesting activity, but all are aware that fame and fortune probably await the person who succeeds in proving the existence of the creatures. Quite a few of the most active sasquatch hunters are gambling on being the lucky winner. It is in their interest to co-operate to some extent with the others, in order to keep abreast of what is going on, but anything that seems like a really good lead will not be shared.

In addition, those who have acquired a store of information through prolonged effort are not keen on too much being made available to newcomers in the field. Some are going to be upset when they see how much is in this book. Even if everything were available for study by everyone there is no indication that gathering and investigating reports would contribute anything significant towards proving the existence of the creatures or making it possible to find one. The blunt fact is that the existence of the sasquatch will not be accepted by science until some part of one is available for study. Only the people who go out with guns to collect one are doing anything that could directly fulfil that requirement, and on an individual basis the odds against them are fantastically high. That such a small and fragmented effort could produce results is really not to be expected. The remarkable thing is that it did produce a motion picture, and had Roger Patterson and Bob Gimlin used their rifles instead of a camera they would probably have brought the hunt to a successful conclusion a decade ago.

-9-

Ivan

Once when I called on Ivan Marx he had a big cage full of magpies on a table in the kitchen. Not everyone I know has a cage full of magpies, in fact I can't offhand recall anyone who has, so I asked him how come.

"Well," Ivan said, "magpies are meat eaters."

He went on to explain that he had been following the activities of a couple of sasquatches. They had hibernated through the winter, and when they first started to move around again they cleaned up all the old cougar and coyote kills they could find. Ivan figured that the racket the magpies made when they were feeding would be helping the sasquatches to find those kills, so with some magpies of his own all he would have to do would be to stake the birds on a hillside in the early spring with something to squabble over and they might attract a sasquatch within range of his rifle.

So, he said, he and Peggy had found a nest of fledgling magpies and brought them home and raised them in a cage.

That sounded sensible enough to me, and in any event, there were the magpies. I asked Ivan how his plan had worked out. It didn't, he said. What, I asked, was the trouble?

"Well," (you could sort of feel his story-telling machine settling into gear) "magpies can talk."

Those magpies, he explained, made such a racket there in the cage in the kitchen that he and Peg got to thinking they sounded like the Farkel family in "Laugh In", so they got in the habit of saying "Hello, Fred Farkel" to the magpies as they went by the cage. After a

160

while the magpies started saying it back to them, and they got pretty good at it.

"It just wouldn't have worked," Ivan went on. "We'd have had those magpies staked out on the hill and here's this sasquatch comes slipping through the trees, and they're chattering away, then he shows up and a magpie looks at him and says 'Hello, Fred Farkel'."

Ivan is a game guide who lives near Burney, California, in a concrete house that is all one very long room, so he'll have a good place to show his wildlife movies. At the time of the magpie incident, however, he was living at Bossburg, Washington, chasing the cripple-footed sasquatch.

Before I ever met Ivan, in 1960, I had heard he was the biggest, well, yarn-spinner, in California. He was also a good bear hunter, with well-trained hounds, and Bob Titmus had hired him for a few months in the off-season to try to get the hounds to run Bigfoot for Tom Slick's Pacific Northwest Expedition. It was after a spell in the back of Ivan's jeepster with the hounds that Rene Dahinden abandoned Bigfoot and headed home for British Columbia. By the time I arrived to spend a month with the expedition Ivan had just signed off the roster to guide some bear hunters, but we saw him around a few times, and an hour with Ivan was better than a show, anytime.

By 1967 Rene had cooled down a little, and when we were making our swing around the sasquatch states we stopped off in Burney to say hello to Ivan. He welcomed us like long-lost bosom pals, and we had an enjoyable visit. Ivan had a big cage near his house, and there was a cougar in it. Not a very friendly one either. I forget how he came to have it, but another time he told me all about how he catches cougars. After his dogs have treed one Ivan goes up the tree after it with a long rope with a noose on the end, and a stick. He crawls out on the limb towards the cougar, and while it is concentrating on him he just slips the noose around its neck with the stick. Then he goes back down the tree and has his friends yank the cougar into mid air with the rope. The cougar lands on the ground and Ivan grabs it by the tail.

A cougar, Ivan explains, always tries to get away if you pull on its tail. It never turns around and claws you. Also, a cougar is short-winded, so if you just hang onto its tail for a minute or so while it tries to get away, it runs out of wind and collapses, and you have your cougar.

For all I know about catching cougars, maybe that's the way to do it, but I don't intend to try it and I don't suggest it to anyone else either.

The next I knew of Ivan, he phoned me in November, 1969, while

161

I was on my way to eastern Canada. When I got the message and returned the call, Ivan said that he had moved up to Bossburg, near Colville, Washington, where there were lots of cougars with no one hunting them. Just lately, he said, a sasquatch had been seen around there several times, and it was coming to feed out of roadside garbage cans. There wasn't anything I could do about it, since I wouldn't be back that way for a month, so I phoned both Rene and Bob, reminding each of them that they knew Ivan better than I did, and should talk to him themselves; as far as I was concerned they were on their own.

That was the kickoff of the great Bossburg sasquatch hunt, which went on until the spring of 1971. There are at least eight imported sasquatch hunters who will never forget it, plus any number of local participants, and it produced some footprints that have greatly impressed at least two physical anthropologists. Whether there ever was a sasquatch around Bossburg I'm sure that I don't know. It depends on whether Ivan took over as producer before the curtain opened or half way through the third act. Whichever it was, he starred in the show from the first.

When I came through Bossburg on my way home a few days before Christmas, Bob Titmus had already written it off and gone home to Kitimat, but otherwise the hunt was in full swing. Rene had acquired two sidekicks, Roy Fardell and Roger St. Hilaire, and was living in a borrowed trailer in Ivan's back yard. Roger Patterson and Dennis Jensen were around too. Ivan was not taking any part in the search himself, which was odd, since he and Rene had together discovered more than 1,000 tracks of the crippled sasquatch in the snow a short distance up the road, and he had a frozen footprint of compressed snow in his freezer. What Ivan was doing was holding court, just about every evening in his living room.

The place Ivan was renting was a hovel. The only water was from a single tap, and the outhouse was so full it leaned like the tower of Pisa. There was a hole right through one wall in the kitchen. The living room, which housed Ivan's expensive radio equipment and the leather-upholstered furniture a pet cougar had ripped up, had knot-holes in the floor and smoke-blackened walls. At one side of the room a bedroom door hung on one hinge. Ivan didn't care. He sat back in his big leather reclining chair and played to the audience, while Peg brought in cups of coffee. She didn't seem to care either. Once I was there while Ivan was skinning a bobcat on the living room floor. He'd sit back in his chair and talk awhile, lean forward and take a couple of whittles at the cat with a penknife, then sit back and spin another tall one. When he finished, Peggy came in and sloshed a bucket of water on the blood, then swept it down a knothole.

Ivan's routine wasn't all extempore. There were some set-pieces in it. He had an old, old dog named Jeremy, who could just manage to sleep and waddle. He also had a special board he called the "Scare the Jeremy Board". Two or three times in the course of the evening Ivan would hold up a hand for silence and announce:

"It's Scare the Jeremy Time!"

He would then take the special board and sneak up on Jeremy, who was dead to the world, and drop the board flat on the wooden floor beside his head. Jeremy would leap up in consternation, and Ivan would make a tremendous fuss over him, sympathizing with tremendous sincerity over the terrible things that people did to him. I think perhaps the regular stimulation was all that had kept Jeremy going for the past couple of years. One thing sure, he didn't have a weak heart.

I don't know just what kind of a dog Jeremy was before he got fat, but he wasn't a hound because he had short legs. The hounds were kept tied outside by their kennels, but the cats had the run of the house. The robin and the crow had their own cages (this was before the magpies) and there were also two Chihuahuas, Finky and his wife. Finky was a hairy Chihuahua, and his wife, who was of the hairless variety, was bigger than he was. So were the cats. Of all the four-legged creatures in the Marx household, Finky was at the bottom of the pecking order. Since the robin was safe behind bars he couldn't pick on anybody. I doubt he could have handled the crow. The only thing Finky could beat up was his own personal rubber tire, solid rubber about six or eight inches across. When he was in an agitated mood, and he was an excitable little creature, he would find that tire and savage it until he felt better.

Finky, Ivan said, was his sasquatch dog. He had trained him to kill sasquatches. And once or twice a night he would demonstrate how Finky would do it. He would gather Finky under his left arm, with his head sticking out at the back and his tail at the front. Then he would crank Finky's tail in rapid spurts with his right hand, hissing the command, "Kill, Finky! Kill!"

Released, Finky would explode like a string of firecrackers, darting back and forth around the room until he found his tire, which he would kill gloriously, to his and the audience's complete satisfaction.

Ivan is one of the few men I have known who seemed to really love the life he led. It didn't matter whether he was loafing around the house day and night or up before dawn to go hunting with the hounds, and those are almost the only things I ever saw him do. Someone asked him once if he had always lived like that, and he explained that, no, when he was a youth he wanted to finish school and get a job just like everyone else. He did it too.

The first day, he said, he found he had to put a card in a clock to be punched before he started work, and again before he went home. The second day, he did it again. The third day he started off to work alright:

"But I just didn't have the strength to punch that card."

From that day, so he said, he had never done anything except hunt with his hounds and his gun and his cameras. And even though Ivan said it himself, as far as I ever saw it might be true.

I was back in eastern Washington in the summer of 1970 while I was doing interviews for the computer survey, and I stopped off to see Ivan. The visiting sasquatch hunters were gone at that time, having seen nothing of interest after the tracks in the snow, although they had found a couple of people who gave excellent sighting reports.

When I was last there early in the year, Ivan was still staying aloof from the great hunt that was swirling around him, but he confided to me that in the spring he and Peggy had started taking trips off into the mountains on the west side of Lake Roosevelt, and they had located the winter dens of a couple of sasquatches, deep in tangled forests of small evergreens, both standing and fallen, where no one could possibly approach without enough thrashing and crashing to wake the dead. It was then that they learned that the animals started the year by cleaning up the previous winter's kills, and after that they hung around the receding snow line, eating the lush plants that shot up as soon as the sun hit the damp earth. He had never seen them, but he had found the tracks of both the cripple and a larger sasquatch often enough to keep track of their movements.

The crippled footprint was one of the biggest ever photographed or cast, 17 inches long and very wide, but the casts of the new print were considerably larger and wider. For a while in the spring he had lost track of them, but early in July a girl had found tracks on a beach at the North Gorge campground. The tracks were both in and out of the water, and speculation was that the creature had been catching carp that spawned there.

That fall I was back again in response to big news. Ivan had a movie of old cripple-foot. He had kept on shadowing the two giants during the summer, and had found and cast huge handprints in the mud where they lowered themselves to drink. He had found bushes they had broken down in feeding on leaves and berries, and even a place they used for a comfort station, leaving their droppings buried deep in a mudhole. He had made color movies of all those things, but the movie of the creature itself had come by accident.

While Ivan was away from home someone had recieved a phone call from a man who said he had seen a sasquatch hit at a place where the highway and the railroad ran side by side. The message was a little garbled, there was the informant's car there and also a train, but it wasn't clear which hit the sasquatch. The man had driven on to Trail, British Columbia, and phoned from there.

Ivan had been doubtful of the message until he went to the place indicated and found scuff marks. He had one hound that he thought would chase a sasquatch—since it once treed a hunter—and sure enough it took the track. Off went Ivan after the hound, while another hunter drove around to a road that crossed up ahead, to listen for the baying if the trail led that way. The rest of the party stayed behind. After a while the other hunter heard a scream off in the distance. Then, some time later, Ivan returned. He kept the movie a secret for quite a while, but eventually he let it be known that he had seen the creature that day, and filmed it.

The animal had been injured in the accident, he said, and had dug itself a hole to lie in while it healed. The arrival of the dog had only annoyed it, but when Ivan got there it heaved itself painfully to its feet and lurched off, all stiffened up, through the bush. Part of the area was open, and the sasquatch couldn't make very good time in its battered condition, so whenever it had to cross an open space Ivan was there with his camera. Once the dog rushed it, but it screamed and the dog took off. Eventually it worked out some of the stiffness and was able to outdistance the hunter. The animal was about 10 feet tall, Ivan said, but was extremely thin. He thought it was very old as well as being hurt. Beyond doubt it was old cripple-foot itself.

The movie showed just the things Ivan had described, but it turned out to be badly underexposed, so that no detail of the black animal could be clearly seen. Its movements were awkward and jerky, but the injuries explained that.

There seemed to be several local people who had been with Ivan when various items of evidence had been found. As the only professional hunter around, and a good one, Ivan was just the man to have done the things he said he had. None of his tall tales had ever involved anything resembling the months of careful stage-managing that would be necessary if all those casts and movies were not genuine. The concensus was that everything was probably okay. I thought so. Ivan flatly refused to show anyone where the movie had been made, but the reason for that was that it was a place the sasquatches had been using on other occasions, and he hoped to find one there again. He told me that next time he would shoot one, so that made things fine with me. Whether the story was true or false, Ivan didn't need any help, and I went back to my own activities, hopeful of further news.

For most of what happened later I have only secondhand

information. Ivan had various offers for the film, but never sold it. He did hire himself out to a man who claimed to head a conservation organization from Washington, D. C. and suddenly the new man was playing the great sasquatch hunter, always front and center, with Ivan somewhere in the background. Ivan does not care for the background. I was disgusted, because I knew quite a bit about the new man, and because I was certain, as I told my contacts in Colville, that there was no way Ivan would find a sasquatch while he was working for someone else.

Sure enough, Norm Davis, who owns the radio station there and who had been extremely helpful to me and various others, phoned one day with the news that he had met Ivan having a cup of coffee in town while everyone else was rushing around in a flap over some tracks in the snow. I am not sure whether those were the tracks that a local man faked with wooden feet. If they were, that might have had something to do with Ivan's lack of interest. But when Norm asked him how come he wasn't out hunting with the others, Ivan, whose contract ran until June 30, replied:

"You know, I've thought about it, and I've thought about it, and I can't think of any way a guy could catch one of these things before the first of July."

One of the things I had been told about the situation at Bossburg was that although Ivan had not sold the film the contract gave his employers control of it while the deal lasted. There was a copy for the great white hunter to show to local dignitaries and such, with Ivan running the projector, but the original was to be kept in a vault in Washington. The film can was sealed, however, and there would be a hefty penalty to pay if the seal was broken. I was in Bossburg once that spring and when I visited Ivan he brought out a film he said he wanted some work done on. I asked him what film it was. His sasquatch film, he said. But I thought that was locked up in Washington, D.C.

"Oh, that," he said. "They can't open it. There's just some scrap film in that can."

Of course I didn't know whether to believe that, and I also didn't know what film Ivan had in his hand, but I later learned it was indeed the original. Then I didn't know whether there really was any film in a sealed can in a vault, but I eventually heard, via the other side as it were, that there was, and that when it was opened the can was found to contain scrap. Later yet, Ivan phoned me to say that one of the financial backers, who was a big-time lawyer, had written threatening a lawsuit. I reassured him that I couldn't see any lawyer from Washington, D.C., going into court to complain that some hick out west in that other Washington had separated him from some money in exchange for a film can he wasn't allowed to open. I expect Ivan already knew that, but he was short of an audience that evening. I don't think he ever sued anyone for breaking the seal on the film can.

Again it is second-hand information, but I am told that the son of a local man involved in the hunt had tried to tell people from the start that he knew where the movie was made, right on their ranch, but no one would listen to him. Finally someone did, and it turned out, as I verified for myself, that the creature in the movie was close to five feet in height, not 10, and that during the shooting of the film it stayed in one small area, although the cameraman had changed his postion considerably to get different backgrounds.

Let that story be a warning to anyone who finds himself on the inside track with someone who has great revelations to share regarding the sasquatch. In this field, as in others, some people will go to lengths other people can't imagine to set the scene for things that were never there at all.

Please note that I do not say that Ivan faked that film, or the one of the white sasquatch in the snowstorm, or any other he has since then or may in the future emerge with. It is my considered opinion that the creature the film shows is a person dressed in black fur, and I know of nothing but Ivan's statement indicating that it is not. As a critic, that is my present opinion of the film, on the basis of the information at my disposal, and it is an opinion I do not expect to have to alter. As to the hand prints, the crippled prints and the even-larger footprints, I am also sceptical, but not to an equal degree with regard to each.

There are people not connected with Ivan in any way that I know of, who say they saw such creatures in that vicinity around that time. There is even one man who claims to have seen the crippled print years before Ivan showed up in that part of the country. There is also the opinion of the physical anthropologists, John Napier and Grover Krantz, that the crippled print is anatomically valid in a manner that no layman could have faked. Grover also says that the position of the bones indicates that the heel projects behind the ankle joint much farther than it would in a human foot, which is just what would be necessary for a creature of such weight, and which is just what the Patterson film appears to show. The hand prints too, Dr. Krantz finds convincingly correct for an animal of such a size with a non-opposed thumb.

How much of this was genuine, and how much was a buildup for Ivan's film? All of it? Or none? At what point do you draw the line? Having no answer, I tend now to write off the whole Bossburg episode to entertainment. It was certainly full value for that, and as evidence for the sasquatch it isn't needed.

I've been told one thing, though, about those nightly performances of Ivan's, and I don't know whether it is true or not. As far as I am concerned, I wouldn't have missed them for worlds, and I met clients of Ivan's who felt the same way. They got value for their money whether their hunt was a success or not. The story in Colville, though, is that if the local audience showed signs of wearing thin, Ivan would approach some sasquatch enthusiast during the day and say he had some hot new information for him, if he would come around to the house that evening. The only thing was, they agreed after comparing notes, he would give the message to more than one person. So they would all show up at the house that evening, and they would wait and wait for the others to leave, until finally they all left together. Ivan, after all, didn't have to get up for work in the morning, and Peg always had the coffee pot ready.

I remember once Ivan leaned back in his chair and fixed his eye on that bedroom door that hung on one hinge, although a single nail would have fixed it so it would swing. He had been living there then for the best part of a year. Ivan stared at the cockeyed thing, and he said:

"Some day I'm going to fix that door. It's been bothering me all winter."

-10-

Twenty Inches of Rain

Information found in old publications has made it clear that stories of people seeing hairy bipeds are part of the heritage of many areas of North America, not just the west coast, but although that information is old most of it did not come to light until fairly recently. In the meantime the many stories from the Pacific states and British Columbia, plus the publicity for the motion picture taken in California, have established a widespread impression that the sasquatch phenomenon is limited to the far west, and have also resulted in investigative efforts being concentrated there. Plenty of people in other places have known of stories in their own locality, and a few have been collecting reports from wider areas for some time, but it is only in very recent years that there has been much of an effort to deal with information on a continent-wide basis. As a result there are many people who take a great interest in the subject of encounters with manlike monsters in the western forests, yet do not realize that there are also reports of similar things going on close to home.

Since no one else seemed to be doing it, Dennis Gates and I took a trip around the United States in the spring of 1976 to meet as many as we could of the people then active in research in other areas, and to make some direct inquiries of our own. And because it has not been done before, I am attempting in the next few chapters not just to outline the situation in various parts of the continent but to present briefly, giving sources, all that I have on record from those provinces and states where the existence of such information is generally unknown. To do so has real drawbacks, because a series of capsule monster reports can't help but be repetitious, and because a list that pretends to completeness has to include stories that are not very substantial—indeed is bound to include some that, unknown to me, have been found to be worthless. I consider the drawbacks worth accepting in order to make it possible for readers to look up their home area or any other that interests them and find out what information is available there.

Looking over the situation in the United States, a glance at the map of reports show immediately that the hairy bipeds don't recognize the regional divisions that humans have set up. The bulk of U.S. sasquatch reports are from the Pacific Coast states, but other than that there is no relationship between them and the various "regions" of the country. Most people associate such stories with mountainous areas, and the Rocky Mountain states of Montana and Idaho do have a lot of them, but the other states of that region have almost none. The Midwest has few reports in its western half, a lot in the east. Southern and Middle Atlantic states are a mixture. New England has few reports state by state, but a fair number in relation to its small total area. Across most of the Southwest there are few reports, but in northeastern Texas and eastern Oklahoma they are quite numerous.

In general the pattern of reports does not appear to have any correlation with the pattern of human population, except that sasquatches are not reported downtown in the cities. Some densely-populated states have a lot of reports, some have few. The same goes for states with little population. Nor does there seem to be a correlation between sasquatch reports and average temperatures. Amount of vegetation is certainly a factor, since there are few reports from areas without some kind of forest cover. Sasquatches don't roam the great plains, or the open mountains either. That could just mean that they now need a place to hide from humans, but there is no indication that they ever shared the range of either the buffalo or the antelope. However a map of sasquatch sightings does not noticeably resemble a vegetation map.

The map that does show some correlation with a map of sasquatch reports in North America is the rainfall map. It isn't by any means a perfect match, but there is a rough outline that fits. Where the annual rainfall is under 20 inches a year there are hardly any sasquatch reports. There is only one major exception to that rule, in western Montana, and three minor ones, north of Los Angeles, at the north end of Lake Winnipeg and in Alaska. Where the rainfall is over 20 inches there usually are reports, although up to now there are more exceptions on the wet side of the line than on the dry side. Why mankind's supposed need to imagine monsters should dry up where it doesn't rain much I will leave to someone else to explain.

It should be borne in mind that this information is only what my files contain at time of writing. When this book is published I may be contacted by someone who has been collecting information for years and has dozens or hundreds of reports from areas where I have only one or two. The one thing that is consistent about concentrations of reports is that they are clustered in areas where someone has been investigating. There has been an extensive exchange of information in recent years, however, and I don't really think it probable that patterns apparent at this time will be greatly altered.

The area of the United States from which there are few reports is shaped something like a very fat, rounded Y, tilted over to the left. Starting in southeastern Oregon and southern Idaho, there is a virtual blank in the sighting report map that sweeps south and east through Nevada, Utah and Colorado. Another all-but-blank area starts east of the Rockies in Alberta and Saskatchewan and covers eastern Montana and all but northwestern Wyoming, all of North and South Dakota, western Minnesota and all but the eastern edges of Nebraska and Kansas. Western Texas is also blank, and the few reports from New Mexico are all from the western and southern borders. Over to the west, southeastern California is blank, but there are a few reports from central Arizona.

As this book goes to press there are reports of sightings and footprints from the vicinity of the Standing Rock Indian Reservation, straddling the border of North and South Dakota. Otherwise all I have from North Dakota is a letter written to Roger Patterson in 1967 by a Mrs. Myrtle Paschen, stating that shortly after the turn of the century a group of people in a sleigh in the Kildeer Mountains saw a large gorilla-like animal run human-fashion towards them through the snow and then turn and run off, leaving immense, humanlike tracks.

The Kildeer Mountains are only half the height of the Black Hills of South Dakota, which have peaks above 7,000 feet and cause a fair amount of rain. Two reports from there were investigated by Tim Church, a researcher whose home is in Rapid City. The first was on a forest service road near Johnson Siding in November, 1973, when three men described seeing in their headlights a 10-foot, shaggy, dirty-white creature near the road that picked up a dead deer under one arm and crashed into the underbrush. The following March two small boys near Nemo described seeing on a hill close to their home a large, heavy-set creature like a huge man covered with orange-brown fur.

A lone report from western Nebraska is also from a high area, at

172

Gering, near hills reaching 5,000 feet. According to the Lincoln *Journal and Star*, July 16, 1972, two housewives were frightened by a hairy creature seven feet high, a neighbour shot at it with a shotgun, and police found footprints like those of a large, barefoot man.

Colorado had the "Lake Creek Monster" in the 1880's, and in the 1960's Roger Patterson received a letter telling of a dozen climbers getting a close look at a seven-foot creature with shiny black fur on Mount Elbert. They were supposed to have followed its tracks for a mile in heavy snow after it walked away into the forest.

More recently there are two more Colorado reports. A boy from Colorado Springs wrote to me that after breakfast one summer day in 1975 he heard a screeching sound and on looking out the window saw about 150 yards away a dark-brown, seven or eight-foot creature walking across a park. Then in the Ouray, Colorado, *Plain Dealer*, February 3, 1977, there was a story about a letter sent to the "Burgermeister oder Sheriff" of Ouray from Heinz Fritz Goedde of Nordhorn, West Germany. Herr Goedde said that while holidaying the previous summer with three brothers who live in the U.S. he stayed at a campsite near Ouray, and on a hike in the mountains saw a big creature that just quietly walked away. It happened on the Fourth of July, and he thought it was some sort of Bicentennial stunt, but next day while trout fishing he had a similar experience. That time he tried to approach the thing to photograph it, but it left him behind. He had no idea what it was until after he was back in Germany, when he happened to see a TV show about monsters in the United States.

It is highly improbable, to say the least, that in an area where no such thing has been reported before a person could arrive for a brief stay and see sasquatches two days in a row. It is also hard to imagine that Herr Goedde would not have discussed his encounters with his brothers and found out what it was that he had seen. On the other hand one can't be at all sure about the accuracy of the details in a newspaper story about a letter translated from another language. If the sightings involved the same creature in the same out-of-the-way place it would improve the probability considerably. I don't suggest that any of the Colorado reports is substantial enough to make any contribution to the basic question of whether there actually are such things as sasquatches, but I think that if a case for the creatures' existence has been established from evidence elsewhere, then those reports do have a bearing on the question of whether there are sasquatches in Colorado.

On the western side of the Y are southeastern Oregon and southern Idaho, both very dry, and Nevada, the only state where my maps show no area that gets 20 inches of rain a year. I have a computer report describing 15-inch, five-toed tracks seen in November, 1960, 7,000 feet up in the Diamond Mountains in Nevada. In addition there is the 1870 report already quoted and three vague modern ones. There is also a letter to the editor from the Reno *Evening Gazette*, August 11, 1973, written by two Murray, Utah, couples, as follows:

EDITOR, the Gazette: We recently visited Lake Tahoe for the first time (a beautiful place) and while on our way up Kingsbury Grade we had a very frightening experience.

It was about 8:30 p.m., July 29, about two-thirds of the way up the grade, which you know is very narrow and steep.

As we came around a turn, we saw something on the side of the road which we thought was a black bear. As we got closer, it was standing on its hind feet in an upright position.

When it saw our car, it went into the bushes. Just before it disappeared, it turned and looked at us.

Its face was more flat, like a gorilla's. It was about seven feet tall, and very shiny.

There was no place to get off the road, so we went on to Tahoe.

There were four people in the car and we all witnessed the same thing.

When we first saw it, my friend began yelling "Blackfoot! Blackfoot! I mean, Bigfoot!" We don't known for sure what kind of an animal it was, but we do know we saw it.

We went to the sheriff's office at Lake Tahoe and told our experience. They said they had had another report that two girls had seen the same thing and were still crying with fright. The excuse they gave us was it was a crazy person in a gorilla suit trying to scare people.

We don't buy this story, after seeing it.

I am a sportsman. I have hunted deer, buffalo, antelope, elk, bear and lion, and I know an animal when I see one. This definitely was an animal.

Until they catch that so-called imposter we still have our own ideas. Did we see Bigfoot?

We are sure some of you will say we are crazy or that we had too much to drink (not one drink). But maybe there is someone else who has seen this animal?

> Mr. and Mrs. D. Cowdell
> Mr. and Mrs. C. Searls
> Murray, Utah.

That report comes from just east of the California border and really relates more to the coastal area. There are several other reports from nearby, on the California side of the line.

From the state the other side of Nevada, Utah, I knew of no reports at all until very recently. Then on August 25, 1977, the Ogden, Utah, *Standard-Examiner* carried a story about two North Ogden men and six children seeing from a distance a black and white beast about 10 feet tall that moved off on two legs, being in sight for about four minutes. Following the route it had taken they later found the remains of a rabbit, skinned and bitten off half way up the body. The incident took place in Cuberant Basin, in the High Uintas, and although most of Utah is very dry indeed the sighting was within a small area that does get more than 20 inches of rain.

After one of the men, Jay Barker, told the story to the newspaper, he was contacted by two couples from Davis County, Mrs. and Mrs. Robert Melka, from Bountiful, and Sgt. and Mrs. Fred Rosenberg, from Hill Air Force Base, who said they had seen three such creatures in the same general area on July 10. The Bountiful *Davis County Clipper* carried their story in full on September 2. They said they were sitting on a ridge half a mile southwest of Elizabeth Lake, about 17 miles from the scene of the other sighting, when two upright creatures, both much larger than humans entered a meadow several hundred yards from them and remained in sight about 10 minutes. The reporter who wrote the story stated that the two couples told him exactly the same things, although they were interviewed separately, and he gave the following quoted account:

"These things, whatever they were, walked and ran on only two legs. They had arms, legs and bodies much like a human but covered with hair. Only their hands and feet did not have hair.

"They were in profile to us much of the time and there is no question about them having only two legs—but their legs and arms appeared to be out of proportion (larger) than those of humans.

"Two of them would romp and play in the meadow while a third stood some 100 yards away at the edge of the meadow near a cluster of pines . . .

"When they would run they would take tremendous strides and would cross a 100-yard clearing in rapid speed . . ."

They agreed that although they had the appearance of humans, they did not act like humans. "They had animal-like characteristics . . ."

The four had not discussed the matter later, even among themselves, until the other story was published. The Melkas had not told their children anything and the Rosenbergs had told their family only that they had seen something strange.

"We thought a lot about it, but didn't want to talk about it and be humiliated by those who wouldn't believe us," said Mr. Melka.

Arizona and New Mexico present a special problem with regard to reports of hairy bipeds, because there are supposed to be "wolfmen" among the Navaho Indians who run about at night dressed in wolfskins. If that is so, it greatly increases the chance of mistaken identification, especially since there are reports from those states of creatures that were man-sized or even smaller.

A school teacher wrote to me in 1969 that one night in the fall of 1965, while driving north from Winslow, Arizona, "I was startled to see what I thought was a charred man climbing over a low fence or guardrail into the road."

He turned his car and went back to the spot, where he saw a figure, "not a burned man at all but a glossy pelted thing that ran like a man mocking an ape. More man than ape in his run, but more ape than man in his looks."

The creature was about three feet high, with a glossy black pelt about an inch deep, no visible ears and great long arms. He saw it for about 15 or 20 seconds. That night he drew a picture of it, and later he found that Navaho and Hopi children at a school where he was teaching recognized it as one of their legendary creatures, a "skin walker."

A woman who wrote to Dennis Gates from Yarnell, Arizona, in 1977 had two sightings to report, more than half a century apart. She said that when she lived near Flagstaff, in 1924, she and her mother saw a creature in the garden on their ranch at the foot of San Francisco Peaks. It was about seven feet tall, heavy and covered with light-colored hair, and it had an armload of corn and was bent over pulling up turnips. They watched it run through a wheat field and jump a rail fence making for the pine forest, carrying the corn and turnips. The more recent sighting was in 1975, when a Yarnell resident told of seeing in his rear-view mirror a huge apelike animal running along behind his car. He stopped and got out and it ran away. Next day he saw it again up on the mountainside, "lumbering along, its arms swinging and extending below its knees."

According to the Flagstaff *Daily Sun*, January 23, 1971, two students from Northern Arizona University reported to Flagstaff city police that about 1 a.m. that day, while parked on a remote side road, they saw a thing with a face covered with hair leaning against their car and looking in the window. He shouted, and it ambled slowly away. The face was apelike, and the creature appeared to be about five feet tall.

In January, 1970, four youths from Gallup, New Mexico, paced a creature running beside their car at 45 miles an hour near Zuni, New

Mexico, according to the January 21 issue of the Gallup *Independent*. They finally shot at it, knocking it down, but it got up and ran off. It was hairy and an estimated five feet, seven inches tall. A letter to the editor, filed by Ivan Sanderson, with no source for the clipping noted, tells of a sighting in the summer of 1968, about six miles north of Anthony, New Mexico. A couple driving north shortly after midnight told of seeing a thing bigger than a dog and looking "like an ape with long black wavy fur" eating an animal that had been killed on the road. The woman who reported that had another story also. She said that a friend in the same area heard scraping at her window, switched on lights in the yard and saw two creatures, one that looked like a bear running on its hind legs and one like a man on all fours with shaggy tan fur.

There are also a number of reports from the Chuska Mountains on the Arizona-New Mexico border. The Navaho sheepherders there are reported in recent years to have been having trouble with "bears that walk upright" that can steal a sheep by reaching over a corral fence. The information was given to me by the people in charge of an Indian mission, who said they had spoken with Indians who had seen the creatures. The descriptions given, in their opinion, fitted the sasquatch, and they had seen 16-inch tracks themselves. I was in the area in 1972, and was surprised to find that high in the mountains, which looked from the highway like bare rocks rising out of a near-desert, there were evergreen forests not greatly different from those on the Pacific Coast.

The Mountain states I have not dealt with are Idaho, Montana and Wyoming. The first two contain areas with heavy concentrations of reports, along the continental divide and on both sides of it. An extension of that area would pass through Yellowstone Park, and the few reports there are from Wyoming are concentrated in that northwest quarter. A New York resident wrote to Ivan Sanderson that in the summer of 1959 a friend of his camping in Yellowstone Park saw a huge hairy hand reach in under the edge of the canvas and pull out a rolled-up blanket. The letter did not say whether the blanket was later found outside.

Two other reports are very similar to each other, one being dated "about 1972" and the other in the summer of 1972, both involving girls about 20, and both being located "20 miles south of Jackson". One involves a girl wrangling horses in a canyon who saw something larger than a person walk upright across a field taking huge strides. The other girl was riding along the Snake River when she saw on the other side a dark upright figure emerge from the timber onto an open slope. "It appeared all covered with hair and strode with arms swinging just like a person." She watched it cover several hundred yards uphill faster than a person could possibly have walked. I do not have either story first hand, so it is possible that they are versions of the same thing, but I don't think so. All three of those reports are inside or very close to small areas where rainfall exceeds 20 inches.

Another story from the same general area involves the killing of a creature by two young men out varmint hunting in the Christmas holidays in 1967. I have recently talked to both of them. They agree that something they took to be a bear was shot, and that it was a real struggle to get to it through the deep snow. One says that they made a close inspection of it, and what he describes would have to be a sasquatch. The other says that he did not even manage to get close to it and doesn't recall that his friend did either, and he still thinks it was a bear. The version that does not describe any unknown animal has to be considered the more acceptable of the two, although one is left wondering what a bear was doing out in the snow at that time of year.

Finally, in the summer of 1972 there was a flare-up of excitement on the Wind River Indian Reserve near Lander after two 13-year-old boys claimed to have seen a tall hairy monster with an apelike head that ran along near them while they were on horseback. It did not get close.

In the Mountain states it is often a question of why there are no reports from areas that seem suitable. Farther east the question is the opposite. Why should there be reports from near Sioux City, or Kansas City? I know of no answers, unless the Missouri River has something to do with it, but the reports are there. A student at Midwestern College in Denison, Iowa, wrote to Roger Patterson in the late 1960's that he and a friend had seen a creature like that in Roger's movie in the woods behind the college at night. Denison is about 70 miles southeast of the Sioux City area, where in the summer of 1971, according to Jerome Clark, writing in *U.F.O. Report*, summer, 1975; three young men claimed to see what appeared to be a huge manlike ape in Stone Park, near Sioux City. The following January, about 1:30 a.m., Jim Britton heard a scream outside his home, about a mile from Stone Park, and on looking out the window saw what looked like a big gorilla about 100 feet away. He went outside with a 30-30 rifle and saw it down on all fours near his barn. As he approached it stood up like a man, with its forelimbs hanging to its knees. It appeared to be about seven feet tall. Britton shot it and it went over backwards, but was up again instantly and fled on all fours at tremendous speed. Questioned by Dr. Laurence Lacey, for the Mutual UFO Network, he said the thing looked like a big man with dark brown hair about six inches long all over its body.

In July, 1974, according to Jerome Clark, four young people at

Oakland, Nebraska, about 50 miles south of Sioux City, went out walking at about 10:30 in the evening and reported seeing a huge, shaggy gorilla-like animal making a screeching sound and moving towards them from about 75 feet away. One of the boys threw a firecracker at it, and it rose on its hind legs and fled into a corn field. There were other sighting reports in the area, plus unexplained screams, causing a good deal of excitement and concern. Then at the end of August a Sioux City resident, who was also interviewed by Dr. Lacey, heard his porch railing rattle, looked out the window and saw close up a "gorilla" on two legs with hair all over its body. His story was reported in the Sioux City *Journal* on August 31, 1974, and the next day a woman called the paper to report that she had seen the same thing the night before the man saw it.

On September 6, shortly after 8 a.m., Jim Douglas was driving a few miles from Sioux City on Highway 77, north of Jefferson, South Dakota. In an alfalfa field beside the road he saw a sandy-colored, hairy creature dragging a red furry object. Douglas got out of his pickup and watched from a distance of about 100 feet. He estimated the creature to be about nine feet tall. Police at Elk Point, South Dakota, refused to pay any attention to his story. He was later interviewed by Dr. Lacey.

I have never been to that area and know next to nothing about it. I would judge from the map that it would be farm country, although there is a reference to nearby woods in one report. The majority of the incidents took place near the Missouri River and the others are near tributary rivers. Farther south on the Missouri, a Kansas resident wrote to Roger Patterson that he and others had seen a seven to eight-foot, light-colored creature in daylight near Leavenworth in the summer of 1968. They had gone to the place where they saw it because they had heard there was supposed to be a "monster" there. Another letter to Roger contained a report of a sighting on the Pottawatomie Indian Reserve northwest of Topeka and two other sightings near Topeka, all about 1960. A man wrote to me that a fellow serviceman had encountered a 5½ foot creature outside his car in a lover's lane in Kansas, but didn't say when or where. None of the reports give much detail, and I have not looked into them myself. Neither do I have any personal knowledge of the reports from Oklahoma, which are more numerous. What I can say from experience is that investigation of a series of reports from an area virtually always substantiates most of them, and it is unusual for even an individual report to prove to be entirely unfounded. That is not to say that the reports can be proved true, but only that when an indirect report is checked there almost always proves to be some person available who will say "Yes, that happened to me."

An organization called The International Association for the Investigation of the Unexplained, in Oklahoma City, has a report of a pilot seeing a sasquatch-like creature after making a forced landing in an isolated area near Nelagony in north-eastern Oklahoma in 1967. While walking out, the pilot said, he saw a man who ran into the woods when he waved. Later he saw this creature standing only 10 yards away as he walked by. He described it as seven feet tall, apelike, with a large head and massive arms and a stale odor. He decided it was not a man after all. Its eyes, he said, were a glowing green. He is reported to have taken a picture of it, but if so I don't know what became of it. The glowing eyes are dealt with elsewhere. The same organization reports that several people saw a hairy, upright creature near Talihina, Oklahoma, in 1970.

Fate magazine in September, 1971, carried a report that several people around Oakwood, Oklahoma, had seen a gorilla-like creature of huge size in the period from November, 1968 to January, 1969. Near Watova, Oklahoma, a creature five to six feet high, built like a man but covered with hair, was reported seen in August, 1974. Then in September, 1975, at Noxie, there was a sighting that resulted in a great deal of publicity.

I can think of no logical reason why the news services should ignore a couple of dozen reports in local newspapers each year, even those in major dailies, but send two or three others to every newspaper in the country, yet that is what happens. I have hundreds of clippings of sasquatch reports that apparently appeared only in the original newspaper. Other reports, not obviously more interesting in any way, will be carried by hundreds of newspapers all over the continent, and will be followed up in television and radio programs. A few even result in investigations and lengthy feature stories by the likes of the New York *Times* or the Wall Street *Journal*. It is perhaps understandable that AP and UPI don't distribute all such stories that they have access to. That would mean carrying one or two every week and might make it look as if they took the subject seriously. But how do they pick the ones they do put on the wire? In any event, the Noxie report was one of those. Here is the United Press International version:

NOXIE, Okla. (UPI) — Farmer Ken Tosh said he saw the monster—seven feet tall, hairy, foul smelling and with eyes that glowed like a cat. It's got the sheriff worried—the people, that is, not the monster.

"I was within 10 feet of it before I saw him," said Tosh, 30. "He growled and ran one way. I screamed and ran the other."

180

Tosh is one of several residents of this tiny community to report sighting what has been dubbed the "Noxie monster".

"The people are scared," Tosh said. "Anything that comes around, they'll shoot. We couldn't capture him."

The reported sightings panicked residents, and armed bands of hunters have been roaming the rolling timber-covered hills along the Oklahoma-Kansas border hoping to get a shot at the creature.

Nowata County Sheriff Bob Arnold is more worried about the hunters than the monster.

"We're going to get some righteous people killed by some idiot," Arnold said Thursday. "I'm telling reporters to stay away from here because they might get shot and I'd hate to have to fill my jail with some of the good people around here.

"I had an unconfirmed report last night there had already been one horse and two dogs shot by these gun-crazed idiots running around drinking.

"A young person in this day and age, with long hair and a lot of stuff on his face, might at night look like a wolfman and one of these guys that are drinking their beer might want to blow him to bits."

"Maybe we'll luck out and find a monster and a UFO both, he joked.

"I've never seen one of those, either."

Interviewed for a network radio program in Canada, Sheriff Arnold said that Tosh had seen the creature on several occasions and shot at it several times. Another man living a mile away had shot at it five times. It was always encountered after dark. He said the men who had seen it had shown genuine fright and some of them were moving away. The sheriff also mentioned that there had been reports of a shorter, but similar creature in Watova 13 months before, and repeated his concern about someone getting shot.

I don't know how many times over the years I have encountered that attitude. A sheriff in Washington state admitted to me that he deliberately lied to the press about what one of his deputies had reported, because of concern about all the people who would be running around with guns. The famous $10,000 fine for shooting Bigfoot was adopted by the commissioners of Skamania County, Washington, because of the same concern, even though they knew full well they had no authority to pass such an ordinance. I have also read over and over about the hundreds of people, armed with both rifles and liquor, who rush to the scene of a monster sighting report. Oddly enough, I have never heard of anyone being shot at, let alone hit by any of the hordes of drunken monster hunters. What's more, although I have probably been on the scene of as many sightings as just about anyone, I have never encountered any of those hordes.

I strongly suspect that there are very few instances when a large number of people have actually turned out to look for a monster, and that in the few cases when it has happened not many of them were drunk and a lot of them weren't armed. In short, I am inclined to the opinion that there is a much greater tendency to tell exaggerated stories about monster hunters, which seem to be acceptable, than there is to tell stories about monsters, which are not acceptable. In any event, I submit that experience has established beyond doubt that the danger of a monster hunter shooting someone is too insignificant to be an excuse for suppressing or failing to investigate a report.

There was another report from Oklahoma in the middle of September, 1975, describing face-to-face confrontations by three persons with a large hairy bulk resembling a monkey, in the vicinity of Indianola, in the southeastern part of the state. The Associated Press carried some details regarding a subsequent unsuccessful hunt for the thing, but neglected to say much about the original reports.

None of those reports go back more than 10 years, but recently I received a letter from a woman from Pryor, Oklahoma, who recalled several incidents from one summer when she was a child, about 1956. Her family lived in a mountainous area, about 13 miles from Wilberton. One afternoon while sitting in the dining room her mother saw a creature, upright like a man but covered with hair and with long arms like an ape, walk along the edge of a pond near the house and pass out of sight. A neighbour saw the same thing hanging over his fence, and the writer's sister had told her that while picking berries she almost put her hand on a huge hairy leg. No further details about the leg were supplied.

All but the western tip of Oklahoma gets more than 20 inches of rain a year, and so does the eastern three quarters of Texas. There was an 1875 report, not particularly impressive, about the capture of a hair-covered boy, near Austin, Texas, and there was the strange case of the "Lake Worth Monster," but Texas has produced quite a few fairly normal reports as well. An interesting story is contained in a letter from a San Antonio man to Dr. Grover Krantz, as follows:

> During the fall of 1975 I was at a private lake some 35 miles north of San Antonio. Roughly 20,000 acres of brushy limestone cliffs, caves, wildlife and very few people. I was sitting on a pier on the lake's edge when I heard a splash across the lake as if something had fallen off the top of the cliff into the water. It's approximately one eighth of a mile across the lake and the cliff is about 150 feet tall at that point. All I could see was something large and grey brown in

182

color. I got my .22 rifle with an eight-power scope and put it on what I had seen. It looked to be about eight to nine feet tall, covered with longish hair, more grey to white in color than I had originally thought. I watched it for a good five minutes and got at least a two-thirds front view of the animal or whatever it was. No distinct features such as eyes, nose, ears, but more of a rounded cat's face with hair spreading outward from the center of the face.

I called the game warden and reported what I had seen but he said it was probably a large Spanish goat or deer. The more I thought about the way it moved around and stooped and moved tree limbs around, I couldn't buy the explanation. Next morning I went across the lake and climbed to the top of the cliff for a look. No tracks because the ground is almost solid rock, but the trees were broken in a few places and small brush trampled where something had come through there. About 20 feet back from the edge of the cliff I found a huge flat boulder that had been turned over and the dirt scraped from the bottom. This rock had to weigh at least 300-400 pounds and I could see no way it could have been overturned by less than three men.

Another letter that I have tells at second hand of two sightings two years apart near Lake Travis, northwest of Austin. In one case two large hairy creatures approached a parked car in which there were several youths, including my correspondent's brother. The other involved a tall, dark-coloured creature seen on a road late at night. The letter did not give any dates but something around 1970 seems indicated. Many other reports come from northern Texas east of the Panhandle, mostly right in the northeast corner. A letter written by Thomas R. Adams, then a university student in Paris, Texas, to Ivan Sanderson in March, 1969, outlined the situation at that time.

I have no way of knowing how much you know, if any, about the monster situation here in northern Texas. There are reports from several points in this area:

The Direct vicinity, in extreme northwestern Lamar County. The residents of this rural area report what they refer to as a "manimal" that makes scheduled appearances in June and October as it "migrates" thru the area. They have reported that it has been seen regularly for the past decade and some old-timers claim they have seen such a thing for the past 50 years. A few years ago, the Paris news decided to run the story and it received moderate regional publicity. Because of the patronizing attitude of the press, though, the Direct residents have become less vocal about the manimal.

However, the opinion of most of the residents appears to be, "It's real."

Descriptions have the creature being about 6'2" as it stood up. It also has been seen on all fours. Several witnesses have reported that it made 8-foot leaps. It has prowled in the vicinity of houses and has been reported looking in windows. Several years ago a game warden was called to take a look at the tracks. He said the tracks were "like nothing he had ever seen."

A man could put both hands in a single pawprint. The cry of the creature sounds like "a man in pain yelling". It is expected through in the last part of June and again in October, just before deer season begins. In June it is always heading west, but in October it is travelling east . . .

The Blue Creek area, approximately five to six miles west of Bells, Grayson County, Texas. I am trying to dig up the news stories on this. Most of the reports were early in this decade. They included that of a travelling salesman who watched an apelike creature bound across the highway in his headlight beams, making the crossing in about two strides. Also, the creature was reported in farmyards and around houses late at night, sending dogs into whimpering, shivering fits of fright.

Denton. An apelike creature was seen in an isolated sighting on the outskirts of Denton.

Caddo, Stephens County. An ape-shaped animal that became known as the "critter" was seen in the area three or four years ago. A rancher named Charlie Gantt, who fired 10 shots unsuccessfully at the "critter" described it as being seven feet tall and four feet wide and covered with hair. Other reports were made and an obscure amateur movie which reportedly showed the creature in action was shown on WBAP-TV in Fort Worth. I remember seeing the film and it was difficult to tell anything from viewing the TV although the film did not seem near as revealing as the Patterson Bigfoot film.

Haskell, Texas. The residents here claim that the "Haskell Rascal" and the "Caddo Critter" are one and the same. Although the Haskell creature has been reported for 80 years, I have had difficulty finding descriptions of it. Residents claim that it spends its summers in the Kiowa Peak area, west of Haskell, and that it prowls in the lowlands during the winter, killing and feeding on livestock. There were numerous reports of cougars in this area around the turn of the century, so this must be considered as a possibility, although I don't see how anyone could confuse the Haskell Rascal with those reported descriptions of the Caddo Critter. The rascal is also reported to wander throughout a 60-mile range.

I have newspaper references concerning the Caddo Critter and the Haskell Rascal, and others describing incidents that took place since the student's letter was written, from Reno, Benbrook and Grand Prairie, every one of which is in the northeast corner of the state. In the same general area, just across the border, are the repeated incidents near Fouke, Arkansas. I haven't visited most of those areas, but I have been to Fouke. There is plenty of wild area around the Sulphur River there, stretching over the border into Texas.

Then there is the Lake Worth monster. If ever there was a story that should be nonsense, this is it. Lake Worth village is a part of the Dallas-Fort Worth metropolitan area, as are Benbrook and Grand Prairie. The lake itself is almost in the city. There is a nature preserve on the lake shore and on Greer Island that is probably big enough for a large animal to hang around in temporarily, but it is by no means big enough to be its permanent home.

The first published monster reports appeared in July, 1969, when six young people told police that a "half man, half goat covered with fur and scales" had jumped out of a tree onto their car while they were parked near Lake Worth about midnight. A police dispatcher was quoted as saying "We've had reports about this thing for about two months and we've always laughed them off as pranks."

Police still considered that a prank was at the bottom of the new report, although they noted that the young people were really frightened. By the next day they had a story that could not be dismissed as a prank. The one road through the Lake Worth nature center had been crowded with cars the night before, and some of the sightseers claimed to have encountered the monster. The Fort Worth *Star Telegram*, July 11, 1969, quoted one of them, Jack E. Harris, of 5537 Terrance Trail:

We were driving around trying to find it, when we heard it squalling. We heard it before we saw it.

I saw it come across the road and I tried to take a picture of it but the flash didn't go off.

I took another picture but I don't know if I got anything because I was too busy rolling up my window.

We watched him run up and down a bluff for a while and other cars arrived. There must have been 30 or 40 people watching him.

Well, some of them thought they would get mean with the

185

thing, but about that time, it got hold of a spare tire that had a rim on it and threw it at our cars.

He threw it more than 500 feet and it was coming so fast that everybody took off. Everybody jumped back in their cars.

Earlier there were some sheriff's deputies there asking us about it and one of them was sorta laughing like he didn't believe it.

But then that thing howled and I think it stood his hair on end. He decided it wasn't so funny anymore.

Those sheriff's men weren't any braver than we were—they ran to get in their car.

Witnesses agreed that the thing was big, hairy, white and walked like a man. It was estimated at seven feet tall and 300 pounds weight.

After that the sightseers and the sighting reports became too numerous for anyone to keep track of them, although one person tried. She was Sallie Ann Clarke, who wrote a book about the goings-on, and achieved nationwide publicity as "The Monster Lady." When Dennis Gates and I visited her in 1976, she told us that at the time she wrote the book she just wanted a subject to write about, and she threw in plenty of fiction to liven it up. The book was published while the excitement was still at its height, and it resulted in many people contacting her with new reports. She told us she spent four months in the area both day and night, as well as returning many times after that, and although she had seen nothing herself at the time she wrote her book, she saw the monster four times afterwards. On one occasion she and another witness saw it break through a barbed wire fence although it had no need to do so, it could just as easily have gone around. It weighed, in her estimation, about 700 pounds. She had interviewed hundreds of people, and we listened to several of her tape recordings, including one on which a witness described being present on the occasion when the wheel was thrown.

One person claimed to have succeeded in getting a picture of it. Allen Plaster, the owner of a local dress shop, produced a black and white, fuzzy picture of the upper half of a hairy, light-coloured creature taken from behind. After the tire incident perhaps the best story was that of Charles Buchanan, who said that while he was sleeping in the back of his pickup beside Lake Worth about 2 a.m. on November 7, 1969, a thing "like a cross between a human being and a gorilla or an ape" picked him up in his sleeping bag and then dropped him. He shoved a bag containing some leftover chicken at it, and it took the bag in its mouth, loped off to the lake and swam out to Greer Island.

186

Did those things really happen? Only the people involved know for sure. Jim Marrs, who wrote most of the stories for the *Star-Telegram*, told me that the police had found some boys with a costume who had started the whole thing by jumping on the car, and that he did not think there was any monster. A drawing made by one of the witnesses, and apparently acceptable to at least some of the others, shows a creature far more human looking that the usual descriptions or pictures of sasquatch. If there ever was a case where mass hysteria took over after the first report this would appear to be the one.

I had seen a story about the Lake Worth monster but did not take it seriously, and had no intention of investigating it. Dennis and I went to the Fort Worth area solely to check on the tracks in limestone in the bed of the Paluxy River near Glen Rose. It was the quantity of clippings that tumbled out of the *Star-Telegram*'s "Monster" file that persuaded us to devote an extra day to calling on Sallie Ann Clarke and looking around at Lake Worth. I learned long ago that police explanations of reports like this are not necessarily correct, and I have no grounds for calling Mrs. Clarke a liar. I am also quite sure that she is not the type to "see" things because of hysteria. There is certainly independent evidence of hairy giants circulating in the northeast corner of Texas, and if that is so there is no reason why one should not have spent some time in the woods of the nature center at Lake Worth, or why it should not still be stopping off there from time to time.

Texas reports don't end with the Lake Worth Monster, in fact they seem to be quite common in the last few years. The following is a quotation from a letter written October 5, 1973 by an editor of the Commerce, Texas, *Journal* to Ivan Sanderson's organization, The Society for the Investigation of the Unexplained:

> You will no doubt recall that last summer (1972) near Texarkana, Texas, a reported incident details the sighting of what has been referred to as the Fouke Monster in this area. Subsequently, the Texarkana Jaycees offered a reward of $10,000 for the live capture of this monster. However, the creature has not been seen in this area since.
>
> Approximately four weeks ago from this date a similar creature was sighted about 7 to 10 miles from here in the South Sulphur River bottoms near Peerless, Texas. Two such sightings were reported, and one of the persons involved supposedly had photographs of this creature's tracks. They seem to correspond with tracks you described in an interview with the Texarkana *Gazette*.

Since that time there have been at least half a dozen newspaper stories. The Antonio *Light*, September 1, 1976, reported that about two weeks earlier Ed Olivarri had seen a brown-haired animal, seven feet tall, run from his back yard near Kelly Air Force Base early in the morning. Some days later his next-door neighbour, Mrs. Rose Medina, reported seeing a brown-furred animal sitting on her back step. It ran off on two legs, but was only three feet tall. In both cases barking dogs called attention to the animals. On November 28, 1976, the Longview *Journal* carried a report of three sightings near Hallsville, Texas the preceding summer, one of which involved a 12-foot, silver-haired creature with a smaller, red-tinged female companion. The big one shucked corn just as a man would. The same paper, on March 31, 1977, reported the finding of a 17-inch footprint near Diana, Texas.

The Abilene *Morning Reporter-News*, July 7, 1977, reported the return of the "Hawley Him". Two teenagers at work clearing brush on a ranch claimed that rocks had been thrown at them by a seven-foot "kind of ape but still a man" with huge arms hanging to its knees. They ran to tell their employer, whose teen-aged daughter returned with them, armed with a 30-30. All three saw the thing and one of the boys shot at it, but missed. It left tracks only a foot long. On July 20 the Longview *Journal* had its third report within a year, this time about three 14-inch footprints found two miles south of East Mountain.

Finally, in mid-September, the Corsicana newspaper reported sightings on August 22 and September 2. The first was a hairy manlike creature seen on highway 274 north of Trinidad by three Corsicana women, who reported the incident to the sheriff. The other was a seven-footer seen by a Corsicana man on Highway 31 east of Corsicana. I have a clipping of the article but it does not have the name of the paper or the date.

-11-

The Mississippi Waterweb

The Mississippi seems to have some significance with regard to the reports of hairy bipeds. There are stories scattered almost everywhere along its length, and along most its main tributaries, and east of the river there are generally more reports than to the west. Once you leave the vicinity of the Mississippi going west there are almost no reports for the next 200 miles, except for a few in Arkansas, which seems to be favorite sasquatch territory.

I have already quoted an Arkansas story from 1851 which referred to reports going back for 17 years before that in Greene, St. Francis and Poinsett counties. There is also a story from the years following the Civil War reported in the chapter on "Lore and Legend" in the book *Ozark Country* by Otto Ernest Rayburn. It tells of a wild man seven feet tall reported many times in the Ouachita Mountains in Saline County. He was said to be of the white race, without clothing, and with his body covered with long, thick hair. He was thought to live in caves, and was seen sometimes in the canebrakes along the Saline River. He was never heard to utter a sound until the time when a hunting party cornered him in a cave and lassooed him, when he "emitted a strange sound like that of a trapped animal." The rest of the story will be told in a chapter dealing with reports of captures.

The next Arkansas report I have comes after a gap of almost 100 years, when things started to happen near Fouke. There, in 1965, 14-year-old James Lynn Crabtree encountered a strange animal while squirrel hunting. He first heard horses running, and heard them plunge into the lake near his home. Then he heard a sound like a dog

189

hollering in pain. He ran towards the noise but as he got nearer it changed to a different sound. Approaching cautiously he saw a hairy animal with its back to him, watching the horses in the lake and acting quite agitated about them. The animal was seven or eight feet high and had reddish-brown hair about four inches long all over its body. It stood upright like a man, but had extra-long arms. Turning, the creature saw the boy and stopped to look at him. Its face was covered with hair, showing only a flat, dark-brown nose. The creature stretched, sniffing the air, and then started walking towards the boy, who shot it in the face with his shotgun, three times. The thing kept coming and the boy ran. He said it showed no sign of being hurt by the light 20-gauge loads.

That was the first report that attracted attention to what has since been tagged the "Fouke Monster", but it turned out that it had been seen at least three times previously, and within a month another 14-year-old boy saw it and shot at it while deer hunting. The following year it was seen by a lady hunter during a deer drive and then a school bus driver saw it cross the road. It was heard screaming in the woods far more often than it was seen.

Then in June of 1971 something left hundreds of tracks in a bean field, and they were very strange tracks. They measured 13½ inches long, 4½ inches wide at the ball, 3¼ inches at the heel, very high instep and only three toes. Smokey Crabtree, James Lynn's father, has this to say in his book *Smokey and the Fouke Monster*:

> Where the sand was one half inch deep or so there was a narrow outer part of the foot touching down, connecting the heel to the front part with a one-inch wide or so strip. Where the dirt was soft there was a full track and very plain ones. They were so plain you could see the imprint of the lines on the bottom of the bare feet. There were only three toes and there wasn't that much difference between the size of the first toe and the third toe. There was no place for other toes. It was not like his little toes had burned off in a fire or frozen off. His foot was designed for three toes only.

Stride measured as much as 57 inches from heel to toe and as little as 26 inches. There was a small toe or thumb about five inches back from the big toe which did not seem to have any bone in it, leaving only a shallow mark in every track where it showed. The tracks wandered aimlessly in a loop, covering several hundred yards and returning to the woods exactly where they entered the field. The creature had avoided stepping on the bean sprouts. He estimated its weight at 300 pounds.

A trapper, hunter and tracker since childhood, Smokey Crabtree terms those tracks "a surprise of my life."

190

There are casts of the tracks to be seen at Willie Smith's service station in Fouke, and also a track itself, lifted in a slab of dirt and preserved in a box. Having seen them and having spent several hours talking to Smokey Crabtree, I am satisfied that the tracks, and the animal, are genuine: the movie made at Fouke, *The Legend of Boggy Creek*, may contain fictional episodes, but there is a Fouke Monster.

The conclusion that there are indeed giant hairy bipeds living in the Sulphur River bottoms in Arkansas creates no problems. It has every appearance of being a suitable place for them. The conclusion that a three-toed track is genuine is quite a different matter. I got involved in this business in the first place by concluding that the "Bigfoot" tracks in California were genuine. That may seem to be no great thing when you read it over quickly, but it involves accepting the idea that mankind shares North America with some other kind of upright primate that despite its giant size has remained undiscovered; and that is a concept so unacceptable that nearly 20 years of effort have been insufficient to persuade the scientific world to so much as look into it. Yet five-toed footprints in themselves are quite ordinary. They mirror just the sort of foot that an extremely-heavy, bipedal primate would be expected to evolve.

Three toes are alright for a bird or a dinosaur and not entirely unknown for a mammal. There is a three-toed sloth, as every crossword puzzle addict knows. Horses have one toe, cows two and dogs four. Rodents are more versatile than most groups. Agoutis have three toes. Pacas have four on the front and three on the back. Several species of rats have four toes on the front and five on the back. The snowshoe rabbit has a remarkably human-like four-toed hind foot. The horse family, which is odd enough with its single toe, has relatives, the tapirs, with four toes on the front and three on the back. Higher primates, however, all have five toes, and the thing described at Fouke has to be a higher primate.

The oldest reference to three toes I know of is in a work of fiction. In *The Hell Bent Kid*, published in 1957, New York author Charles O. Locke puts in the mouth of an old Texan the following words:

> If we leave her down here they are feared at home she
> might marry the first three-toed bush ape who comes along.

I have been unable to learn where the late Mr. Locke picked up the term.

It is entirely unacceptable to have to assume that there is not one unknown giant bipedal primate, but two, one of which has evolved for no apparent reason, a very unprimatelike foot; but it isn't much better to have to assume that there are three-toed individuals in a five-toed population. The problem is considered at more length later on. For me personally, it is the tracks at Fouke that establish that the problem has to be faced.

Not all the sightings were made by Fouke residents. May 25, 1971, the Hope, Arkansas, *Star* quoted two Texarkana couples describing a large hairy creature that crossed Highway 71 in front of them as they were driving home at night from Louisiana. It had long dark hair and ran upright in a stooped position with its arms swinging. Nor is Fouke the only place in Arkansas where there have been recent reports. In September, 1968, according to the September 26 *Arkansas Gazette*, city and county authorities were flooded with phone calls from the Sweek Gum residential area at Hamburg where people reported seeing "a thing that looks like a man but has a gorilla head." Hamburg is in the southeast corner of the state.

November 27, 1973, the Little Rock *Arkansas Gazette* carried a report of another sighting near Fouke by a farmer named Orville Scoggins, and his son and grandson. This time it was a little creature, about four feet tall with long, pitch-black hair, upright on two feet. I had a phone call reporting another sighting at Fouke in December, 1973, and a letter reporting new tracks there, identical to those described by Smokey Crabtree, seen in May, 1975.

On November 2, 1974, the *Arkansas Gazette* reported several sightings in Dallas County, the most recent being at Fordyce. One man, who shot at the creature in his front yard at night, said it was seven or eight feet tall and looked like a gorilla, although at first he thought it was a bear. He fired at it twice, using a 30-30 by flashlight, and it turned and walked away. In March, 1975, the Camden *News* reported the sighting of a seven or eight-foot creature completely covered with hair in that community. Camden is southeast of Fordyce, and both are in the south central part of the state.

From Louisiana a news service carried a fascinating story in 1973. It quoted a television news report that an experienced swamp guide ran over something in a bayou in Honey Island Swamp, northeast of New Orleans, causing his outboard motor to kick up. Looking back he and his party saw coming out of the water "an animal about five feet tall. It was black. I couldn't tell if it had hair or it was skin. It was running on two legs." The thing went up on the bank and disappeared in the swamp, leaving fairly large tracks with five webbed toes.

There seems to have been quite a bit of interest in the sasquatch generated in Louisiana in the past year, and I have seen references to no less than eight sighting reports in the last three or four years, but I don't have adequate information about any of them. The best source I have is a tape recording of a conversation in which a youth named

192

Chris Denaro tells of seeing a nine-foot creature while parking with his date in some woods in the summer of 1977. He says he reported it to the East Baton Rouge sheriff's office and footprints were found, but no size is given.

Moving north into Mississippi, I have no definite report on file until November, 1967. The editor of the Winona *Times* received an unsigned letter posted in Atlanta, Georgia, of which I have a photocopy. Excerpts from it follow:

This letter I am writing will be hard to write. But being it concerns an object and in an area east of Winona on the highway to Europa I feel like someone in that area should know what me and my brother seen.

The date was about 7th of November, the time one or one thirty appx. A.M. We were travelling east to Marietta, Georgia, in a Chevrolet pickup when my headlights picked up an object running down a steep hill on my left on its two legs, as if to run out and stop our truck. Then my headlights was on the creature. Its size was like that of a huge Kodiak bear but it was running on two legs not four. Its eyes were bright red appx. 2 inches in diameter. On its body was hair appx. 1½ long. Its weight was 5 to 700 lbs. Its height was appx. 7 ft. tall or more. Its left arm was held up like waving goodby or giving a stop signal to us. The expression in its eyes was like a human in mortal terror. And my brother and I both agreed it (the expression) was like a person saying, "Please help me."

The face of the object was like a person gone wild or crazy. Its shoulders were appx. 4 ft. wide with narrow waistline.

The object then tried to hide behind the shoulder of the road. If the object was scared, which no doubt in my mind it was, well I was a helluva lot scardier.

On June 22, 1971, the *Delta Democrat Times* at Greenville, Mississippi, ran a story quoting Mrs. Mae Pearl Young, of Greenville as having seen beside her daughter's home at night a black creature over six feet tall with "a great big head . . . broad shoulders and its hand on its hips."

In the mid 1970's there have been quite a few Mississippi reports. The Corinth, *Daily Corinthian*, March 16, 1976, reported that deep four-toed tracks 15 inches long and 6½ inches wide had been followed for 200 yards from a road north of U.S. 72 up a bank and into the woods. Later stories told of more such tracks five miles away and mentioned that there had been reports of a large hairy creature in Alcorn County two years earlier.

In early 1977 there were a sighting report and two more footprint reports, all in the southwest part of the state. The Natchez *Democrat*, on January 20, told of sightings of a huge hairy "Bigfoot" type creature that resulted in Natchez police being deluged with calls. "Upright" footprints were also mentioned, but no description was given. On February 16 the Brookhaven *Leader* carried photographs of one of several five-toed, 15-inch tracks, extremely deep, found two days before beside the Buie Mill Road west of Loyd Star in Lincoln County. On April 17, the Tylertown *Times* also had pictures, of one of six 18-inch, three-toed prints found by Magee's Creek near Holmes Water Park, Walthall County, by five boys cutting through the woods to a swimming hole. Not at all like the three-toed Fouke prints, these were eight inches wide, with wide rounded toes.

From Missouri comes an exciting story about a thing that had been killing sheep and goats at Piney Ridge in 1947. When hound hunters tried to go after it, it killed the dogs, then doubled back and overturned their jeep. I have read the story, which was written by Glenn Payne, of Sedalia, Missouri, but I do not have a copy of it. There are two other reports of sightings in Missouri, right on the Mississippi, the second of which is perhaps the most publicized item of any in the history of sasquatchery except for the Patterson movie. The first report, on the other hand, I know of only from a reference in the January-February, 1973 edition of *Flying Saucer Review*. It says that Joan Mills and Mary Ryan stopped for a picnic lunch beside Highway 79 north of Louisiana, Missouri, in July of 1971. They smelled a bad odor, then saw the top part of a creature standing in a thicket with some weeds in front of it. Miss Ryan is quoted:

It was half-ape and half-man. I've been reading up on the abominable snowman since then and from articles you get the idea that these things are more like gorillas. This thing was not like that at all. It had hair all over the body as if it was an ape. Yet the face was definitely human. It was more like a hairy human.

The creature made a gurgling noise and proceeded towards the young women, who dashed to their car. Continuing to gurgle, the article states, the thing patted the hood of the car, and tried to open the doors. Having left the keys outside, the women could not drive away, but succeeded in making it back off by sounding the horn.

It stopped where we had been eating, picked up my peanut butter sandwich, smelled it, then devoured it in one gulp. It started to pick up Joan's purse, dropped it and then disappeared back into the woods.

Joan Mills retrieved her purse with the keys, and they fled. Back in St. Louis they reported the incident to the Missouri State Patrol. Miss Mills wrote:

. . . all you have to do is go into those hills to realize that an army of those things could live there undetected.

To me that seems a more interesting incident than the one on July 12 the following year at Louisiana, Missouri, that was so well publicized. Major newspapers in the eastern U. S. sent reporters to do feature stories, and "Momo, the Missouri Monster" almost became a household word. At 3:30 on Tuesday afternoon, July 11, 15-year-old Doris Harrison heard her young brother scream outside the house, looked out and saw a thing that stood like a man but was covered with long black hair. It was six or seven feet tall, had hair all over its face, and seemed to have no neck. It smelled awful, and it left one possible footprint 10 inches long and five inches wide, with three toes. After the fuss started several other people claimed to have seen something similar, generating even more excitement, and a lot of people spent time monster hunting, but nothing came of it.

Things have been quiet since, apparently, except for a couple of reports from Pacific, on the outskirts of St. Louis, reported in the Union, Missouri, *Franklin County Tribune* on May 25, 1977. A woman had reported to police that about 8:30 at night on May 18 her husband had seen an eight or nine-foot creature walking like a man but covered with hair that came out of a creek and walked across the road. She said he had seen a similar, but smaller one at the same location in October, 1975. The paper says that the local name for such creatures is "Brush Ape", and that other people in the Pacific area have reported seeing them over a period of several years.

Another story came to light in the St. Louis *Post-Dispatch*, April 28, 1977, but it is the oldest Missouri report of all. Sue Hubbell, writing from Mountain View, Missouri, said that she had found in the June 26, 1925 edition of the Mountain View *Standard* an account of three sightings of a peculiar beast near Alton. One man who got within 50 yards of it described it as "an animal that walks upright like a man, rather brown hair all over and had a face something like a monkey." The paper also published, on August 13, 1925, a supposed explanation for the sighting, after a sheriff at Chillicothe interviewed a transient who was found to be covered under his clothes with a six-inch growth of matted hair. The story says that the man admitted having passed through the Alton area in July. However it does not say that when he was at Alton he was wearing no clothes, and nothing is said of the fact that the Chillicothe man's hair was not brown, but black.

Farther up the Mississippi in Iowa there are a couple more reports from fairly close to the river. One from Maquoketa, Iowa, was published in 1970 but goes back a long way. I have a clipping of the story, but no note of the paper that printed it:

'It' is the mysterious creature folks around here have heard about for 50 years—a furry, scary thing that stands like a man.

. . . A woman who lives near Baldwin said hunters and trappers told of animals that walked like men along the Maquoketa River near Canton when she moved to the area 50 years ago.

Gary Koontz, 21, of Maquoketa, says he saw a strange animal four years ago while hunting the dense woods and brush land north of the state park.

Koontz said he spotted the creature from a distance of about a hundred feet.

He said it stood upright, was four or five feet tall, was covered with dark fur or hair and had a flat, frightened face.

"I knew it was something I had never seen before," Koontz said. "I was startled but decided to take a shot at it with my shotgun and I'm sure I hit it. It let out a high-pitched, woman-like scream and disappeared into the brush."

He said that his uncle, on leave from the army, had seen the same thing just recently.

The other report carried by Associated Press from Lockridge, Iowa, in October, 1975, tells of several people seeing a thing with long, shaggy, dark hair that was usually on all fours but sometimes stood on its hind legs. It had an apelike face. Lockridge and Maquoketa are each within 40 miles of the Mississippi and each is beside a tributary river.

I have already referred to several reports on the side of Iowa bordered by the Missouri, and there are also a couple of recent ones well removed from either river. A Dumont, Iowa man wrote to Grover Krantz that on August 10, 1976, he had found a half mile of big tracks going along a gravel road. Color pictures that he sent showed that the tracks were 12 to 13 inches long and five inches wide. Stride was about six feet. The tracks came from open country but headed towards a wooded area paralleling the West Fork river. His own tracks, 230 pounds on a size 10 shoe, hardly showed. The other report, published in the Sibley *Tribune*, August 26, 1976, told of a 13-year old boy seeing a six-foot, hairy black creature standing by the Ocheyedan River near Ocheyedan, Iowa about 9:30 on the morning of August 22. It was holding its hands as if drinking from them. Although he was in a car with his family he said nothing about it until late in the day when they wanted to know what was the matter with him. When they went back next morning they found 14-inch footprints, narrow in the heel but seven inches across the five toes, in the dirt along the river bed.

The situation in Minnesota is an unusual one. It has a lot of the sort of country where sightings could be expected, and I have several good reports from just north of the Canadian border. There are seven reports in my file from Minnesota too, but almost all are brief references in "monster story" books and magazine articles, lacking in specific detail that could be checked. One that is a bit more complete, from Eric Norman's *The Abominable Snowman* says that at LaCrescent, Minnesota, in the fall of 1968 a man in a duck blind thought he heard a friend approaching, rose to tell him to be quiet, and saw a tall black hairy thing so high he came up to the middle of its chest. He dropped his shotgun, which discharged, and the thing ran off screaming.

Most detailed of those accounts is a story from an article in *Saga* magazine, June, 1969, which states that "several months ago" an Iowa student named Larry Hawkins saw what he thought was a person in trouble crouching beside Highway 52 south of Rochester, Minnesota. He pulled over on the shoulder and then realized that the thing was apelike, with thick shoulders and covered with hair. It leaped up and ran up a steep hill into the woods, leaving behind a rabbit that was dead but had no blood on it. As the student was examining the rabbit he heard a harsh roar from the woods, so decided to leave. The story states that he reported the incident to police but they refused to take him seriously.

There is also a letter from a deer hunter to Roger Patterson in the fall of 1968 in which he says that as he was sitting on a stump in the woods 10 miles north of Floodwood, a creature about 4½ feet tall jumped down from a tree about 125 feet away and walked into the woods "on its back feet." He says that he thought it was a child at first but then realized it was "something I had never seen before," but he did not specify that it was covered with hair.

From Wisconsin I have seven reports, three of them from 1976. An Oregon resident wrote to *Argosy* magazine that when she was about 10 years old, in 1910, she met a large animal in the woods near a resort her father owned at Mirrow Lake, Wisconsin. Coming home at dusk she encountered a creature covered with fur that walked like a man and had a man's eyes. It followed her most of the way home but did not seem menacing, and she thought it was in need of help in some way. She tried to get her father to go and see it, but he just patted her curls and said she had met a friendly bear. Fishing with her father that day was Al Ringling, then head of Ringling Bros. circus, who perhaps missed a chance at the greatest sideshow attraction of all time. The next day, the lady wrote, she went back and found footprints in the snow, twice as long as her father's, that went back up into the deep woods behind the house. No one would look at those either, and that night a blizzard buried them. I don't think the letter was published, but there is a copy in Ivan Sanderson's files.

A Wisconsin resident wrote to me that in July, 1964, shortly after midnight, he saw a sasquatch run across state highway 89 near the Illinois border, jumping the fences on either side. In Benton, Wisconsin, according to the Dubuque, Iowa, *Telegraph-Herald*, about 15 people saw a hairy, white "monster" in August, 1970.

A compilation of descriptions and conjectures puts the tale like this: a white furred, apelike 300-pound creature about seven feet tall with claws and pink eyes has established a lair near here. He—almost everyone agrees it's a male—has been around about three weeks.

Benton is close to the Mississippi.

Most of the reports given in this chapter and the preceding one have not been investigated by anyone, but one Wisconsin report was given a thorough shakedown by Ivan Sanderson and Dr. Bernard Heuvelmans, who were in the area the month after the incident. They interviewed six of a group of 12 men who had encountered a strange creature on a deer drive in Deltox Marsh, near Fremont, on November 30, 1968. The creature had appeared in front of the left end of the line of deer hunters as the men went through the swamp. Word passed down the line, and the men tried to approach it. Some of them got very close before it finally disappeared in some woodland. In an article in *Argosy*, Ivan Sanderson summarized what they told him:

> The composite description of the creature that emerged was that of a large and powerfully built man covered with short, very dark brown or black hair and with a lighter and hairless face and hairless palms. The head appeared smallish, also with short hair, but the neck appeared to be enormous and so short as to be almost non-existent.
>
> The shoulders were very wide and large and the torso barrel shaped. In a six-way discussion at our interview some time was spent on the proportionate length of the arms, body and legs. Analyzing this exchange (from the tape), it seems that while the body seemed to be very long, this was due to the absence of any noticeable waist. All of them said that it tapered from the shoulders right to the hips. As for a description of the legs, they could only guess since the creature was standing in grass which they estimated to be between three and four feet tall. Some at first said the legs were short; others that they were long—but this was before they decided to speak of their length in proportion to the

body rather than in comparison to a man or an ape. Then they agreed that they would be of about average length for a tall man, since the grass did not reach to the crotch. But it was concerning the arms that all seemed agreed, feeling that they were exceptionally long for a man.

The creature had not seemed to be afraid and was obviously observing the men, even moving towards them when they backed up, but keeping at a minimum distance. It walked like a man but slightly bent forward. The investigators were told that the creature had been seen twice previously that fall, once in Deltox Marsh and once in Lebanon Swamp, and had been seen on a road at night the day after the deer hunters saw it. Everyone agreed that no man would have dared masquerade in a fur suit during hunting season, it would have been suicide. Yet no one, apparently, had shot at the thing.

The LaCrosse *Tribune*, October 28, 1976, states than "an unidentified creature is stirring interest in the Cashton area" and tells of an unnamed farmer seeing a seven-foot, upright creature covered with dark hair in mid-August. It had a very strong smell. The same paper, on November 17, 1976, had a follow-up story on the Cashton situation which mentioned that four 18-inch footprints, seven inches wide but showing only four toes, had been found by three people walking in the woods late in October. In the meantime, United Press International carried a story out of St. Croix Falls telling how 12 teenagers went out shining car spotlights at deer in Sterling Township on the night of October 8 and saw a seven-to-eight-foot creature covered with hair. The story said that reporter Pete Jensen went to the scene and found tracks measuring 19 inches in length and nine inches in width, and followed them for half a mile, "finding along the way several branches broken off trees at the seven and eight-foot level and a dry stump that appeared to have been freshly uprooted". The editor of the St. Croix Falls *Standard-Press*, Frank Zaworski, was quoted as saying that he did not think the boys were lying.

Over in Michigan, old reports go back as far as 1891. The Colfax, Washington, *Commoner* carried the following story on November 6 of that year:

WILD MAN OF MICHIGAN

AN UNTAMED GIANT, SEVEN FEET TALL, SEEN IN GLADWIN COUNTY

Gladwin, Mich. Oct. 26.—George W. Frost and W. W. Vivian, both reputable citizens, report having seen a wild man near the Tittabawassee River, in Gladwin county. The man was nude, covered with hair, and was a giant in proportions. According to their stories he must have been at

199

least seven feet high, his arms reaching below his knees, and with hands twice the usual size. Mr. Vivian set his bull dog on the crazy man, and one mighty stroke of his monstrous hand he felled the dog dead. His jumps were measured and found to be from 20 to 23 feet.

Another story that would date about 1915 to 1920 was told in a letter to Ivan Sanderson. Two hunters somewhere in northern Michigan were said to have followed what they thought might be bear tracks and come upon a thing resembling a human giant with long arms, and short light hair covering most of its body.

In the Folklore Archives at Indiana University Library there are the following entries:

> In the year 1937, Saginaw, Michigan, a fisherman sitting on the banks of the Saginaw River is reputed to have seen a manlike monster climb up the river bank, lean upon a tree and then return to the river. The man suffered a nervous breakdown.

> In Charlotte, Michigan, the whole town was upset one fall (1951) when there were reports of a strange monster loose in a swamp just outside of the town to the west. The people of the town couldn't identify it but one of the men decided it must be a gorilla because it stood on two feet. Because of this strange creature, seen on six or seven occasions by many different people, the area where it was seen is known as Gorilla Swamp.

As if having a wealth of old stories weren't enough, Michigan also had two of the most widely-publicized modern incidents. The "Monster of Sister Lakes" first made the headlines in June, 1964, but there were reports that it had been seen off and on for two years before that. Cass County deputies were dispatched to find the monster after Mrs. John Utrup, of Dewey Lakes, reported that a thing nine feet tall and weighing 500 pounds had chased her into her house at night. Other descriptions had it covered with black hair up to its neck and with eyes that glowed like an animal's when a light hit them. The deputies were kept too busy directing traffic to have time to hunt the monster. So many curious people turned up that men had to be called in from the next county to help keep them moving. According to one report police did find barefoot, humanlike prints six inches wide across the ball in the mud in the Utrup farmyard. There were apparently other serious reports, but within a couple of days the whole thing had turned into a carnival. Every place of business had a Monster Sale on, the theatre had a double-bill horror show, restaurants were featuring Monster Burgers, and a variety store was advertising a Special Monster Hunting Kit—a baseball bat, a net, a flashlight, a mallet and a spear, all for $7.95. All over the place, according to a magazine article, teenagers were draping themselves in old fur coats to do monster imitations on busy highways. That sounds suicidal in an area overrun by men with rifles, but despite the usual dire predictions from the lawmen, no one seems to have been hurt.

The following year the spotlight shifted east from Cass County across the state to an area north of Monroe, where something big, black and terrifying was rumored to have been seen in the thick woods and swamps. On the evening of August 13, Mrs. Ruth Owens and her 17-year-old daughter, Christine, were returning from a visit to Mrs. Owens' parents, when they saw something moving where the road passed through a wooded area. Christine, who was driving, panicked and stalled the car just as it brushed past the creature. Mrs. Owens said she saw a "huge hairy hand" on top of her daughter's head, and the girl's face was slammed against the door post. Their screams brought men running from a farmhouse, who found the lady hysterical and the girl unconscious. Next day it was like Cass County all over again, with hordes of hunters in cars with guns, not to mention bow hunters and a motorcycle gang. The Detroit News has a large file on the two incidents, but the best over-all account was in True magazine in June, 1966.

There hasn't been that much excitement in Michigan since, but reports have kept coming in, several of them in personal communications. One of my correspondents, from Sault Ste. Marie, reported being told by the chief of the police at a nearby Air Force base in 1967 that he had seen and chased an apelike thing but lost it in the woods. Another said that he and his young brother saw a huge, hair-covered creature in the woods while camping at Bay City near Saginaw Bay in the summer of 1973. There were other sightings reported from Port Huron, in April, 1969, and New Buffalo in the fall of 1973.

Wayne King, at Millington, Michigan, has started the Michigan Bigfoot Information Center, and has investigated a number of recent sighting and footprint reports in Benzie County and elsewhere. He passed on to me an intriguing account of a sighting, on the evening of June 4, 1977. A young couple whom he had interviewed said they saw three creatures they estimated to weigh 300 to 500 pounds run across U. S. 27 in Clare County, taking tremendous strides.

Finally, the Traverse City Record Eagle, September 8, 1977, carried a long story by Outdoor Editor Gordon Charles about tracks made on the nights of September 2 and 3 behind the home of Becky and Bob Kurtz in a wooded area in Mecosta County near Barrytown. The tracks were 16½ inches long and 9½ inches wide, with five wide toes, and two odd bulges on the balls of the feet. They indicated a five-foot stride, and sank to four times the depth of the men's prints. Whatever made them had ignored a garbage barrel, but cleaned up some partly-spoiled peaches. Mr. Charles made clear that he and a couple of his bear-hunting friends are no longer scoffing at the idea of sasquatches in Michigan.

Up to this point I have tried to at least mention almost every report from each state, but in Illinois, there are too many. This results, in my opinion, not from any preference of the hairy creatures for Illinois, but from the efforts of one of the most dedicated researchers, Loren Coleman, who now lives in New England but was active in Illinois for a decade. One of the items Loren collected is a letter to the editor published in the Decatur *Review*, August 2, 1972, from Mrs. Beulah Schroat, of Decatur, which takes the hairy creatures back to the early years of the century in Illinois. She wrote:

In reference to the creatures people are seeing, I am 76 years old. My home used to be south of Effingham. My two brothers saw the creatures when they were children. My brothers have since passed away.

They are hairy, stand on their hind legs, have large eyes and are about as large as an average person or shorter, and are harmless as they ran away from children. They walk, they do not jump.

They were seen on a farm near a branch of water. The boys waded and fished in the creek every day and once in a while they would run to the house scared and tell the story.

Later there was a piece in the Chicago paper stating there were such animals of that description and they were harmless. This occurred about 60 years ago or a little less.

My mother and father thought they were just children's stories until the Chicago paper told the story.

The earliest specific report Loren uncovered was in the Carbondale, Illinois, *Free Press*, on March 30, 1942:

MASS HUNT FOR ANIMAL FAILS TO SOLVE MYSTERY

Mt. Vernon, Ill. (AP)—The mystery of Gun Creek bottoms is still a mystery. Not even 1,500 Southern Illinois hunters could flush the mysterious animal with a "wildcat's scream" and "paws like a huge raccoon" yesterday although they combed the bottom with shotguns and rifles on the ready.

The object of the mass hunt was blamed last October for an attack on a squirrel-hunting Mount Vernon minister. Screams like those of wildcats have been reported heard frequently at night. The animal was declared responsible for the death of a dog and was said to make tracks similar to a raccoon "but four times as large."

Raccoon tracks look a lot like human hands and are usually something over three inches long. The animal was referred to elsewhere as "a large animal that looked something like a baboon", and as "a half-man beast."

In Centerville, Illinois, in May, 1963, the description given was "half man, half horse." Centerville police had more than 50 calls reporting sightings in a few days, according to the Chicago *Daily News*. In 1965, according to a letter they wrote to Roger Patterson, two boys camping on a farm outside Trenton, Illinois, woke to find barefoot prints 14 inches long and 5 to 6 inches wide going across the field and into the woods. They sank two inches deep where the boys sank only a quarter inch. The boys said they sent casts of the prints to the University of Illinois and never got them back.

Four youths reported seeing a black manlike monster near Decatur in September, 1965, according to the September 22 issue of the Decatur *Review*. On August 19, 1968, the Carbondale *Southern Illinoisan* reported that two young people encountered a 10-foot black creature with a round hairy face that threw dirt through their car window. In July, 1970, according to an article by Loren in the March, 1971 issue of *Fate*, four youths camping a mile south of Farmer City heard something moving in the tall grass, turned on their car lights and saw a creature squatting near their tent. It ran off on two legs. Early in the next month, in the same general area, three youths from Rantoul said they saw a thing as big as a cow and covered with hair ambling on its hind legs along the shore of Kickapoo Creek. They said they found half-eaten minnows and scooped-out clam shells where it had been walking. That story was in the Champaign-Urbana *News-Gazette*, August 10. There were other sighting reports August 11 and 16.

In July, 1972, the hunt was on again, this time in the woods outside East Peoria. "Two reliable citizens" were reported by the July 27 Pekin, Illinois, *Daily Times* to have seen on July 25 a 10-foot thing that "looked like a cross between an ape and a cave man" and smelled like "a musky wet down dog." The same evening, Leroy Summers, of Cairo, Illinois, reported to Cairo police that he had seen a white, hairy creature 10 feet high standing near the Ohio River levee. From Cairo to East Peoria, which is on the Illinois River, is three quarters of the length of the state.

Early the following May the action moved to Enfield, United Press International reported that four men saw a grey-black apelike creature five to 5½ feet tall standing in the doorway of an abandoned barn. Loren Coleman was on hand the next week to investigate. A month later a light-colored seven or eight-foot creature was reported, seen by a parking couple near a boat ramp at

Murphysboro, and the next night a seven-foot white-haired creature covered with mud and river slime, seen by two teenagers there while they were sitting on the front porch about 10 in the evening. Dubbed "the Big Muddy Monster", that one achieved considerable notoriety. It was reported seen three or four times later in the summer, and even the New York *Times* considered it worth a feature article, although they didn't get around to it until November 1. On October 17, 1973 the Champaign-Urbana *Courier* reported that four youths in a car saw a five-foot "gorilla like" creature near St. Joseph. January 26, 1974, the Elgin *Daily Courier News* reported a dirty, hairy white monster at Aurora, where it was called "Big Mo." In August, a seven-foot shiny dark brown number with 13-inch footprints six inches wide, and four toes, was at Carol Stream, where it was eventually reported about a dozen times, according to the Chicago *Tribune*, October 6, 1974. Aurora and Carol Stream are close together and both are close to Chicago. In the meantime, according to the Centralia *Sentinel*, July 9, 1975, "The Big Muddy Monster" had been seen back at Murphysboro in July, 1974, and it was reported again in July, 1975. That is the most recent sighting report I have from Illinois.

Michigan City, Indiana, was the scene of an 1839 "wild child" report, but there is nothing more in the file for 110 years. On July 17, 1949 International News Service carried a report from Thorntown, here reprinted from The Richmond *Palladium Item*:

> State conservation department officers are heading four posses searching for a big hairy beast who has been terrorizing fishermen and keeping Thorntown residents indoors at night for some time.
>
> Witnesses say that the animal weighs about 250 pounds and is brown, hairy and has protruding teeth.
>
> Nearly 30 Thorntown residents allege that they have seen the beast, including Charles Jones and George Coffman, fishermen, who fled Thursday night when the monster pursued them. Jones ran through the woods and Coffman waded the creek to escape.

The creature, referred to as a "gorilla" was being sought by the posses in "the Sugar Creek wilds."

In August, 1962, and later, a group of young people claimed to have seen at an Indian graveyard at Blue Clay Springs, near Richmond, a seven-foot creature "made like an ape with long arms" that stood on two feet but ran on all fours, had white hair all over its body, four toes all the same size, and eyes that shone red when the lights were on them. By September, according to a report sent to Ivan Sanderson, they had identified two different creatures, both

204

light-colored, but not the same shade and one larger than the other. One was supposed to have torn the door off a car. Shot at with a high-powered rifle, a .22 automatic and a shotgun, it only flinched.

That story is basically unbelievable, but I have been involved with the investigation of a couple of similar ones out west that seemed pretty substantial. At least the young people had told their parents and the police what they were doing at the time and they definitely were cruising around at night heavily armed, whether or not there was actually anything to shoot at. On the other hand I can't say anything in support of a story in the Indianapolis *News*, March 15, 1965, about youngsters at French Lick, Indiana, seeing a 10-foot tall monster that was green in color and had glowing red eyes. They called it "Fluorescent Freddie"

In May, 1969, near Rising Sun, according to an article by Jerome Clark and Loren Coleman in the January-February, 1973, issue of *Flying Saucer Review*, George Kaiser, a farmer's son, saw in the farmyard a muscular creature of human height but covered with hair that made a grunting sound when it noticed him, leaped a ditch and ran off down a road at high speed. Later, four-toed footprints were found in the dirt by the ditch, and cast.

August 13, 1970, the Petersburg *Press Dispatch* reported that teenagers were searching for a "Booger Man" around McCord's Ford. The Evansville *Press*, August 15, said that those who had seen it described a "ten-foot tall creature covered with hair, appears to walk on hind legs, top speed 60 miles per hour." It had been reported on two successive nights south of Winslow. Then in the summer of 1972 the action shifted to Putnam County, with a series of reports from Roachdale that were repeated in papers all over the country. The fuss was about something humanoid in form, fur-covered, about six feet tall and capable of walking and running on its hind legs. It was reported to have a musty scent and eyes that reflected light "like balls of fire." Roachdale is the only Indiana location I have visited. It is in flat farm country that appears totally unsuitable for a sasquatch. Even the nearest river of any size, the Wabash, is more than 30 miles away.

Fate magazine, August, 1973, had an article on the Roachdale incidents, mentioning specific sightings by Mrs. Lou Rogers and by Carter and Bill Burdine. The latter two saw a large hairy creature in the doorway of their chicken coop and went after it with shotguns. They were reported to have lost at least 170 chickens, not eaten but drained of blood, over a period of time. The story also states that in Parke County, which is west of Putnam County, the sheriff, Gary Cooper, had broadcast a warning on September 20, 1966 for people to be on the lookout for a 10-foot monster covered with hair that left tracks 21 inches long, and had reportedly killed two coon dogs and a pig.

I didn't know it at the time when I was there, but Indiana has a resident researcher, Don Worley, at Connersville. Like a lot of the active people in the eastern U. S., and some of those in the west too, he is also an investigator of U.F.O. reports and is of the opinion that there are probably two kinds of hairy bipeds, one a normal, Earth animal while the other, or others, arrive in spacecraft or else are not real creatures in any ordinary sense. It is my opinion that trying to use one unexplained phenomenon to explain another one is not at all helpful, and that so far there is no evidence for it anyway, except for people claiming to have seen something in the U.F.O. line at about the same place and time as a "creature" report. Such connections are often made, however, by the people who do the investigating of reports in the Midwest and in Pennsylvania, and I hope they will not take too much exception to my printing only the information about the hairy bipeds and leaving out the lights hovering in the sky, and so on. So that the reader will understand what I am talking about I will give my condensed version of two of Don Worley's Indiana reports, and then repeat one of them in full as he originally wrote it.

Don tells of a series of incidents at Sharpsville that began in June, 1971, and went on until the next year. It began when a young man went out in his yard to see why his dog was acting up and saw about 25 feet away a great apelike creature which the dog was lunging at. The creature was growling in a deep rumbling manner and taking slow swings at the dog without touching it. Calling off the dog, the man got his shotgun and chased after the creature as it moved away down by the creek, firing two shots from a distance. The creature returned to the home five more times in 1971 and 1972. On one occasion the man found his wife and her girl friend cowering in terror. A powerful smell of rotten meat filled the air. They said the creature had tried to enter the home by pulling loose part of the aluminum storm window. The man, an ex-marine, attempted to hunt the creature and three times pursued it back to the wooded creek after it appeared about 11 p.m. It never ran, but kept ahead of man and dog taking quick giant steps. Once the man's mother saw it follow him back towards the house after a chase.

The other report is from Galveston, in October, 1971. Jim Mays, fishing one evening at a spot called The Pits, felt a hand placed on his shoulder. He turned and saw a towering creature's hand, legs and face before it turned and fled. The creature ran down the road over a bridge, leaped a ditch and went into the woods. Two nights later he and four other men saw the same thing by flashlight, standing at the edge of the woods. It was eight to nine feet tall, sandy colored and

206

apelike. They threw rocks at it, but were not sure if they hit it. An approaching car interrupted the sighting and then the thing was gone.

Here is the way that story was told by Don Worley:

On September 11, 1975, the Kokomo *Herald* published an article about the dozen or so UFO's seen since 1965 by John Grey, a middle-aged factory worker. In one of his first experiences, he felt a strange sensation forcing him to go outside, and an unspoken command to shine his flashlight up in the sky. Suddenly an orange glow, which he had seen at other times, appeared in the sky. He flashed his light three times and the glow blinked three times. Grabbing his binoculars, he saw in the sky a round black disc as big as a house with a curved lighted section. As it sped away he heard a humming sound. Then three jets came swooping in from Grissom Air Force Base.

Mr. Grey's soon-to-be son-in-law, Jim Mays, a relative skeptic about the unknown, was curiously enough destined to introduce Grey to the creature. At twilight one October evening in 1973 Jim sat fishing at a spot called The Pits, when a creature approached from behind and placed its hand on his shoulder. The startled witness got a quick look at the towering spectre's hand, legs and face before it fled in a strange leaping fashion up a blacktop road and across a bridge. Jim could hear the slap of its feet on the blacktop. The creature leaped a ditch and went into the woods, from which a glowing bronze light source soon ascended into the sky, shooting up and fading away so quickly that it seemed almost instantaneous.

Two days later an uneasy Jim, with a bayonet stuck in his belt, returned to the scene with John Grey and three others. A glowing aerial light seemed to follow them part of the way and then disappeared near the bridge. The creature was spotted standing at the edge of the woods from which it had disappeared before. Later, by comparing the height of the weeds with that of a human, it was established that the creature was eight or nine feet tall. Two witnesses retreated in fear to the car. The creature was illuminated by two flashlight beams and questions and a few curse words were yelled at it. One witness believed the flashlight beams seemed weaker when on the creature. A few rocks were hurled, but it could not be determined whether they missed, bounced off, or went through the creature. The giant, sandy-colored apelike thing did not move at any time. An approaching car

207

made it necessary for the witnesses to move their car, and when they returned the creature was gone.

Undoubtedly my version is the less interesting of the two, but with so many reports to cover, most must be drastically cut down. There are just three more in my records from Indiana, all 1977 newspaper stories. The Warsaw *Times-Union*, March 4, tells of two nine-year-old boys finding several sets of 18-inch tracks 10 to 12 inches wide near State Road 15 just south of Warsaw after a snowfall. They followed one set across some railway tracks and another into the back yard of a home, but "lost the trail" both times. How the trails stopped is not explained, but the boys agreed to a suggestion that the tracks might have been faked. On April 20, the Cincinnati *Post* carried a story from Rising Sun, Indiana, scene of a 1969 report. Mrs. Connie Courter, of Aurora, was quoted as saying that on the preceding Tuesday night when she and her husband drove home to their trailer about 11 p.m. they encountered a 12-foot creature right beside the car, which bumped into the car and dented it. They drove away and called the sheriff, but the deputy found no sign of the thing. The next night about 11:45 when they arrived home they saw it perched up on a hill, and her husband fired 15 shots at it with a .22 It jumped up at the first shot, but then crawled to get away from the shower of bullets. Aurora is right on the Ohio River, as is Rising Sun.

August 11, 1977, the Anderson, Indiana *Herald* carried a somewhat more prosaic sighting report. Mrs. Melinda Chestnut, of Elwood, had told state police that while driving east on County Road 1100N around 2 a.m. she saw an object eight to 10 feet tall walking near the county landfill. It was upright and completely covered with hair. She slowed her vehicle and turned on the high beams, at which it fled into the woods. The final paragraph of the story is intriguing:

> State police officials at Pendleton said creatures of this type were reportedly seen quite often in this same area some 80 to 100 years ago.

With things so busy elsewhere in the Midwest, Ohio would hardly be left out, and although it was late getting started, the Buckeye State has yielded more reports that any of those dealt with so far except Illinois and Texas. The earliest report in my file came to light only this year in the August 29 edition of the Miami, Florida, *Homestead City News*. It quotes Dean Averick, now a Florida resident, as saying that he had seen a sasquatch when he operated Deans' Boat Landing near Padanaram, Ohio, in 1954. He said he saw a hairy creature that "had a snub nose, peaked eyes, very chesty and light-brown scraggly hair, over six feet tall, and had a long straight back." It walked out of the brush about 125 yards up the shoreline and waded into about three feet of water, then, "I figured it must have seen me because it headed for a small island off shore."

The next report I have is from George Wagner, Fort Thomas, Kentucky, via Loren Coleman. Quoting from the Cincinnati *Post and Times Star*, on January 30 to February 3, 1959 and from personal recollections, he says that a trucker reported seeing a hulking creature climb out of the Ohio River onto the shore at Cincinnati. It was a cold night and a violent wind was whipping the river into six-foot waves. The trucker refused to describe the monster to police, but more calls came in, and across the river at Covington, Kentucky, a motorist reported seeing a thing on two legs, three or four times the size of a man and much bulkier, on a bridge over the Licking River. On March 28 that year and again in 1963, according to John Keel's *Strange Creatures from Time and Space*, a seven-foot "thing" covered with grey hair and having large luminous eyes was reported near Mansfield, Ohio.

In 1966, according to Tom Archer, an Ohio researcher now in the army, a seven-foot ape-man was reported roaming Wayne National Forest at the southern tip of the state. In April, 1968, John Keel reports, an eight-footer covered with hair was supposed to have been seen in the woods behind the Cleveland zoo, at the north end of the state. It turned up again at that location in August, 1972, described in the August 14 Cleveland *Plain Dealer* as seven feet tall and covered with black hair. Wayne E. Lewis, of 3861 West 39th Street, said he encountered it at 9:30 p.m. while looking for a kitten. It was standing behind a fence and was a lot bigger than he was. A six-footer, Mr. Lewis gave his weight as 360 pounds. A 19-year-old who also saw it said it looked like a gorilla except that it stood straighter.

Meanwhile something had been seen on Fox Road south of Huron, Ohio at night in October, 1970. A Lorain, Ohio, businessman wrote to Ivan Sanderson that he had talked to a girl who almost hit it with a car. She described it as seven feet tall, covered entirely with blackish-brown hair, with a pointed head, deepset eyes and arms longer than average. Huron is about 50 miles west of Cleveland. In November, 1972, according to the November 30 *Plain Dealer*, a cab driver reported to police at Ironton, Ohio, that he saw a large white apelike thing dragging a dog or deer. Ironton is on the Ohio River at the south edge of Wayne National Forest. Sometime around that period two security guards reported seeing an eight-foot hairy monster at night on a golf course under construction at Dublin, Ohio. It left a foot-long print which appeared to have three toes. I checked that story in 1976 with the agency that had employed the guards.

August, 1973, was a busy time in Ohio. Just north of Mansfield, according to a report investigated by Albert Hartman of that city, a man saw an eight-foot creature, very wide and with very long arms, standing by his barn at about 2 a.m. He shot at it with a shotgun and it fled. He then called the sheriff. At approximately the same time of night, at Oberlin, Rudy Reinhold and five other coon hunters saw an eight-foot, shaggy-haired, stinking animal with red glowing eyes. The dogs crowded it into a cornfield, walking backwards on its hind legs, but when the men were returning to their cars it chased them. A police sergeant who had experience as a park ranger and a zoo operator found some large tracks in the vicinity, not clear enough to identify but bipedal, and with a five-foot stride. Oberlin is about midway from Cleveland to Huron and a bit inland. The story was told in the October, 1973, issue of *Cleveland Magazine*. On October 27, 1973, the Akron *Beacon-Journal* reported that a seven-foot hairy monster in need of a strong deodorant had been seen at Massillon by several residents, who had notified Massillon police.

Tom Archer passed on a report of five-toed footprints 14 inches long and six inches wide, seen and photographed in a field close to some woods near Westerville, in September, 1974. Don Worley investigated a report by three children on a farm in Preble County, Ohio, about July 10, 1975. Playing in an open space at the edge of a cornfield, two 10-year-olds saw something watching them from 15 feet away over the top of the six-foot corn stalks. They ran and got a 12-year-old friend. Returning, all three saw it. They tried to get the older child's father to come, but he paid no attention, so they climbed up on a shed roof and watched the thing from there. After a while it ran off, very fast, towards some woods and disappeared over a hill. They described it as being upright but leaning forward. It was covered with long, smooth brown hair.

In 1977 there have been two reports in Ohio newspapers. The Jackson, Ohio, *Journal Herald*, February 7, carries a story of tracks, 20 inches by six inches, found in snow beside a mobile home and proceeding up a steep hill with strides "monstrously" long. The tracks were followed almost a mile into brushy woods. On March 10, the Revenna *Record Courier* reported that Mrs. Barbara Pistilli, from Nelson Township, Portage County, had phoned the sheriff's office on the evening of March 8 to say that two teenagers had seen and shot at a giant furry creature. They told deputies that it was eight feet tall and weighed about 500 pounds. It had walked out of the woods and started across the field, but fled when they shot at it. Mrs. Pistilli said it had been seen in the area before.

In the last 23 chapters I have dealt with 25 states, and aside from the traditional sasquatch areas in the west there are 20 states, provinces, and territories to go. To take a break, let's examine a couple of trends that are evident in the reports already covered.

The one thing that stands out above everything else is that the reports from central areas of the continent relate primarily not to mountainous areas but to water. In 1976, I checked all the reports in Wisconsin, Michigan and Indiana, and noted their distance from the nearest creek, river or lake showing on a small topographical map. There were 18 reports for which there was a pinpointed location, and for 16 of them the mileage to the nearest water was zero. In the same states I checked on the same map every town in each of two counties selected at random—six counties and 65 towns in all. Of those towns, 36 were beside water, 29 were not. To be in the same proportion, eight of the 18 places where sasquatches were reported should have been away from water.

Loren Coleman, in a paper written in 1973 on *The Occurence of Wild Apes in North America*, makes this point in his opening paragraph:

> Throughout the Mississippi Valley and the valleys of its tributaries, a vast network of closed-canopy deciduous and mixed forests exists. The gallery forests of the Mississippi Waterweb consist mainly of oak, gum and cypress in its southern portions and elm, ash and cottonwood in the northern branches (U. S. Department of the Interior, 1969). These Bottomlands, as they are technically termed, cover a good deal of the south, and are more or less unexplored, ignored, and generally unknown by man (Sanderson, 1961:91). For several years, it is from these Bottomlands that reports of large apes have been emanating.

If the bottomlands do indeed provide the habitat for hairy bipeds it is not at all necessary that all the sightings should take place there. It would be when they ventured out into inhabited territory or crossed roads that they would be most likely to be seen, and that is in accord with the reports. I have done an analysis of the locations where the creatures have been reported and what the people making the observations were doing. In the areas east of the Mountain states, two categories, "seen by a driver on the road", and "seen in the yard by someone at home", account for more than one third of all the reports, yet there are almost 300 other categories. Road reports are the most frequent kind everywhere but in British Columbia, but people west of the mountains do not often see monsters in their yards, or on their farms. Perhaps I should not make such comparisons, since I am working with detailed information for many of the west coast reports while for the eastern ones I often have only brief newspaper stories. One thing that has to be significant, however, is the fact that fully 10 percent of the creatures reported in the Midwest, East and South were seen in swamps and ponds. In the far west sightings in such locations total less than one percent.

Another thing that is apparent is that the descriptions vary considerably. Heights range from a good deal smaller than a man up to 10 feet. Colors vary from black to white, with plenty of shades in between. About the only things that are entirely consistent are that the creatures seen are hairy and that they are up on two legs at least part of the time. Actually, that isn't significant either, since those are my own minimum requirements for filing a report. There are undoubtedly reports of strange creatures that lack one or both of those attributes, but I don't file them.

Does that mean that there is actually an unlimited range of queer things of all kinds being reported and that I have just made an arbitrary selection, like a philatelist specializing in the stamps of a single nation? Yes and no. There probably isn't anything imaginable that someone hasn't reported seeing, but not often. A lot of the people who collect sasquatch reports do so as part of a general collection of all types of oddities. People seldom report giant snakes, or dragons, or tigers, elephants or kangaroos, or any of the infinite variety of out-of-place or imaginary animals available to them. They don't even report unicorns or fairies at the bottom of the garden. They do report unidentified flying objects—far more of them than sasquatches—but in the animal kingdom the hairy biped is the only strange thing to be reported over and over again, all over the continent.

The variations in the descriptions given are considerable, but they aren't anything like what they could be. People don't say they see sasquatches that are blue, or purple, or even yellow, and the only story of a green one doesn't sound very serious. There are rare reports of something very big or very small, but almost none over 13 feet or under four feet. There are very few mentions of claws or fangs, both of which would surely be expected in made-up descriptions of monsters. People don't describe big ears or long necks, or short arms. Why not? They are as likely as the alternatives. The idea of long arms could be copied from gorillas, but if so, then why not fat abdomens, and why not short legs? There could be long flowing locks on the head, or long beards. There could be tails, and drooling mouths, they could make deep bellows, and low growls and pound on their chests. Clubs could be carried, skins used for clothing. They could be seen crouching over their campfires—the embers of a fire would be easy evidence to fake—and they could be seen swinging from trees. That is not to say that none of the above items has ever been mentioned, but most of them have not and the others only once or twice in 1600 reports. Considering the wide range of possibilities, the descriptions stay within fairly narrow limits. Assuming that the extremes of size are miscalculations and that the small ones are the young of the big ones, they could all be a single kind of creature.

Does that mean that there is only one species of animal behind all the reports? There is no answer. Considering that there are sure to be many reports that are not true, and many honest ones containing errors, it would be quite possible to discard the anomalies and bring everything within a range of one species with rather varied individuals. All higher primates do tend to show great individual variations. There is, for instance, only one species of Homo sapiens, to include all the sizes, shapes and colors of men.

The most outstanding argument against a single species is the difference in footprints. Most show five toes, but about 20 percent of reports describe either four toes or three toes. Probably the proportion with less than five toes is not actually that great. The number of toes often is not mentioned in a footprint report, and it seems likely that when prints show three or four toes that would usually be remarked on, while five toes would be taken for granted.

There are also a couple of tracks with a sort of thumb on the side. One of those, 10 inches long, was found by Loren Coleman near Decatur, Illinois, and he has a drawing of one of a similar shape from Northumberland County, Pennsylvania. If there were just five-toed tracks and three-toed tracks and each type was of consistent shape, I would accept that as a clear indication of two different species. Since there are four-toed tracks as well, and the three-toed kind are very inconsistent in shape, I don't think such a conclusion would help much.

Supposing that there are two species, is there one kind on the Pacific slope and a different kind in the Mississippi waterweb? I don't think so. Five-toed tracks have been reported in the Midwest, East and South, as often as three-toed ones. There aren't actually very many reports of either kind east of the Rockies. Most stories involve only sightings, without tracks. There are also a few three-toed tracks reported from the Pacific area, so if two species are required to explain the difference in the tracks, then both inhabit both areas. The most convenient explanation is that all the tracks with less the five toes are faked, but some of them would present as many problems for a faker as any of the "Bigfoot" tracks, and it doesn't make sense that so many people would trouble to fake something with such an obvious flaw.

Basically, I would like to make the point that there is enough evidence of something in all areas to justify devoting some resources to finding it. The sorting out can wait until something is found.

-12-

Injun Devils and Woolly Boogers

There are plenty of mountains in the South, but not many sasquatch reports come from them. Aside from Arkansas and Florida the pickings in the southern states so far are rather slim. There is some material indicating that reports are there alright, but not much that is specific. There has been no Loren Coleman spending patient years there digging out information, and no heavily-publicized incident has stirred up enough interest to bring other reports to light. Much of the information available to me is in the form of clippings and letters, largely from Ivan Sanderson's files at the Society for the Investigation of the Unexplained, and from the files of other investigators, but I have had direct communications from people involved in a few cases.

A story in the Birmingham, Alabama, *News*, December 29, 1974, quotes Robert Gilbreath, owner of a bookstore in Guntersville, as having a dozen sighting reports from Alabama, most of them in the Guntersville-Albertville area. The one witness mentioned in the story is an Albertville woman who saw a creature six or seven feet tall walk away as she was taking out the garbage one night. No date is given. There is also a Portland, Oregon, man who told John Fuhrmann that he and four other boys saw a "Bigfoot" in a swamp near Mobile in 1948, when he was 12 years old. They estimated the height at 11 feet.

The Red Bay *News*, May 6, 1976, had a story about a sasquatch seen in the 1880's that apparently made a habit of disappearing under the bank of a creek. The full story is told in a later chapter. That's not much of a collection but Don Worley comes to the rescue with a letter from a notary public in Dothan, Alabama, recounting what he

214

was told by a friend, regarding an incident in Dothan on the night of May 23, 1976. The friend, a Baptist minister, was driving on Old Taylor Road, in a wooded area within the city limits, when something dark jumped out on the side of the road just where it crosses a stream. He stopped abruptly, and got a good look in his headlights at a creature slightly under six feet tall, very heavily built and covered with dark hair. The face resembled that of a human.but the eyes were red.

The creature started towards the car, and the minister, frightened, floored the gas pedal. As he accelerated the creature ran beside the car, along the edge of the road, and it was still there when the car reached 70 miles an hour. Just past Hatton Road, approaching the crossing of another stream, the creature dropped slightly behind the car, crossed the road in one leap and disappeared in the woods again. It had run beside the car for about a mile.

The writer said that he had heard of a couple more people in the area that had seen such a creature in the past year but gave no further information.

From Georgia I have nine reports, but I don't have much information about most of them. Ramona Hibner, whose main area of investigations is in western Florida, obtained two reports from a woman in Valdosta, Georgia, concerning incidents near Boston, Georgia, about 1951. The woman said that she went outside because her dogs were barking and saw a giant of a man, upright and covered with hair, cornered by the dogs on the front porch. Her husband shot at it and it ran off. She also said that her stepfather had found 20-inch tracks outside his cabin in the morning after seeing what he thought was a black man seven feet tall looking in his window at night. He had gone outside with a pistol and fired at the intruder, which ran into the woods.

In 1955 there were some reports that were publicized. An Associated Press story, as it appeared in the Evansville, Georgia, *Press*, August 4, 1955, read as follows:

Dawson, Ga. — A "hairy ghost" described by countless quaking witnesses has finally "attacked" a youth and state crime officials hope soon to put "the thing" where it "belongs". That would please frightened people here and at Edison, 40 miles away, where reports of the marauding monster with claws and blond hair, running and walking upright, have authorities in a tizzy.

But none of the people who fled southwest Georgia woods and fields with the monster in pursuit gave clear accounts until forestry worker Joseph Whaley, 20, produced claw marks as evidence of his encounter.

215

"It reminded me of a gorilla," Whaley said. "It was six feet high, gray in color and built something on the order of a man."

Major Delmar Jones, head of the Georgia Bureau of Investigation, instructed his field agents to go over the reports with the authorities here and at Edison.

"We have no idea what the thing could be," Jones said, "but sooner or later we hope to put it where it belongs and where it can do no harm, whatever it is."

Whaley told Forest Ranger Jim Bowen he was clearing brush from around new forestry signs when he "heard the bushes rattle" and "saw a strange creature coming toward me."

Whaley said he swung his grass blade against its paws and chest but the creature kept coming. He raced to his jeep to radio for help.

"Then something hit me on the left shoulder, tore my shirt and left scratches on my shoulder," Whaley said. He said he scrambled from his jeep, circled back again with the thing in pursuit and drove away.

Sheriff Ivan Jones and Police Chief W. B. Lawrence at Edison said the stories first centered near the farm of Wayne Dozier, vocational agriculture inspector, who found a tuft of hair clinging to the top of one of his fences. "The hair was sent to the state crime laboratory for examination. It was found to be human."

A letter from Wayne Dozier is in the S.I.T.U. files, stating that a young man referred to as "Tant", while mowing a field near a dense wooded area, was alarmed by the sight of a little hairy man about three feet tall that walked out of the woods and along the edge of a fence. Mr. Dozier says he found some tracks at the place "Tant" had indicated, but in the letter does not describe them. Two other men the next day found in the fence at that place a three-inch white curly hair that was later pronounced to be human. The letter refers to a "fantastic" article that later appeared in a newspaper.

The S.I.T.U. files also contain another version of the Whaley story. It quotes him as saying that the creature had claws and long pointed ears, and says, without attributing such an observation to anyone, that it had "fangs as long as a man's fingers." The story says that a shaggy, white-haired creature over five feet tall chased Tant King at Edison, and that it left tracks "indicating a paw with four claws, that curved outward on each foot." I would think that the claws, long pointed ears and fangs make up the "fantastic" article but have no proof.

216

Roger Patterson had a letter from a man in Hollywood, California, who said that a number of people had encountered a seven-foot, upright creature on a particular stretch of road near Columbus, Georgia, around 1956. Dawson and Edison are within about 50 miles of Columbus, so this may be a slightly altered recollection of the same incidents, or it may be an extension of them.

Gordon Strasenburgh, a researcher who lived for a time in Atlanta, Georgia, received two vague Georgia reports from people who phoned him on a radio or television program in 1975. One said that a man had seen a sasquatch near Tarrytown about 1965. It was over eight feet tall and "appeared mangled". The other said that a sasquatch with shining eyes had been seen recently in north Georgia.

There is a report of a sasquatch being killed in Georgia in 1943, which will be dealt with elsewhere. All the other reports I have come from a publication called *The Yeti Newsletter*, which was published at St. Petersburg, Florida, by the Yeti Research Society, and edited by L. Frank Hudson and Gordon R. Prescott. I don't know them personally, or know anything about what standards they set for inclusion of reports in their publication, but I certainly have no reason to believe that they made any of them up. They usually did not give much specific information, but that does not establish that they did not have any. Some of their sources may have been confidential, and they may have had other reasons for not wanting to give specifics as to dates and places. One of their reports quotes two farmers who were hunting deer as having seen a sasquatch cross the Southern Railroad near Waycross, Georgia. It was about 7½ feet tall, covered with greyish-brown hair. For a time it stood and looked at them, then wandered into a swamp. No date is given.

Two other reports have fairly recent dates. One attributed to H. G. Pilcher, Lakeland, Florida, states that a friend of his and several other people saw an eight-foot, brownish-black sasquatch emerge from a swamp by a cotton mill somewhere in Georgia. It left footprints in red clay, and took shiny objects from a pickup truck. Date was July 29, 1974. On September 2 of that year, the newsletter states, Les Alexander, Bob Martin and Chris Stevens, of St. Petersburg, Florida, saw a sasquatch near Dahlongea in northern Georgia. They were camped in Blackburn State Park and at 4:30 a.m. saw an eight-foot creature feeding out of garbage cans. It made a high-pitched howl. The story reports that rangers at Henry Johnson State Park said the creature had been seen before by local residents and was called a "Billy Holler Bugger." Dogs would not track it.

For South Carolina all I had before 1977 was an article from the St. Petersburg *Times*, July 8, 1974, which states that although no one claims to have seen Bigfoot in South Carolina, huge tracks have been

217

found in fresh condition. The author, Red Marston, quotes Dean Poucher, former executive secretary of the Beaufort Chamber of Commerce, regarding two experiences he had. One was on a hunting trip to a privately-owned island, the interior of which was covered with impenetrable undergrowth. The first thing that happened was that the deer hounds refused to leave the boat, much to the embarrassment of their owner. Then Poucher and another hunter separately had experiences which they kept to themselves until a chance conversation led to their comparing notes:

At opposite ends of the island we had come across tracks. I had no opportunity to see these tracks other than those buried up in a foot of blue marsh mud, but I will never forget their size, nor the depth to which they were sunk. My boot alongside, a size 11, was hardly half as large as the track. To compare, I put my full weight on one boot but I sank less than an inch in the mud.

My guess is that our big-footed friend would tip the scales at three to four times my weight, placing him somewhere between 600 and 800 pounds. The tracks on the north end of the island, where my hunting companion came upon them, were in hard sand. They were about 18 inches long, seven to nine inches wide. There was no instep, the animal was flat-footed. The tracks were likewise packed so deeply into the hard sand that our weight estimates checked out.

Three years later Poucher and his wife walked out on a dock on another island, this one inhabited, at a place where they had a view across the water to the beach where the hunting dogs had cowered in the boat:

I happened to look down at a mud bank alongside the dock. It was low tide. There were the tracks! They came up out of the creek and disappeared into the thickness of the island."

The two recent reports from South Carolina also deal with tracks. From February 14 to 17, 1977, the Rock Hill *Herald* ran stories and pictures about a single, deep, flat-footed, five-toed print found in a vegetable garden at Fort Mill. It is a good-looking footprint, with clear detail, but only 12½ inches long and 5¾ wide, which isn't much to get so excited about. What is rather more interesting is the explanation offered by the head of the South Carolina Game Commission after he had seen a cast of the print. If the Rock Hill *Herald* is correct, he said, "It could have been created by an air pocket in the ground created when the ground was frozen. With nice weather the ground thawed and dirt fell in where the air pocket was." No comment seems required.

About when the Fort Mill fuss was ending, on February 18, the Rock Hill *Herald* had a story of some more tracks, about a mile of them this time, found on a farm in Saluda County, 35 miles west of Columbia, and about 75 miles from Fort Mill. The new prints were apparently older, having been found on February 6, and they were also larger, 14 inches long and seven inches wide, and a bit deeper. They took a four foot stride. All that was impressive, as were the pictures, but the route they took, as described in the paper, seemed strange. They went from a pond down a little-used road, over a sawdust pile, along a creek bank, across a pasture and back to the pond. The paper said that loud screams had been heard in the area for years, and also noted that "County police are searching for the barefoot phantom, believed to be responsible for killing several cows reportedly found half-eaten in Saluda."

North Carolina is the scene of several recent sighting reports, but no old ones. The Durham *Morning Herald*, August 6, 1976, reported that Roger Hoffman of Mount Holly, on July 30, saw a seven-foot hairy creature cross a path leading to a creek in the woods in front of his home as he sat on his porch at night. He took after it with a rifle and shot at it.

A September, 1976, story carried by United Press International, states that Jim Hollingsworth, a Goldsboro mental therapist, has spent the past six years trying to find out what makes 18-inch, three-toed tracks in Chatham County; and that he has learned of several sightings in the Cape Fear River area of a seven-foot-tall, apelike creature with black hair, a slumping gait and a "deathly" scream. Brody Parker, a Chatham farmer, is reported to have watched a figure seven to eight feet tall, with black fur, for 20 minutes in the spring of 1976.

A September, 1975, Associated Press story indicates that 18-inch footprints were found in the vegetable garden by Mrs. Brody Parker at Farmville. Presumably those are the same footprints and the same location that the UPI report refers to, although it is plain that no sighting by Brody Parker had been reported in 1975.

In 1974 a man from Charlotte, North Carolina, phoned to tell me that he had seen hairy giants four times over a 20-year period on South Mountain, in the Blue Ridge Mountains of eastern North Carolina. The most recent occasion was approximately two months prior to the phone call, at which time he had shot at one of the creatures. I was in no position to follow up a report in North Carolina, so I turned it over to Gordon Strasenburgh. Here is an excerpt from a letter from him:

> As for his latest encounter, he was camping near a stream
> in South Mountain. Around twilight after he had finished

eating but with a fire still going, his dog first barked and then curled up and whined. He saw something bent down, which when it straightened up was, "I know it was seven feet tall, because I'm 6'2".

It had no neck, smelled like a really bad dump, and made a sound which began with a growl and went to a howl or whistle. He took it for granted that it was a male. He had a 30-30 Winchester with him and fired at it after it began to walk away. It let out another howl or whistle and went off into the woods.

He said he almost fainted when it first stood up, and that it was like a big human, with long arms and shoulders "wide—huge—maybe four feet across. The arms were as big as telephone poles."

On later contact, the man told Gordon he no longer wished to discuss the subject.

There was also an Associated Press report, January 14, 1972, in which a deputy sheriff in Burke County said that he and a volunteer deputy, while parked on a roadside near Drexel, saw a creature cross the road in front of them that was huge and kind of grey, and appeared to have no head. It looked like a six-foot man with a fur coat pulled over his head, but it was not that, and it was not a bear. They turned on their headlights, and it turned around and walked down a path into the woods.

That story prompted a follow-up January 26, in which it was explained that the deputy was now satisfied that he had seen a bear that covered its face with its paws as the lights of the moving car hit it. (Can anyone really imagine a bear doing that?) The story also referred to several other sighting reports, and to police stations being "flooded" with calls about "The Woollybooger." I also have a letter from a man who went to see the two deputies the day after the sighting. He says that they were satisfied that the thing could not have been a bear, but appeared to be a sasquatch, and that they told him the sheriff and two other deputies had seen a sasquatch years ago "but thought it was a ghost."

The name "Woollybooger" rather intrigues me. It is obviously a description as well as a name, and I have talked to several people who come from rural areas in the South who are familiar with the term, some of them having come west a good many years before westerners ever heard of "Bigfoot." The "Woollybooger" also crops up in print from time to time, and it seems to me to indicate quite definitely that there is a widely-distributed tradition of hairy bipeds in the South. Whether much documentation for this exists, I don't know. Very few old published reports from the area have come to light so far. The proprietor of a Spokane bookstore told me that when he was a boy in Tennessee there was a well-known tradition of hairy giants in the Great Smokeys. I haven't heard of that, however, from any other source.

Aside from that, the oldest Tennessee report I know of is contained in a letter from a James Meacham, which Ivan Sanderson reproduced in *Abominable Snowmen*. Meacham wrote that in 1957, in a marsh near Jackson, Tennessee, he put 13 long-rifle .22 bullets into a thing that looked like a small ape or tailless monkey that was reddish in color. He said it did not even move when it was shot, but when he started towards it the creature rushed off through the trees. It was in a tree when he first saw it, and it swung by its arms, but he does not say how it moved as it departed. It sounds more like an orangutan than a sasquatch, but no real animal could be expected to absorb that kind of a barrage without moving.

Two years later, September 24, 1959, the Knoxville, Tennessee, *Journal* carried a story that two Clapps Chapel Road residents had fired shotguns at a creature described as being eight feet to 10 feet tall and weighing more than 200 pounds. (I would hope so.) One of them had seen the thing come to the window of his front porch. It ran when he went outside, so he and a neighbour sat under a tree and waited to see if it would return. They didn't see it, but they heard something thump around the side of the house. Going to the spot, they saw the creature running towards the woods, and fired after it. "I don't believe it was human, it was too big and moved too fast and quiet," one of the men is quoted as saying.

The next report is from the Nashville *Tennesseean*, August 24, 1965, concerning sightings in Rutherford County. Roy Hudson and Terry Ring reported seeing the thing at seven in the evening on Brown's Mill Road. The deputy who was sent to check that report was stopped on the way back to Murfreesboro by Dorris Barrett, who, with Terry Lester, had seen the creature later in the evening, when they shined a light in its face. Hudson told a newspaper:

We saw this thing walking along one side of the road to another. We drove down the road and turned around . . .we were sure scared . . . and when we came back we passed about three yards from it. It jumped over a five-foot fence and tore the fence down.

He is reported to have said that the monster was about seven feet tall and four and a half feet around the waist. It was covered with long reddish-brown hair and had teeth about three inches long, and a pug nose, like an ape.

Don Worley, in an article in *Argosy UFO*, July, 1977, reported that in October, 1975 a farmer in Giles County saw in his barn a

221

black shaggy-haired monster with no neck, and that it fatally injured a calf by lifting it overhead and throwing it down.

In April, 1976, Tennessee produced perhaps the most widely publicized story from any Southern state, and for once there was something special about the actions of the creature to account for the media attention. Mrs. Jennie Robertson, who lives near Flintville, in Lincoln County told the story. The following quotation telling of what happened to her and her four-year-old son is from the May 18 issue of a paper called *The Star*, but the clipping does not say what town the paper is from.

It was about 9 o'clock in the evening and Gary had walked out the back door after supper. I was standing just behind him and I heard him cry out.

I rushed out the door and saw this huge figure coming around the corner of the house. It was seven or eight feet tall and seemed to be all covered with hair.

It reached out its long, hairy arm toward Gary and came within a few inches of him before I could grab him and pull him back inside.

I called to my husband but by the time he got out the back it was gone.

Mr. Robertson then drove to Fayetteville and placed a notice in the *Elk Valley Times* warning that the creature was around and that it might be dangerous. He told the interviewer that he had seen it previously, near the woods, but didn't say anything because he thought people would consider him crazy. The Robertsons' closest neighbour, Houston Smith, told *The Star* he had been seeing the thing around the woods for three years but had not wanted to tell anyone. "It walks upright, but kind of bent over," he said. "It's not a bear and it's certainly not a person. A lot of this country's never even been explored. Anything could live there, especially if it only comes out at night like this does."

James Stevens, of Shelby Forest in Shelby County, said he had been tracking the monster part-time during the past three years and it once chased his truck down a gravel road "at up to 40 miles an hour."

Just over the border in Allen County, Kentucky, there is a story that goes back 100 years about a group of creatures that lived in a forested valley called "Monkey Cave Hollow." Ivan Sanderson has on file a letter from Harold Holland, of Scottsville, Kentucky, including the following statement:

About 20 years ago one old man who had moved from this area but returned for a final visit to his home, told me that when he was a boy of about 7 or 8 years, he saw the carcass of the last "monkey". He stated that a hunter came by his father's house and displayed the dead beast. He said that he could not recall exactly what it looked like (after all it had been 80 years or thereabouts) but that the creature had hands and feet "like a person" and was about the same size he was himself, had no tail and was covered with brown hair.

In June, 1962, the Louisville *Courier-Journal* had a story about Owen Powell, in Trimble County, seeing a gorilla about six feet tall, walking on its hind legs. Its arms hung to its knees.

The Nashville *Banner* on July 11, 1977, published a feature article about James Vincent, of Hendersonville, who has investigated numerous reports in Tennessee but did not give specifics of any of them. It indicated that Dr. Richard Young and Charles Denton, of Clarksville, Tennessee had seen a sasquatch near Murray, Kentucky, in 1968. It was apparently on the road, since one man is quoted, "I haven't travelled that route since then."

Another investigator, Michael Polesnek, was in Henderson County, Kentucky, investigating the "Spottsville Monster" that had been seen there in late 1975, according to the Owensburg *Messenger-Inquirer*, February 9, 1977, but again no details were given.

James Moorhatch, of Louisville, investigated and wrote a comprehensive report on an incident near Pembroke, Kentucky, in the spring of 1976. There were three night sightings of a bipedal creature six feet or shorter but with massive shoulders and glowing green eyes, in the vicinity of three isolated houses. Three of the five witnesses lived in one or other of the houses. I also have letters from Mr. Moorhatch and another Louisville resident, Michael Johnson passing on sighting reports from Simpson County, Kentucky, in January, 1977. One witness was a policeman, who said he had seen a large hairy animal crossing a road at night.

Stories in my file from West Virginia go back only to 1960, but that year there were three, a couple of which are very strange. John Keel, in his book *Strange Creatures from Time and Space*, quotes from a personal letter concerning an incident in the summer of 1960 near Davis. His correspondent stated that while cutting wood for a fire at night he felt someone poke him in the ribs, turned and confronted "a horrible monster."

> It had two huge eyes that shone like big balls of fire and we had no light at all. It stood every bit of eight feet tall and had shaggy hair all over its body. It just stood and stared at us. Its eyes were very far apart.

The creature shuffled off into the darkness. Next morning the men found gigantic footprints where the creature had been seen, but didn't feel like following them.

Eyes that shine when there is no light to reflect from them are bad enough, but according to *Flying Saucer Review*, January-February, 1973, issue, a hairy creature seen in Monongahela National Forest near Marlinton, West Virginia, in October, 1960, could stall the engine of a car. The story quotes W. C. "Doc" Priestley as saying that while driving in a car following a group of friends in a bus he saw an eight-foot, hairy, apelike monster "with long hair standing straight up." At the same time his car stalled. After a while the bus backed up to where he was and:

> It seemed the monster was very much afraid of the bus and dropped his hair, and to my surprise, as soon as he did this, my car started again.

A little further on the car quit again, throwing sparks under the hood as if there was a bad short circuit. There by the road stood the creature. Again the bus backed up and the creature fled.

I would greatly appreciate it if stories like that did not exist but unfortunately those are not the only ones. To the north, around Pittsburgh, there are several equally strange, and I will make my comments after outlining the Pennsylvania situation.

On December 31, 1960, the Charleston, West Virginia, *Daily Mail* carried a story headed "Is Braxton Monster Back?" Mrs. Charles Stover had told the newspaper that when her husband was driving home about 11 p.m. on a backwoods road near Hickory Flats, between Braxton and Webster Counties, he rounded a curve and saw "the monster, standing erect, with hair all over his face and body." The thing was only six feet tall, which should have convinced the reporter that it couldn't be the original "Braxton Monster." That was described in September, 1952, by a woman from Flatwoods, who said she had seen a "monster about 10 feet tall, with a face of fiery red, a green body and a headpiece which looked like the ace of spades." The 1952 report is supposed to have drawn 10,000 to 15,000 visitors to the area. Another report was in *Flying Saucer Review*, July-August, 1968:

> There are no bears in Mason County, but in November, 1966, Mr. Cecil Lucas saw three bear-like creatures sniffing around an oil pump in his field. His farm is located a few hundred yards from the McDaniel home and is on the banks of the Ohio. When he came out of the house to investigate, the dark, hairy forms ran off erect towards the river, disappearing into the thicket.

A different flying saucer publication, *UFO Report*, in December 1976 had an article stating that two teenagers had reported to the Braxton County Sheriff's office that while heading for a drive-in theatre at Weston, West Virginia, on August 17, 1976, they encountered an eight-foot, black apelike creature crouched in the road. It stood up and watched them for about 30 seconds, then walked into the woods.

From Virginia I know of only three reports, all in the 1970's. One is a letter from John Lutz, of Odyssey Research Association in Baltimore, to Sabina Sanderson. He repeats a story told to him by a resident of Atlanta, regarding a skunk hunt on June 29, 1972, in mountainous terrain somewhere in Virginia. (A skunk hunt?) The man said he and two relatives were walking up the side of a mountain about 8:45 p.m. when something huge was heard coming down, and they scattered out of the way in panic. He described the thing as seven to eight feet tall, 500 pounds, running on two feet, covered with hair and with a face like an ape.

About the end of April, 1977, according to the Roanoke *Times*, a set of 16-inch footprints were found crossing a strawberry patch on the farm of Rupert Williamson in Dinwiddie County. On May 17, the Danville *Bee* carried a story that a Pittsylvania County deputy was checking two Bigfoot reports. A woman who preferred to remain anonymous but "whose reputation for truthfulness and veracity is irreproachable" said the thing she saw was hairy and seven feet tall, but after she calmed down she thought it must have been a person in a gorilla suit.

For a teaser, the Dumfries, Virginia, *Potomac News*, June 17, 1977, published a long article about the sasquatch that contained the statement, "Much had been written about the monster at the ammunition storage area at Quantico Marine Corps Base . . ." but doesn't say what. There is a reference to one Marine claiming to have seen a brown thing walking on two legs and another reporting something that looked like a cross between an ape and a bear covered with very long hair. Several people had apparently heard loud screaming:

> We called the ASA (Ammunition Storage Area) sergeant of the guard to determine whether there had been any more sightings or sounds in the area but were told that all information regarding the ASA monster is considered classified.
>
> When asked why that is so, the guard answered that he was not allowed to answer that question either.
>
> He later said that the information is not really "classified", but everyone at the compound has been ordered not to talk about the monster at all.

The Washington *Post/Potomac* magazine, October 10, 1976, had a lot of fun with a couple of Maryland monsters, the Snallygaster and the Dwayyo, which are apparently well established in local folklore. It says that a Dwayyo, described as "apelike," was reported seen by

Middletown residents in 1934, while one reported in Carroll County in 1944 left huge footprints.

Stimulous for the article apparently was a series of sighting reports from the Whitemarsh area northeast of Baltimore in swamps around Bird River. The only one referred to specifically was by Richard and Elva Stewart, who said they were cutting brush near a grove of trees when they heard branches breaking, smelt an odor like swamp gas and saw a creature about eight feet tall with big red eyes. It ran away. John Lutz was deeply involved in the investigation of a series of reports in Maryland, near the small community of Sykesville, just west of Baltimore, in the summer of 1973. Those events received wide publicity, and stimulated an unusual amount of investigative activity, but not very much seems to have happened. The first report to police was the night of Tuesday, May 29, according to the May 31 Sykesville *Herald*. Anthony Dorsey went to find out what was upsetting his dog and saw by flashlight two large glowing eyes seven feet off the ground. Police responded, and although nothing was found the officer was apparently impressed by the sound of something big moving in the underbrush. Next day Police Chief Omer Herbert joined the search and found a 13-inch footprint and two partial ones showing a six-foot stride. On Sunday, three people told of seeing something moving in a thicket. It had short fur-like hair and an egg-shaped head, and appeared to be about eight feet tall. Next day a man reported seeing it, describing it as large, hairy and apelike. Besides the local police and Odyssey Research personnel; conservation officers, state patrolmen and the assistant director of the Baltimore Zoo participated in the search.

In a letter to Grover Krantz, written July 1, John Lutz said that more than 30 people had seen the creature and descriptions were almost the same although some of the witnesses were tourists who had never heard of the thing. It was an estimated seven to eight feet tall, 350 to 400 pounds, covered with 1½-inch black hair. It walked upright like a man, and sometimes was covered with mud on its legs. Dogs refused to take its trail. There were no reports of it harming livestock, but wildlife in the area seemed to have disappeared. The area where it was most often seen was heavily wooded, part of the Patapsco State Forest, which is about 20 miles long and three or four miles wide, along the Patapsco River. By July 22, Mr. Lutz had six more reports. He told Dr. Krantz that Odyssey Research had all the witnesses on tape. Despite all the activity, nothing definite was ever found—and despite the usual concerns expressed by the police, none of the armed amateurs shot anyone.

Three other Maryland reports were published in *The Aegis*, Bel Air, Maryland, apparently in late 1975. My clipping is not dated.

Things started with a report by a Bel Air motorist to state police in April, 1975, that he had hit a seven-foot upright creature that crossed the road in front of his sports car. He said he knocked it down but it got up and limped off, still on two legs. The 50-mile-an-hour collision did considerable damage to the car. Police studied some hair taken from the car and announced that it came from a bovine. Several weeks later a woman from the Darlington area, a few miles downstream on Deer Creek from the collision scene, reported that she had found huge human-like footprints in her garden. They were not photographed or cast before they were destroyed.

Publication of those accounts resulted in a call from a former resident of Harford County who said he had seen a sasquatch in Churchville 62 years earlier. The man said that in 1914 he was eight years old and living at "Will Finney Corner" near Route 155 and Glenville Road. One evening while looking for his cap out behind his home by lantern light, he saw a creature sitting on a log between two trees. He was holding the lantern high in the air and it was at eye level with the seated creature. The thing was close enough to have grabbed him, but just watched as he walked by. It had a definite nose, a sharp chin and was hairy all over. He didn't tell his parents, for fear he would get a licking.

Most recent report I have from Maryland was in the Annapolis *Evening Capital*, September 1, 1977. Ronald Jones, of West River, told Anne Arundel County police that on Route 258 about 11:30 p.m. August 30, he saw a creature seven to eight feet tall, 400 to 500 pounds, erect and covered with long dark brown hair, and with a strong, pungent odor. When he first saw it he thought it was a person's body near the road and got out of his truck. The paper was not able to interview Jones, but his wife said he had gone into a tobacco field after it and thrown a tire iron at it, and that it had hit him on the back of the head and had held onto his truck until he backed up into it. She said that he had been reluctant to call police, but that a neighbour, "recalling former stories about a Chalk Point monster reported in the area several years ago", had called them for him. Her husband does not drink, she said.

Even tiny Delaware has been the source of one newspaper report of a sasquatch. The copy I have does not show the name of the paper or the date, but seems to be fairly recent. It tells of a variety of teenagers claiming to have seen a half-man, half-animal running on two legs in the woods of Cedar Swamp, near Selbyville.

There is only one mainland state for which I have neither a specific report nor a suggestion that there may ever have been any, and that is Rhode Island. Across the border in Canada, Nova Scotia and Prince Edward Island are also completely blank, as is the island of

227

Newfoundland. Maine has an ample supply of woods, water and mountains, and apparently has sasquatches too, In the S.I.T.U. file there is a letter from an Air Force sergeant who is also a newspaperman, and a native of Connecticut. He says that as a boy he spent most of his summers at a cabin in eastern Maine, near Topsfield, and that he knew people there "who almost take the existence of this animal for granted just as a New Yorker may believe in the fact that taxi-cabs exist." He says the creature has been called "Indian Devil" for longer than anyone can remember, and is described in the same terms as are used in other parts of the country. Specifically, he refers to a "seven-foot-tall, reddish-brown, fur-covered, almost-human-looking creature."

Another man wrote to Ivan Sanderson with what could be the oldest sighting report of all, but a very indirect one. As a boy, 40 years ago, he had read what he believed to be a non-fiction account of incidents involving "Pomoola, the 'Injun Devil' " in Maine. The book, entitled *Camping Out*, was about three young men on a prospecting trip just after the Civil War near Mt. Katahdin. They had as a guide an old trapper named Cluey, who told them of an adventure he had when a young man, trapping in the same vicinity. That ought to take the story back to the early 1800's.

Talking about "Pomoola," the guide said that while camping at night by a lake he had seen a manlike creature covered with hair approach the camp, then walk away. Next day at dawn he and his friends saw the same creature standing on a rock by a small island in a lake. The Pomoola was supposed to eat fish alive, and to kill Indians, but not white people. Even if the book was actually fiction, it probably indicates that the tradition of a hairy, half-human "Injun Devil", existed in New England in the mid 1800's.

There are also three recent reports from Maine. The Portland *Press Herald*, July 28, 1973, has a story from Brunswick about Mrs. Neota Huntington, of Durham, and four children seeing what she described as a member of the ape family, possibly a chimpanzee, five feet tall, weighing 300 to 350 pounds and covered with shaggy black fur. When she saw it, at about 7 p.m. on a Thursday, it was daylight and the animal was out in the open, just at the side of the road. It ran away on its hind legs. A previous encounter by the children, the afternoon before, was written up as follows:

> Coming upon the animal unexpectedly, 13-year-old Lois Huntington took a spill when she braked her bike on loose gravel. The chimp, if that is what it is, was only a few feet away from the place where she fell. Also in the group were Tammy Sairo, 12, George Huntington Jr., 10, and Scott Huntington, eight.

"If he had been vicious he would have done something to her then," said Mrs. Huntington. "He just stood there and looked."

The sighting was reported to police, "but no one believed us," she continued. "I believed them, because I know my daughter wouldn't lie."

Driving past the spot on Thursday evening, Mrs. Huntington said she cut the car's engine and let it coast in hope of seeing the animal. It was standing beside the road but ran into the woods when it heard the car approaching.

Summoning neighbours, she returned to the place and sat in the car while the others scoured the woods. Then she saw it again, about 20 feet away, peering at her through the crotch of a tree. It disappeared in the woods when she called to the others.

A later search by about 20 police and volunteers turned up nothing but a rotten stump that had apparently been torn apart by some animal looking for grubs. Radio and television reports stirred up a fair amount of excitement, and police expressed the usual fears of someone getting shot. As to the animal itself, Mrs. Huntington commented, "I never said it was a monster." In fact she had taken some overripe bananas to the area on Friday in hope that the animal would find them.

Finally, there are two reports from Manchester, Maine, in the fall of 1975. Brent Raynes, who interviewed the people concerned, writes that at 5:30 p.m. on September 22, a 15-year-old girl and her younger sister were walking on Camp Road when they saw a hair-covered, manlike creature watching a man chopping wood. It was swaying back and forth as it watched. The girls screamed and the creature turned facing them, then ran across the road into the woods. They said it had long brown hair of a bushy appearance about the jaw and upper head, and a white patch on the chest area. Its face was apelike.

Two months later a woman reported that she sat in a parked car at a neighbour's residence from 8:15 to 8:30 in the evening, and watched a shadowy figure like a man six to seven feet tall that was standing by a nearby tree. It had long arms and legs and at one point crouched down with one arm around the tree. She sounded the horn lightly hoping her husband would come out of the house. The creature rose and started toward the car. She then blared the horn and it stepped back. It was hairy and black, but with a light-colored face.

In New Hampshire U.F.O. investigator John Oswald talked to two people who described seeing a dark-furred creature at night near

Freeport, in July, 1974. They were turning their car when the lights hit a six-to-seven-foot, muscular animal with large shoulders, a prominent nose, a pointed chin and a triangular face. It was only a few feet away. As they watched, it backed into the woods.

Ramona Hibner passed on to me a summary of an interview by Brent Raynes with a Central Sandwich, New Hampshire, man. He said that one summer when he was about 14, around 1942, he was cutting spruce late one afternoon when he was frightened by a six to seven-foot "gorilla-looking" creature "nearly upright on long legs" that hung around and followed him for about 20 minutes. No one that he tried to tell at the time would believe him. Then in February, 1977, he found footprints 16 to 17 inches long and six to seven inches wide that led to a small shed, then came back down by a small stream, crossed over and went 2,000 yards to a highway and crossed that.

On May 16, 1977, the Boston *Herald American* carried a story about a creature that shook a camper truck at Hollis, New Hampshire on the night of May 7. Gerald St. Louis was quoted as saying that he and his two boys were sleeping in the rear of the truck on the grounds of a flea market where there was to be an auction next day. About 10:30 p.m. he felt the car shaking so badly he thought it would tip over.

I opened the door of the truck, threw on some lights and started to step outside when I saw it face to face. It was all hairy, brown colored and eight or nine feet tall with long arms, long hair. Thank God for the lights, for it apparently startled the creature and it ran toward a fence, about four and a half feet high and jumped over it with ease. I could see it standing there in the distance—just looking at us.

"I got out of there so fast I left everything I was going to sell the next day behind on the ground.

The two boys confirmed that the truck had been shaken. Next day the *Herald-American* had a story referring to tracks 16 inches long and eight inches wide being found near the Flea market. It also quoted Mrs. Regina Evans as saying that her son Stanley had a similar experience about the same time while spending the night in a camper truck with a friend, Jeff Warren. They did not see what did the shaking.

Over in Vermont there was a story that made the front page of the New York *Times* almost a century ago. Loren Coleman sent me a transcript of the story, from the issue of October 18, 1879, as follows:

A WILD MAN OF THE MOUNTAINS
Two Young Vermont Hunters Terribly Scared
Pownal, Vt., Oct. 17 — Much excitement prevailed among

the sportsmen of this vicinity over the story that a wild man was seen on Friday last by two young men while hunting in the mountains south of Williamstown, Mass. The young men describe the creature as being about five feet high, resembling a man in form and movement, but covered all over with bright red hair, and having a long straggling beard, and with very wild eyes. When first seen the creature sprang from behind a rocky cliff, and started for the woods near by, when, mistaking it for a bear or other wild animal, one of the men fired, and, it is thought, wounded it, for with fierce cries of pain and rage, it turned on its assailants driving them before it at high speed. They lost their guns and ammunition in their flight and dared not return for fear of encountering the strange being.

There is an old story, told many years ago, of a strange animal frequently seen among the range of the Green Mountains resembling a man in appearance, but so wild that no one could approach it near enough to tell what it was or where it dwelt. From time to time hunting parties, in the early days of the town, used to go out in pursuit of it, but of late years no trace of it has been seen, and this story told by the young men who claim to have seen it, revives again the old story of the wild man of the mountains. There is talk of making up a party to go in search of the creature.

The only recent Vermont report that I have is third-hand, and apparently was not published. My informant states that she was told it a week after the event by a relative of the couple involved. In July, 1974, near Rutland, they reported to police that they had seen an eight-to-ten-foot monster, hair-covered, in a meadow after midnight. When police went to the area, they said, the creature ran across the road in front of them.

From Connecticut for a long time I knew of no specific reports at all. A man did phone me about eight years ago to say that he was investigating stories of people seeing a sasquatch-type thing in a lovers' lane near Trumbull, Connecticut. It was supposed to be big, white, and able to run 35 miles an hour; and it left humanlike footprints twice as deep as he could make by jumping off a three-foot bank. He weighed 225 pounds. The man said he would write to me with full details, but he never did. At that time I had no information indicating that such creatures were known in New England, and I made no attempt to follow up the story.

Recently Tim Church has sent me a copy of a story from Colebrook, Connecticut, that goes back to the 1890's. It appeared in the Hartford, Connecticut, *Courant*, August 21, 1895, as follows:

A WILDMAN — he appears to selectman Smith and scares his Bull Dog (Winsted *Citizen*)

Last Saturday, Selectman Riley W. Smith went up to Colebrook on business. Mr. Smith, while there, went over into the fields and began picking and eating berries from the low brushes in the field.

While he was stooping over picking berries, his bull dog, which is noted for its pluck, ran with a whine to him and stationed itself between his legs. Mr. Smith being in a bent over position picking berries. A second afterwards a large man, stark naked, and covered with hair all over his body, ran out of a clump of bushes, and with fearful yells and cries made for the woods at lightning speed, where he soon disappeared.

Selectman Smith is a powerful, wiry man, and has the reputation of having lots of sand, and his bull dog is also noted for his pluck. But Riley admits that he was badly scared and that his dog was fairly paralyzed with fear.

If any of the readers of the *"Citizen"* have lost a wild hairy man of the woods, six feet in height, and want to find him, they can go up to Colebrook and when near the "Lewis place" wander around in the woods and fields and perhaps recover their lost property.

My Massachusetts information is even less explicit than that from Connecticut. In the S.I.T.U. file there is a letter apparently quoting police reports and a clipping from the *Herald-Traveler*, April 9, 1970, regarding a "thing" seen several times near Bridgewater in December, 1969, to April, 1970. The December report indicated that it was walking upright in the vicinity of Bridgewater State Teachers' College. In March the "thing" was reported near Route 28, and "large footprints" were found in mud. The news report said that on the night of April 8 "a seven-foot bear" was almost run down on the road by "one of our more responsible citizens" and heavily-armed state and local police were out hunting it with a pack of dogs. They found only tracks proclaimed to be those of "a very large bear." The story says that no bears had been seen for many years in the Bridgewater area, but the large Hockamock Swamp area does support some 40 deer.

In every case the police had been called out, but never saw anything until one April night when they were twice called to the same house without finding the "thing" there when they arrived. An officer then remained in his patrol car in the dark, "when without warning something began to pick up the rear end of his car. The officer spun his car around and got his spotlight on something like a bear running around the corner of the house."

New York has produced no old information, but quite a lot has come to light in the 1970's. Bob Jones, who was then with the Society for the Investigation of the Unexplained, followed up in 1975 a letter written to Ivan Sanderson shortly before his death about incidents in January, 1971, in Dutchess County. One young man reported seeing a large, hairy, bipedal creature on three occasions—once while snowmobiling, once while leaving a lovers' lane area, and once while looking for the animal. On the last occasion, the sasquatch hunters were surprised by an eight-to-nine-foot creature with a four-foot stick in its hand, approaching the back of their pickup—which is a little hard to swallow. There were other reports as well, including one of the animals running through a barbed-wire fence and breaking off three posts.

Two other reports, from Staten Island of all the unlikely places, were covered in full by Robert C. Warth, writing in *Pursuit*, in April, 1975. On December 7, 1974, in Richmondtown, two boys, aged 11 and 12, were climbing a wooded knoll behind the parking lot of St. Andrew's Episcopal Church at about 4:15 p.m. when, they said, they heard a "loud roar" behind them. Looking back they saw a "big furry thing" standing erect facing them. It was about six feet tall, black, and had its forelimbs raised in a gesture that the boys interpreted as threatening. They fled. Police were called but could find nothing. Both boys were interviewed two days later by Mr. Warth. Their parents confirmed that they had obviously been seriously frightened. Then on January 21, 1975, a nurse on her way to work at 10:45 p.m. had to slam on her brakes to avoid hitting a large animal on Richmond Road in from of St. Andrew's Church. She stopped her car within six feet of it and watched it walk in front of her from the church parking lot across to the entrance in the stone wall of the churchyard. It was, she estimated, 5'8" to 5'10" high, and walked upright with its arms hanging slightly forward of its body. It was covered with long black hair. The creature paid no attention to her car. There is a large swamp area and a garbage dump behind the church. Two four-toed footprints about 10 inches long were found, and it was reported that on a preceding morning just after midnight a young couple had seen a bearlike figure in the same parking lot. However investigators could not find those people.

The Watertown *Daily Times*, August 2, 1977, refers to an incident in January, 1975, when Steve Rich, Jerry Emerson and a boy from Massachusetts say they saw a five-foot animal at the edge of the woods on State Street hill in Watertown, "just walking and swinging its arms." The boys were interviewed by Milton LaSalle, of

233

Watertown, who also passed on to me some more recent reports.

In January, 1976, he had talked to a Watertown man who told him that the preceding June he and a friend had seen a very large, hairy animal squatting on the shoulder of Route 3 near Saranac Lake, at night. As their car came around a curve it stood upright and walked into the bush. Mr. LaSalle also interviewed two boys in their late teens who said that on the early morning of August 11, a few miles east of Watertown, they had twice seen a huge black, hairy creature. They told him they had decided to try to stay up all night, and were walking down Overlook Drive about 5 a.m. when they heard pounding noises and screams from a bushy area. The noises went on for about 15 minutes. Half an hour after that they saw a huge black thing standing in the road about two blocks away. One of them yelled and the thing ran off, upright. As it turned to run they could see that it was covered with hair. After that they went to a friend's house, and from his kitchen they saw it again, walking across an open field near where they had first heard the noises. They said it was about eight feet high, had very wide shoulders and was heavily built.

According to the Glen Falls, New York, *Post-Star*, August 30, 1976, there were also reports of a "huge ape" at Whitehall, near Lake George, in the latter part of that month. Marty Paddock, of Whitehall, and Paul Gosselin, of Low Hampton, said they saw a huge shape by the side of the road and returned two or three times more to get another look at it. On the last occasion they had a third man with them, and after that they reported it to the police, attracting Whitehall police, New York State police and a Washington County deputy sheriff, who searched the area but "were only able to spot the creature from a distance."

Although descriptions vary somewhat, the creature has been widely described by both police officials and civilians as between seven and eight feet tall, very hairy, having pink or red eyes, being afraid of light and as weighing between 300 and 400 pounds.

It reportedly makes a sound that has been described as a loud pig squeal or a woman's scream, or a combination.

The creature also walks upright, rather than on all fours, which has resulted in the eye-witnesses ruling out the possibility that the mysterious creature was a bear.

Whitehall Police Sgt. Wilfred Gosselin said all he was able to see of the creature was an "awful tall shadow" in the field off Abair Road.

He said based on the descriptions given by his son and friends as well as other policemen, "I'm not saying this is a monster or anything else, but there is something out there, and it's no animal that belongs in the northern part of this state!"

The sergeant said footprints much wider and three times the length of a man's were found in the area. The footprints have no claws.

The sergeant reported that the creature has not hurt any humans but said it was possible it had killed a deer found in the meadow there.

Brian Gosselin, the sergeant's son and Paul's older brother, a patrolman on the Whitehall force, said he saw the creature on Wednesday.

He said the creature came within 25 feet of his squad car before a state trooper flashed a light in the creature's eyes. Gosselin said the light caused the creature to cover its eyes and run away screaming.

Milton LaSalle also passed on a second-hand story about a man from Oxbow, New York. He had told a friend that when he was coon hunting near Oxbow in the fall of 1976, he saw a seven-foot, very heavy, upright creature covered with dark brown hair run across a clearing in the light of a full moon.

Then on August 2, 1977, the Watertown *Daily Times* reported that two pairs of railroad workers near Theresa, New York, had seen a large hairy animal standing on the right of way at a distance. The sightings took place June 27 and 28 and were not at the same place. In each case after it walked into the bush large human-shaped five-toed footprints were found. Again, Milton LaSalle was on hand to investigate.

That completes the summary of information from those states, 41 in all, that do not have a major role in the sasquatch investigation up to the present time. The remaining Canadian provinces and territories will be dealt with in another chapter. Of the other nine states, Washington, Oregon, Idaho, Montana, California and Florida all have far too much information for this kind of presentation. So does Pennsylvania, but whether most of it refers to the sasquatch seems doubtful. New Jersey probably could have been covered in full as far as the information I have on record goes, and so could Alaska, but for different reasons they fit in better elsewhere.

The common denominator of most of what I have reported in this section of the book is that I know very little directly about any of it. I have spent a few days in New Mexico and Texas and made a few enquiries in other places, but mainly I am dependent on information from printed sources, from letters and phone calls, and from records kept by other investigators. About all I can say with absolute certainty is that these reports do exist, but I think that there are enough of them, with enough of a pattern to them, so that at the barest minimum they demand an answer to the question, Why?

Why so many reports of hairy bipeds, from so many places?

There are those easy answers about mankind's need to imagine monsters, and about pranksters in gorilla suits, and about the tendency people are supposed to have of seeing something ordinary and making something strange out of it—but do they really provide any explanation for the phenomenon I have documented? The existence of all these reports in all these different places is a fact. It has to have a factual explanation or explanations. Even on the basis of that perennial fear of "the sheriff" and his ilk, that the reports are going to lead to some trigger-happy monster hunter shooting another human, isn't there enough here to justify the spending of some time and money to find out what is behind it all?

Beyond the single certainty, there are a few more things of which one can be all but certain. One is that not all the reports are true. There are sure to be some that result from hoaxes; some that are sensationalized news reports; some from people too eager or frightened after reading publicity of earlier reports. I fully expect that there will be some readers who already know the fallacies in one or more of these reports and will be ready to dismiss them all as equally inaccurate. But on the other side of the coin, it is just as nearly certain that the incidents I am aware of represent only a small fraction of what is out there.

All the old stories that never got into print are gone forever, and there must be lots of other stories in old publications that no one has yet stumbled on. In modern times, the places where there are a lot of reports are mainly the places where someone is making an effort to find them. There are almost certainly a comparable number to be found in similar localities where no one is doing anything about it. It is a fair assumption that there are not hundreds of such stories unrecorded, but thousands. Is it reasonable to assume without a real investigation that all of them are wrong?

In considering that point there is a Catch 22 situation. You can disprove or discount all the reports you like, but if just one remains that is true, then the existence of a hairy biped is established. But if hairy bipeds do exist, then there is no reason to doubt that most of the people who think they have encountered one have actually done so. In other words, if one report is true it is almost certain that most of them are. To dispose of the subject, every single story must be wrong, and every footprint faked. If you find yourself forced to accept any of the reports as valid, you lose your basic reason for questioning the validity of the others.

Of course it is not reasonable to expect anyone to try to disprove each and every report. They spring up faster than weeds, and a lot of them simply aren't subject to proof, one way or the other. That being the case, it is hardly reasonable to expect anyone to devote great effort to what can't be done, and there is a valid reason for all those who put no stock in the reports to ignore them. Only those who nourish some hope that they are dealing with a genuine phenomenon have any reason to do anything about it.

That would be fair enough if things were divided up a little more equally. As it is, the vast majority of the people in a position to assign significant resources to an investigation are sure that there is nothing to investigate, and those few who would like to do something are held back by the weight of contrary opinion, if only through fear of being discredited in the eyes of their peers. That is not, by the way, an imaginary fear. It might seem to the outsider that the world of science would not operate that way, and that every interesting question would provide its own motivation for adequate efforts to find an answer, but such an assumption is not correct.

In spite of the billions of dollars spent each year on inquiring into the nature of our universe and ourselves; in spite of the tens of thousands of people employed in the pursuit of such knowledge; there is in a normal year not one person and not one penny from all the vast establishment of science devoted to assessing the possibility that man has a living relative in North America. When you consider the very marginal value of many researches for which men and money are available—with millions of dollars and lifetimes of work expended on projects known in advance to be insignificant even if totally successful—it seems unbelievable that anything with possibilities so basic, and so exciting, could be not just overlooked, but outlawed.

Assume for the sake of argument that no such animal actually exists. You then have a situation in which hundreds of people are finding reasons to claim encounters with an imaginary animal, and other hundreds are going to the trouble of counterfeiting tracks that such an animal might make. Consider that this phenomenon is not just a passing fad, but has been going on at so many times and in so many places of which we have record that it can be assumed to have existed everywhere, throughout man's history. Surely such a phenomenon should be considered worthy of study. Surely it would be studied by any society that wished to understand itself.

Or is there something about this subject that activates a prejudice so deep as to make it untouchable? If that is the case, and it certainly seems to be, then that in itself is a phenomenon that demands investigation.

237

-13-

The Canadian Scene

With two partial exceptions there is hardly any information about sasquatches in Canadian provinces east of British Columbia. The exceptions are Alberta, which shares the Rocky Mountains with B.C., and Manitoba, right in the middle of the continent. Oddly enough, although recent reports are fairly common, I know of absolutely no information from either of those provinces before 1968. In that year something interesting was going on in the uninhabited upper valley of the North Saskatchewan River. At Kootenai Plains a group of Indians had set up camp in tents and teepees determined to make a life for themselves and their families more like that of their ancestors and less influenced by the undesirable aspects of the white man's culture, particularly alcohol, than was possible on their home reserve near Edmonton. Before long they found themselves encountering something from the old days that they hadn't counted on, the hairy giants of their legends.

The creatures they saw didn't cause them any trouble. One man who saw one watching him at close range as he was cutting brush told an *Edmonton Journal* reporter in August 1969, "I didn't know what to do, so I just went on chopping wood." After a while the creature went away.

In September they found some huge tracks in the sand near where the Cline River joins the North Saskatchewan. George Harris, who then lived at Nordegg, took pictures of tracks 17 inches and 13 inches in length. That same summer, according to a letter he wrote to Roger Patterson, a Michigan man and his family saw a large black figure walking upright with long strides on a ridge across the valley from the Columbia Icefields. It was a mile or more away, but they felt sure it was too large and walking too fast to be a human. On the other side of the ridge are the headwaters for the Cline River, which is the area the Kootenai Plains Indians believed their giant visitors were coming from.

In 1969 preliminary work was being done for construction of the Big Horn dam on the North Saskatchewan about 20 miles below the Kootenai Plains. On August 24 about 9:30 in the morning, a man working on a pump installation down beside the river noticed a dark figure on the high bank over half a mile away and 300 feet above him. It looked like a large man standing watching what was going on. The figure stayed there for about half an hour, sitting down part of the time, and walking a considerable distance along the edge of the bank before it disappeared in the trees. Before it left the number of men watching it had increased to five. Although it was on the same side of the river as they were there was a loop of the river between them, so they could not readily approach it. They did try waving to it, without result. They thought it was too big for a man, but did not appreciate its real size until two of them went over where it had been. Using the small trees in the background for comparison the other men, still watching from below, estimated that the creature was at least 12 feet high and more probably 15 feet. No footprints were found.

I interviewed three of the five men, and looked over the area of the sighting. There was no doubt that they were in a position to make a very accurate estimate of comparative heights, and no reason to think that they were making the story up.

The following month, according to a letter I received from one of the men involved, three amateur prospectors eating lunch near the Ribbon Creek Road, east of Banff, saw a "gorilla" sitting on its haunches watching them. It stayed there for about five minutes, then got up, "sort of chattered its teeth like it was cold and also moved its arms in an up and down movement," and walked off. They agreed that it was at least seven feet tall. It had longish arms and droopy large breasts, and was covered with long brownish hair. They could find no tracks that day, but the following day the man who wrote to me and another friend returned to the area, and saw on the road some tracks that they estimated to be about 15 inches long. I have a second-hand report passed on by Jon Beckjord, of another sighting a few miles east of that area, just at the edge of the mountains, by a Wisconsin resident and a friend in July, 1971. He described what he saw as a moose-sized, upright animal. It was in a stream bed at the bottom of a canyon, about half a mile from the observers.

In the summer of 1972, according to a youth who wrote to me from St. Albert, Alberta, three people from Rocky Mountain House were reported to have seen the head and shoulders of a hair-covered man eight to ten feet tall in a berry patch in a valley not far from the town.

All those reports were either in the Rockies or close to them in foothill country, but in the fall of 1972 and in October and December of 1973 there were sightings of large upright creatures reported far out on the prairie, near Seven Persons, in southeast Alberta. The reports were passed on to me by L. Edvardson, at Medicine Hat. Some tracks 15 inches long and seven inches wide were also found on the ice of Seven Persons Creek. They showed a stride of six feet. One man who had watched an animal for several minutes in the dusk drew several pictures of the creature, which was carrying something in its arms. One shows it moving through some trees. The outline was like that of a heavy woman carrying a child. He said it was dark reddish-brown, and this seemed to be fur. The thing disappeared into some bushes. All the sightings were along Seven Persons Creek, which presumably runs below the level of the surrounding land, probably with a good deal of brush around it. It is a long way from the Rockies, but Seven Persons is close to the Cypress Hills, which rise to nearly 5,000 feet, and the creek has its headwaters there.

Aside from the sighting by the dam workers, none of those reports came to public attention, but in 1974 and 1975 there were two incidents that were widely publicized. The first story, carried by Canadian Press May 12, 1974, took place on the David Thompson Highway close to the lake now formed behind the Big Horn dam. Ron Gummell, a 42-year-old Calgary resident, said that on Saturday morning, May 11, 1974, he rounded a bend in the road and confronted two creatures standing in the middle of the blacktop. He estimated them to be 12 feet tall.

"They were so big they could have picked up my car and thrown it in the lake," he told a reporter. His car stopped about 30 feet short of them, and they stared in his direction for about 10 seconds before jogging away. He described them as being apelike in appearance, and covered with dark brown hair.

The next sighting was reported by a man with 45 years experience as a trapper. According to the Grande Prairie *Daily Herald*, Bob Moody, of Woking, Alberta, said that as he was driving towards Sexsmith about 8:30 in the morning of October 2, 1975, he saw at the edge of the trees beside the road a creature about seven or eight feet tall covered with hair about an inch long, and two smaller ones about three and four feet high. They were behind the large one, so he did not see them well. They all disappeared in the trees as he began to pull his truck over. The area is in the northern part of Alberta, a long way from the mountains, but it is forested. A game guide named Art German went to the site as soon as he learned of the report. He told me that he found marks showing three toes dug into the bank by the road, five to six inches wide and with a four-foot stride. He also found three small piles of fibrous droppings, one of which he sent to me. I sent it to Dr. Vaughn Bryant at Texas A. & M., who had offered to analyze such things, but all I learned was that it contained grass.

The game guide's description of the area was interesting. He said the sighting took place just where a jungle of poplar and willow merges with an expanse of pine timber. There is a river nearby, and in behind, a hill and then a swamp. It is completely wild country, with no livestock there. He confirmed that Mr. Moody was a lifelong woodsman who could not possibly have mistaken some other type of animal.

A Calgary man wrote to me about 10 months after the event that in September 1, 1976, he saw a black, eight to nine-foot animal that looked like an ape standing in a fast-flowing river in the forest west of Caroline, Alberta, where he was hunting. As he started to run towards it from a quarter mile away it waded to the far side of the river, climbed out on the bank and looked at him, then headed into the bush. On October 31 that year, according to the November 3 issue of the Athabaska *Echo*, 11-year-old Daryl Lange and Paul Schleier, a grade 12 student, said they had seen a big dark animal about seven feet tall walking on a country road 10 miles east of Rochester, Alberta, at midday. Indistinct tracks were found at the scene. As a result of the story the paper received a report that a forestry tower man in the Slave Lake Forest had entered a sighting of an unidentified two-legged animal in his log, and that there had been a sasquatch sighting earlier that summer near Cold Lake.

The province of Saskatchewan comes pretty close to being a blank on the sasquatch map. Rene Dahinden was told of 16-inch tracks with a 48-inch stride being seen near Birch Lake in the early 1960's, and a man named Al Sumerfelt was supposed to have a cast. Also, a sasquatch was supposed to have been seen somewhere north of there. Both reports were from the same man and have not been checked. In 1969, a Cree Indian named John Tootoosis, from Cut Knife, Saskatchewan, wrote to me to say that the word in his language for the "mountain giants" was "Mistapawe", but he did not say that they were known in Saskatchewan. On June 11, 1977, Canadian Press carried a report from Saskatoon about a conservation officer studying some 18-inch tracks near Theodore. There were five in a farmer's garden and some more on the road, but none were really clear.

To the north the pickings are even slimmer, one specific report and a few vague ones from 1,300,000 square miles. A man from Yellowknife in the Northwest Territories wrote to me that he had been making inquiries and had learned that Indians along the McKenzie River knew of a "bushman". Specifically, a friend had heard stories of such a creature in the vicinity of Fort Good Hope, where it was reported to be on both sides of the McKenzie. There had also been a report of one seen near Fort Norman, in November, 1973.

A Vancouver woman told me that when her grandfather had a trapline near Lansing, 300 miles northwest of Whitehorse, in the Yukon Territory, he had run into huge five-toed prints and the Indians had told him they had seen a big hairy man. That was in the early 1900's. Besides that, I have a letter stating that in April, 1970, the owner of the Keno Hill Hotel and an Indian came to a place while hunting where several holes in the snow seemed to be emitting steam. The Indian got very upset and insisted on leaving, but as they went they looked back and saw what looked like an ape stand up from under the snow. It's a fine yarn, but I have heard a slightly different version of it from somewhere else.

In another letter, a Gary, Indiana, man tells of seeing a creature 10 feet tall in the Yukon in the 1940's and shooting at it with a 30.06. He said its tracks were 18 inches to 22 inches long. A somewhat more restrained report, and the only recent one from the Yukon, was printed in the Atlin, B. C. *News Miner*, on December 17, 1975. It states that Ben Able, of Atlin, while driving near Jake's Corners in the Yukon on October 4 of that year, passed a small figure by the road at night and backed up to offer a ride. Whatever it was moved off the road and didn't reply when spoken to. It was about five and a half feet tall, with fur all over the body, and it looked bluish in the headlights. The face was grey. He reported his experience to the R.C.M.P. at Atlin.

Construction of a dam on the North Saskatchewan was involved in a couple of the Alberta sasquatch reports. In northern Manitoba damming of the Saskatchewan River has been indirectly responsible for almost all the reports. A dam near Grand Rapids flooded out an Indian community and a new town called Easterville was built for the displaced people on the south shore of Cedar Lake. To provide access to the highway system, a new road had to be built running east to meet Highway 6 south of Grand Rapids. It is on that new road that nearly all of the sightings have taken place.

The first incident occurred on August 23, 1968, when three men had stopped in a car about four miles west of the Highway 6 junction. They said they saw something walk out of the bush about 100 yards away and onto the road. It walked like a man, but was much too large and covered all over with short hair. The men left hurriedly. A few days later, near the Easterville end of the road, two other men saw a similar creature. They too drove off in a hurry. Those two stories were reported in the Winnipeg *Free Press* not long after, but weren't taken seriously. No one bothered to keep in touch with what might be going on at Easterville. After all, the road was used to get to the liquor store at Grand Rapids. What more need be said?

In January, 1974, Brian McAnulty, police reporter on the *Free Press*, was doing some checking around after someone had shot a cougar, a rare occurence in Manitoba, and he stumbled on something much more interesting. Just a couple of months before, the Manitoba Museum of Man had received a remarkable letter from a wildlife conservation officer at The Pas, which is well north of the prairie country, in the timbered, lake-strewn Canadian Shield. The subject of the letter was a set of footprints 21 inches long and seven inches wide. It ran as follows:

On September 29, I found some human footprints which I believe you might be interested in.

These tracks were near Landry Lake, 20 miles west of The Pas in Tsp. 56, Rge. 23 WPM. The enclosed photo shows the size of the print in relation to an 18" ruler. The depression was approximately 1¼ inches deep. My footprints in the same soil sank less than ½ inch. There were imprints of four large round toes almost in a straight line across the front. There was a slight depression in the ground before the print was made so that the two toe marks, across from the ruler in the photo, do not show up well. The print is generally flat from heel to toes. Distance between prints was approximately 20 inches.

The area in which I found the prints is a limestone base ridge covered with spruce and jackpine surrounding the north and east side of Landry Lake. The prints, three of them, were in an old ant hill that had been spread. Aside from this one soft spot, the ground cover in the area is mainly light moss and vegetation over a base of clay-gravel mixture and does not lend itself to tracking animals.

I was moose hunting at the time I found the three tracks. My impression at the time was that they were not man made and that they fitted no animals in the area. They were sharp and clear at the time and were made after the rainfall on September 25. I did not carefully compare all three prints at the time but they appeared identical except for the toe marks. On October 1, I returned to the area with R. J. Robertson,

A 21" footprint found and photographed by Conservation Officer Bob Uchtmann near The Pas, Manitoba.

Wildlife Biologist, and B.E. Jahn, Wildlife Technician, to look at the prints again. That morning a moose hunter had walked through the area and stepped in two of the prints, making it impossible to compare the three. A light rain was falling and had washed the sides of the impression in so that plaster of paris casts were not practical. This, also, resulted in poor photography.

There is the possibility of someone having made the prints but this strikes me as being highly unlikely. The prints were approximately ¼ mile from a bush trail passable to vehicles. Back along the road 200 yards is a dike across a creek on the north end of Landry Lake. It seems more likely that someone creating a hoax would have placed the prints on the dike where there would be more chance of seeing them. As for the other hunter on Monday walking through the area, I believe this to be pure coincidence. I, also, had placed a red ribbon by the tracks to which he may have been attracted.

The letter concluded with some points to note in studying the photos, and was signed: *R. H. Uchtmann*, Conservation Officer.

Nothing ever came of the tracks at The Pas, but because of them Brian McAnulty decided to check whether anything more had happened at Easterville. It turned out that things had never stopped happening. Not only had there been further sightings by Indian residents of the town, there had been three incidents in which the witnesses included non-Indian teachers from the Easterville school. All these sightings had been while driving the road at night, but they told Brian that there was not any doubt about what they had seen. A woman who had taught at the Easterville school for five years had seen something twice, in the summer of 1969 and in November, 1970. On the first occasion an Indian couple was with her, and all of them saw a dark manlike figure run from the road, hurdling over willows and small bushes with long strides. She was no longer at Easterville in 1974 when I went there to get computer reports, but one of the Indians told me the thing was the color and size of a moose, yet jumped over things like a man.

On the second occasion a dark, heavy, shaggy form beside the road which they thought at first was a man turned out to be seven or eight feet tall and covered with hair. It had an extremely short neck, was leaning forward a little, and had long, hanging arms. They backed up to try to get another look, but it had vanished in the trees. The other witnesses that time were the principal of the school and an Indian who had died by the time Brian learned of the story.

The third sighting, in April, 1973, involved yet another teacher, in fact there were two of them in the car but one had dozed off. The driver had to slam on the brakes to avoid hitting a dark thing that he estimated at nine feet in height and extremely broad, which was walking on the road in the same direction that he was driving. As he braked it turned and he got a look at a flat-profiled face. The car went into an uncontrolled skid and was travelling sideways when the thing leaped to the edge of the road and vanished in the bush.

All of those teachers had left Easterville by the end of 1973, but Brian got in touch with them and did a full-page feature story in the *Free Press* on January 26, 1974, about the creatures the Crees in Manitoba called the "Weetekow" and the Saulteaux the "Wendego." He also told the story of a man and woman driving from Werner Lake, Ontario, towards Winnipeg in December, 1972, who saw a tawny creature "sort of like an ape" run across the road upright in front of their car at night. They estimated its height at seven feet. When they backed up they saw tracks in the snow too far apart to be made by a man, and the tracks were also seen by other people the next day.

"It's like you walking down a back alley and bumping into a Frankenstein monster," the young man told Brian. "Everybody knows there is no such thing, but you've just seen him."

By the time I went to Manitoba in the spring of 1974, Brian had

245

also found two people near Reynolds, Manitoba, southeast of Winnipeg, who had seen a hair-covered creature larger than a man near their homes during 1970, one in daylight in the summer, while he was walking down the road. The witness in the second incident had a nerve-shattering experience. Hearing a crackling sound from the direction of an old beaver house near the road as he walked along alone at night, he shone his flashlight over there and saw a dark creature seven feet tall suddenly stand up. He fled.

Brian and I and another man went to Easterville and spent some time driving the road as well as talking to some of the Indian witnesses. He has since uncovered several additional reports, including some more on the border of western Ontario. Sightings on the Easterville road have continued, the most recent I know of being in April, 1976, and there have also been two sightings near Grand Rapids, in one of which nine members of a family saw the creature swim up the Saskatchewan River past their house and then come out and walk into the bush a quarter mile away.

Brian also investigated two reports from the vicinity of the Long Plains Reserve near Portage La Prairie, Manitoba. In July, 1975, two teenage boys reported seeing a dark eight-foot creature of human form at night near a house on the reserve. No one took them seriously when they rushed into the house with the story, but next day three five-toed 20-inch footprints with a five-foot stride were found and photographed in the clay bottom of a nearby ditch. The prints had a decided arch, with the ball of the foot seven inches wide and the rear half of it only four inches wide. Brian found a farmer who had seen a similar creature in a field where he was working eight miles southeast of Long Plains a few days later. Its footprints were only 15 inches long.

The other area of activity in Manitoba is from Beausejour to the Ontario border northeast of Winnipeg. In July, 1974, a man from Lac du Bonnet saw what he described as "an overgrown ape or monkey" about six foot six, with dark hair or fur, at night on the road to Pointe du Bois. That story was published in the *Manitoba Beaver* at Beausejour, and Brian subsequently talked to three more people with reports from the same area. One youth saw a huge, hairy thing thumping on the trunk of his car as he made a U-turn on a road near Beausejour about 4 a.m. on a night in June, 1975. In December of that year his cousin, who had not been told of the previous incident, saw a seven or eight-foot creature approaching his stopped car on the road between Lac du Bonnet and Beausejour. He went back the next day and found 15-inch footprints in the snow which he and others followed for more than seven miles. They had five toes like a man's and were flat.

"It could be a prank alright," he told Brian, "but where do you find someone around Beausejour between seven and eight feet tall who walks around barefoot in the winter time?"

The previous October, farther east in the Agassiz Provincial Forest, a man reported seeing a heavily-built, brownish animal running through the jack pines on its hind legs. It was less than six feet tall, and had long arms swinging. The witness, who was walking around looking for mushrooms, told Brian, "It took me an hour and a half to get there and 15 minutes to come back."

The most recent report I have from Manitoba reached me via the Royal Canadian Mounted Police, at Vancouver, who had received a telex from their Norway House detachment as follows:

Norway House to Vancouver Lower Mainland Division, attention Sergeant Doane.

It was reported to our office on the 26 July 76 by the chief of the Poplar River Indian Band that many of his people have sighted on the reserve many times a large hairy animal that walks on two legs. Poplar River is located approx. 76 miles to the south of Norway House. An investigation was conducted and the results are as follows:

Several people were interviewed and they all stated that the animal was approximately seven to eight feet tall and was very broad at the shoulders. It had the general body structure of a man only many times larger. A foot cast was taken of the foot impression that was left behind by the so-called monster and is held at this detachment. It measures 16 inches by five inches, and has only three toes. Its fur is a glossy gray color and it has white hair on its head. They stated that it was very powerfully built and one man reported that he saw it swimming. To date there have been no further reports of sighting in our area. It should be noted that this so-called monster seemed very inquisitive towards the people and would come around the houses on the settlement and look in doors and windows.

This is about all we have so perhaps you could pass this on. Thanks in advance.

Downing, Norway House detachment.

I turned this over to Brian McAnulty, but Poplar River is not an easy place to get to—even Norway House is far from any highway—and I have not heard anything more about it. Neither do I have any particulars of a sighting and some 14-inch footprints that Brian has learned about just over the border in Ontario. The most significant thing about all this material, I think, is that but for the coincidence of one man running across a report of big footprints while enquiring about something else, and deciding to follow it up, no more than three or four of these stories would be public knowledge at all. The same sort of thing has happened in so many other places where someone has made a substantial effort to obtain information that it seems to me almost certain that there are many other areas where just as much is going on but nothing is heard about it.

There is little likelihood of such creatures existing in open, inhabited areas, but such areas are usually broken up at least by wide, wooded river valleys, if not by patches of actual brush or forest. In northern Canada there is nearly two million square miles of forested country not greatly different from that around Easterville or Lac du Bonnet, yet the number of sighting reports from all the rest of that area can be counted on your fingers. One factor to consider is that most of the sightings in Manitoba are road sightings, the majority of them at night. Most of the northern country has no roads whatever, and not many people are abroad at night. Still, I am sure, and the Indian traditions confirm it, that information is there to be found if someone should ever look for it.

Moving over into Ontario, there is almost 800 miles of country of that type to cross, all within the same province, to get to the next place from which I have sighting reports. That is in the vicinity of Sudbury and North Bay, an area that anyone looking at a map would say was in the southeastern part of Ontario, but which most of the residents of that province, crammed in near the Great Lakes and the St. Lawrence River, would consider to be in the north. From Cobalt there are four reports covering a time span from the early 1900's to 1970, all published in the North Bay *Nugget*. Bill Davis, who has recently checked into some of the Ontario material (he is not the Ontario prime minister of that name) reports that he was told at the *Nugget* office that the 1970 report was considered a hoax. Perhaps it was, but it seems unlikely that the same hoax has been going on for nearly 70 years, and since all the stories are short I will reproduce them here. The first was published in late July, 1923. My photo copies, courtesy of Rene Dahinden, are not entirely legible, so that I might, for instance, have a name spelled wrong:

COBALT—July 27, 1923—Mr. J. A. MacAuley and Mr. Lorne Wilson claim they have seen the PreCambrian Shield Man while working on their mining claims North and East of the Wettlaufer Mine near Cobalt. This is the second time in seventeen years that a hairy apelike creature nicknamed "Yellow Top" because of a light-colored "mane" has been seen in the district.

The two prospectors said they were taking test samples

248

from their . . . property when they saw what looked like a bear picking at a blueberry patch. Mr. Wilson said he threw a stone at the creature.

"It kind of stood up and growled at us. Then it ran away. It sure was like no bear that I have ever seen. Its head was kind of yellow and the rest of it was black like a bear, all covered with hair.

The first report of the creature was made in September, 1906, by a group of men building the headframe at the Violet Mine, east of Cobalt. It has not been seen since that time.

COBALT—April 16, 1946—Old Yellow Top, the half man, half beast that is supposed to be roaming the wilds around the Cobalt Mining Camp was reportedly seen again, this time by a woman and her son, who live near Gillies Depot, while they were walking the tracks into Cobalt.

The woman, who did not want her name made public, said that she spotted a dark, hairy animal with a "light" head ambling off the tracks into the bush near Gillies Lake. She said she did not get a clear look at the thing but said it walked almost like a man.

The sighting is the second such report to be made since 1906 or 1907. A search party may be formed to try and find "Old Yellow Top."

COBALT—August 5, 1970—Twenty-seven miners on their way to work the graveyard shift at Cobalt Lode could have lost their lives when the bus which they were riding went out of control and almost plunged down a nearby rock cut. Bus driver Aimee Latreille, who has been doing the route for the past four months, said he was startled by a dark form which walked across the road in front of him.

"At first I thought it was a big bear. But then it turned to face the headlights and I could see some light hair, almost down to its shoulders. It couldn't have been a bear."

Although no one was hurt, Mr. Latreille said he did not know if he would continue to drive the bus. "I have heard of this thing before but never believed it. Now I am sure."

One of the miners at the front of the bus said he just caught a brief glance of the creature. Larry Cormack said it "looked like a bear to me at first, but it didn't walk like one. It was kind of half stooped over. Maybe it was a wounded bear, I don't know."

Mr. Cormack added that he did not believe the PreCambrian Shield Man existed anyway. "My father used to talk about it, but I've seen it close up," he pointed out.

249

The story started back in the early 1900's when a group of construction workers claimed to have seen the creature. There have been two reports of a strange hairy creature scaring bus passengers in the area. The creature pops up out of the bush and waves a club in the air. One theory is that he is a pre-Cambrian man routed by visitors from some long-lost cave. The other theory is that he is a reveller from the Open Cut Saloon working off a big head.

The last paragraph makes it clear that the writer of the story did consider it a hoax, and the initial report could easily have involved mistaken identification. Note that none of the actual reports say anything about clubs or caves.

In August and September, 1965, there were reports from the heavily-populated area in the southern part of Ontario. The Hamilton *Spectator* in early August carried a story about a Lakeview trucker who claimed to have seen a half-human, half-animal beast six or seven feet tall on a side road near Smithville. He estimated the thing to weight 500 pounds. He went back to look for tracks but found none. I have a photocopy of another clipping, from the Beamsville *Express*, August 25, 1965, that refers to the Smithville monster having turned up in Campden, about seven miles to the northwest, apparently a few days later after it was in Smithville. No less than seven people were reported to have called the provincial police to say that they had seen the monster, which all of them described as big, black and furry. Wayne Beach said that the thing he saw a yard behind his car "just looked like a great big gorilla." Two 16-inch prints were reported found near a garage where Manfred Berg saw something he couldn't identify. Hector McDonald said he got a back view of the creature at night in his barn. Sergeant Bass of the Ontario Provincial Police had no problem disposing of it.

"It's not a monster," he said. "There's no such thing as a beast half human and half animal. Somebody is playing a hoax."

Then on September 4, the Kitchener-Waterloo *Record* carried a story that a mysterious monster reported in the Tillsonburg area had been "identified" by the provincial police. Several persons were said to have sighted the monster in the tobacco fields, although no description of it was given. Tracks were plainly visible and measured 18 inches long. Police reported that the tracks in the sand were those of a tired tobacco worker crawling on his hands and knees between the rows.

It's a little puzzling how the tracks of a person on his hands and knees would look like anything but the tracks of a person on his hands and knees, but not having seen the tracks, I can't comment.

I have a letter from Vince Lefebre, at Sudbury, who said that in mid-July of 1974, in a sparsely-overgrown area, he saw three inch-deep impressions in moss, 13 to 14 inches long and 6 to 7 inches wide, indicating a six-foot stride. Human feet left no permanent mark. Nearby he found another track in dried mud, 14 inches long by seven inches wide and an inch deep, without any arch, showing four toes shorter than human toes, with the big toe much larger than the others. He attempted to get someone from a university to look a the tracks, but no one would go.

Wayne King, from Michigan, sent me an account of a sighting by a 15-year-old boy at Ruthven, Ontario, near Pelee National Park. He said that on June 4, 1977 from the verandah of a house overlooking a gorge, he saw a tall upright black creature almost hidden in the foliage. The boy took two pictures showing what could well be a black head among the leaves, but details are not discernable.

There certainly should be reports from Quebec, but I have never come across any. Ivan Sanderson, on the other hand, had received three before he wrote *Abominable Snowmen* some 15 years ago. They were contained in letters that are not now filed with his sasquatch material, but Sabina Sanderson says that she has seen them fairly recently, so they should turn up. At this point all I can do is quote what he wrote in his book:

Those from Quebec have puzzled me for years. I have constantly heard about them but have only three pieces of paper to show for my exhaustive and prolonged inquiries and appeals. These are all letters from American summer visitors on serious hunting and camping trips by canoe, guided by professional Amerindian trappers and hunters. All three are substantially identical and all give somewhat similar accounts of events in widely separated places. One is from a lone man, a business executive from Chicago; one is from a party of four men of assorted professions who have hunted for years on their annual vacations together; the third is from the father of four—three grown sons and a (then) teenage daughter.

In each case a tall, very heavily built, man-shaped creature with a bullet-head and bullneck, and clothed all over in long shiny black hair, with very long arms, short legs and big hands, is said suddenly to have appeared on the bank of a river in which the party was quietly fishing. On one occasion, the creature is said to have carried off some fish left on a rock on the bank; on another it chased the Amerindian guide out of the woods and into his canoe and then waded some distance out into the water after him. The family party seem. to have become fairly familiar with two of the creatures over

251

a period of several days. They say they constantly prowled around their camp, and showed themselves among the trees whenever they went out in the canoes. One seems to have shown signs of chasing the girl on one occasion but, the father told me, they gained the impression that this seemed to be more through curiosity than menace. Two of the Amerindians are said to have asserted that they and their people knew the creatures quite well and that there were quite a lot of them in those forests. The other guide, who was chased, appeared to be scared almost witless and swore that the thing was some form of spirit or devil. However, it smashed branches and hurled stones, it is reported.

I am frankly stymied over these reports. Two of the writers asked that I withhold their names in perpetua as they did not want the reports to become known to their business associates. The third man I never traced. It was many months before I could get to the places from where these people wrote and although I traced two of them, they all stopped answering my letters and I am left with nothing to follow up.

From the Maritime provinces I have only the one vague old report from New Brunswick, already quoted, despite the fact that Bruce Wright, the late director of the Northeastern Wildlife Station at the University of New Brunswick, was a sasquatch enthusiast. He told me that his interest stemmed from the fact that his godfather, Bruce Irving, had seen a sasquatch somewhere in central British Columbia while he was with the provincial police. He was told the story in 1924, when he would have been 12 years old, and remembered only that his godfather "simply stood and watched it for a brief period until it was out of sight. He was greatly impressed with its size."

Professor Wright also investigated reports of two creatures that were seen at Traverspine, in Labrador, in 1913, which do not match the desciption of sasquatches, but were certainly most unusual and interesting. The story was first told in a book titled *True North* by Eliot Merrick, who lived in the community in the 1930's. Traverspine is not on most maps, but is, or was, at the west end of Lake Melville, near Goose Bay. Merrick's account was as follows:

> Ghost stories are very real in this land of scattered lonely homes and primitive fears. The Traverspine "gorilla" is one of the creepiest. About twenty years ago one of the little girls was playing in an open grassy clearing one autumn afternoon when she saw come out of the woods a huge hairy thing with low-hanging arms. It was about seven feet tall when it stood

252

erect, but sometimes it dropped to all fours. Across the top of its head was a white mane.

She said it grinned at her and she could see its white teeth. When it beckoned to her she ran screaming to the house. Its tracks were everywhere in the mud and sand, and later in the snow. They measured the tracks and cut out paper patterns of them which they still keep. It is a strange-looking foot, about twelve inches long, narrow at the heel and forking at the front into two broad, round-ended toes. Sometimes its print was so deep it looked to weigh 500 pounds. At other times the beast's mark looked no deeper than a man's track. They set bear traps for it but it would never go near them. It ripped the bark off trees and rooted up huge rotten logs as though it were looking for grubs.

They organized hunts for it and the lumbermen who were then at Mud Lake came with their rifles and lay out all night by the paths watching, but with no success. A dozen people have told me they saw its track with their own eyes and it was unlike anything ever seen or heard of. One afternoon one of the children saw it peeping in the window. She yelled and old Mrs. Michelin grabbed a gun and ran for the door. She just saw the top of its head disappearing into a clump of willows. She fired where she saw the bushes moving and thinks she wounded it. She says too that it had a ruff of white across the top of its head. At night they used to bar the door with a stout birch beam and all sleep upstairs, taking guns and axes with them.

The dogs knew it was there too, for the family would hear them growl and snarl when it approached. Often it must have driven them into the river for they would be soaking wet in the morning. One night the dogs faced the thing, and it lashed at them with a stick or club, which hit a corner of the house with such force it made the beams tremble. The old man and boys carried guns wherever they went, but never got a shot at it. For two winters it was there.

After visiting the community to enquire into this incident, Professor Wright included this version of the story in his own book, *Wildlife Sketches*.

About 1913 the little settlement of Traverspine at the head of Lake Melville was visited in winter by two strange animals that drove the dogs to a frenzy and badly frightened the people. They left deep tracks about twelve inches long indicating great weight and they rooted up rotten logs with great strength and they tore them apart as if searching for grubs.

They sometimes stood erect on their hind legs (at which time they looked like great hairy men seven feet tall, and no doubt from this description Merrick got his title of the Traverspine Gorilla). but they also ran on all fours. They cleaned up some seal bones "too big for the dogs", and what is too big for a husky is really big, and many dogs followed them and did not return when they came around the settlement at night. This was a serious loss as dogs were the people's sole means of transportation.

These two strange animals, which the inhabitants called the man and the woman because one was larger than the other, stayed about the settlement despite attempts to trap them or drive them away. One day Mrs. Michelin was alone in her house with her young daughter playing in the edge of the bush behind. Suddenly the child rushed in crying, "It's following me Mummy, it's following me!"

Mrs. Michelin reached for a shotgun loaded with buckshot which she always had near when her husband was away, and stepped out the back door.

"All I could see was the moving bush and the shape of a great animal standing seven feet tall in the alders. It seemed to have a sort of white ruff across the top of its head, I could not make out the rest. I fired into the bushes and I heard the shot hit. I went back into the house and bolted the door. It never came back and there was blood where it had stood when the men from the sawmill came back."

. The sawmill closed down and the men turned out in force to look, but they never found it. Similar animals have been reported since and their tracks have been found at intervals, the latest being about 1940.

I asked Mrs. Michelin point blank if this could have been a bear. "It was no bear Mr. Wright. I have killed 12 bears on my husband's trapline and I know their tracks well. I saw enough of this thing to be sure of that. I fired a shotgun at it and I heard the shot hit."

No explanation is given of how there came to be two creatures in the second account, and Bruce Wright has no comment to make on the two-toed tracks. I have never known what to make of this story. It seems to me to have the ring of truth, in both versions, but they are not quite the same and neither of them fits either the sasquatch or any known animal.

254

-14-

Eastern Action

The oldest sighting report in print that I have is that of the "whistling wild boy" in Pennsylvania in 1838, following which there is a gap of 120 years before the next definite report from that state in my file. Grover Krantz received a letter from a woman in Miamisburg, Ohio, early in 1974, saying that in the late 1950's when her husband lived at Cambridge Springs, Pennsylvania, a friend of his complained that some creature was banging on their farmhouse at night. A group of young people went to the house and sat out on the porch in the dark. Sometime after midnight a creature approached and one of the men shot at it, at which it ran off through a cornfield and into a wooded area. They tried to pursue it in a jeep and then to track it, but it got away. It was eight to 10 feet tall "and looked somewhat human, except that it had fur over its entire body."

In September, 1964, or 1965, according to a clipping Ron Olson showed me, Glen Varner, of Kingsport, Missouri, saw a giant, hairy, apelike creature peering in the window of his mother's home at Garrison, Pennsylvania, at dusk. It then took off, "leaping five to six foot fences." A story in the West Chester *Daily Local News*, January 24, 1976, tells of footprints found in snow in Pike County earlier that month that measured 16 inches in length and eight inches across the ball, were flat, had very wide heels, and went five or six miles before being lost in the deep forest in Promised Land State Park. They were found by 11-year-old Anthony Torriero Jr. near his parents' cabin in the Ponoco Mountains. That is just a few miles across the Delaware River from the area where there have been many recent reports in

255

northern New Jersey. The prints are described as closely matching a picture of a "Bigfoot" cast, so presumably they had five toes.

Such reports are nothing out of the ordinary, but in Pennsylvania the ordinary reports are the exceptions. Starting in 1968, but particularly from 1973 on, there are apparently hundreds of Pennsylvanians who claim to have seen huge hairy bipeds, and in the composite descriptions that have been assembled from those reports there are several things that differ decisively from descriptions elsewhere. There are also a few very strange incidents, like nothing I have heard from anywhere else, and some thoroughly odd footprints. Artists' representations of the descriptions show pointed ears high at the top of a head that is both wide and flat at the top. One version has long pointed fangs. Arms are supposed to hang below the knees. Eyes are supposed to be very large and to glow brightly even when not reflecting light from another source. Even though they show something hairy and upright the pictures don't look at all like the creature in the Patterson movie or like drawings made from descriptions given by witnesses in other areas. By contrast they serve to emphasize how consistently apelike the usual sasquatch descriptions are, and how different they easily could be.

Complicating all this is the fact that most of the people investigating the Pennsylvania reports are convinced that the creatures have some link with unidentified flying objects. Sasquatches are obviously Earth-type animals. Walking like men, looking like apes and living like bears, they fit right in as terrestial creatures, and they are hardly what anyone would expect to find flying a space vehicle. I greatly doubt that anything should be made out of the fact that a hairy biped might be reported seen at about the same time and place that a U.F.O. was reported. If there actually are craft from other places flying around (a subject on which I have no knowledge) and Earth is being studied by some kind of intelligent beings, it seems reasonable to me that they might be having a bit of a problem figuring out why the two-legged creatures that obviously dominate the planet seem to have direct contact with every other kind of creature on it except the one that somewhat resembles themselves. They might very well find the sasquatch a fit subject for observation. Also, if there actually are space vehicles being parked here and there from time to time, I see no reason why the occasional sasquatch might not feel an urge to look one over, as they have done at times with man-made machinery.

Even if a hairy biped were reported seen inside a mysterious craft, or entering or leaving one, I would still not accept that as proof that the creature came to Earth in that craft. It would seem to me at least as likely that such an observation would indicate that the actual operators of the craft were studying the hair-covered fellow—perhaps to find out why he didn't behave like his smaller, cloth-covered relatives. If someone claimed to have been for a flight aboard a U.F.O. and to have observed sasquatches operating the controls, I would have no answer, but so far I haven't heard of that.

My reactions are influenced, I realize, by the fact that space ships and space visitors have no reality for me. I don't have any theoretical objection to them. It seems to me the height of conceit to assume that we Earth creatures are the only beings in the universe clever enough to fly around in space. Since we are now able to do it, I would think it not only reasonable but statistically almost inevitable that there are a variety of other intelligent beings out there who are a lot better at it than we are. They may well be able to travel here, and they may have been dropping in from time to time for thousands or even millions of years, keeping an eye on developments. A time when we are starting to set off atomic explosions and to shoot things out of our atmosphere would certainly be a logical time for such surveillance to be stepped up. But if such is the case, they appear to have done us no harm up to now, so their intentions are presumably not hostile. Even if they were there would be nothing I could do about it, so I simply don't concern myself with the subject, and I have no inclination whatever to consider U.F.O.'s as part of the explanation for anything on Earth. To people who are convinced of the reality of space craft and space visitors, I suppose it is natural to think of them as a possible answer to the problem of something seen on Earth that isn't supposed to be here.

The same goes, I presume, for people who are convinced that there are other forms of reality on Earth with which we normally have no contact. They may well find it entirely satisfactory to assume that a creature that man can see but hasn't been able to catch or kill just steps across some borderline that takes it beyond the contact of man's sight, touch and hearing. Having no way of understanding the universe, or the existence of life, or so many other things that do appear real to me, I am certainly not about to argue that reality contains only the things that register on my five senses. But with the sasquatch I see no need to fall back on either extra-terrestrial or extra-sensory explanations. Considering the consistent resistance our society maintains to any exposure to information on the subject, our failure to have stumbled on proof of their existence, while considerably against the odds, doesn't seem to me to require any unearthly explanation.

But what about this business in Pennsylvania? The January-February, 1973, issue of *Flying Saucer Review* contains a reference to a sighting of a tall, evidently anthropoidal figure by five people in a car parked on Presque Isle beach, at Erie, Pennsylvania, on July 31, 1966. It was said to have left large footprints in the sand, and the same five people were supposed to have seen a U.F.O. land shortly before seeing the creature. That was just a curtain-raiser.

The main event began at the end of July, 1973, in Westmoreland County, just southeast of Pittsburg. What went on there has been reported extensively in newspapers and magazines but perhaps the best basic source of information is a paper presented to a U.F.O. symposium by Stan Gordon, director of the Westmoreland County U.F.O. Study Group, who was the moving spirit of the on-the-spot investigation. The paper is too long to quote verbatim, even if I had permission to do so, but I will attempt a summary.

Stan Gordon states that his organization investigated about a dozen reports in the summer of 1972 of creatures like gorillas and baboons seen on highways, many of them around a large cemetery surrounded by dense woods about two miles out of Greensburg, where he lives. The creature at that location was described as being five feet tall, hair covered and much broader than a human. Many residents in the area complained to police of a night prowler that made a sound "like a man in pain screaming at the top of his lungs." There had been "a major U.F.O. flap in Westmoreland County" during April and May of that year.

The events of the following July began when a young woman called to say that her stepfather had seen two large glowing red eyes at a window eight feet above the ground. His attention had first been caught by an overpowering stench like rotten cucumbers. The location was on Rodabaugh Road, near the Greengate Shopping Mall at Greensburg. It then turned out that the man's 14-year-old son and several other boys had seen a gorilla-like creature walking on two legs while they were taking a shortcut through the woods behind their houses towards the mall. Looking around in the wooded area, Stan Gordon found one perfect three-toed footprint in soft ash, the rest of the ground being hard. The print was photographed and cast. It was 13 inches long and eight inches wide, with very fat toes, not at all like the three-toed tracks at Fouke, Arkansas.

That same day a message was received from an investigator in Beaver County that at 1:40 that morning a man had seen an eight-foot, gorilla-like creature with glowing red eyes staring at him through a window, and police had later found a footprint 18 inches long. That had happened 50 miles from Greensburg.

Information was given out to the news media, and the U.F.O. center, the police and the newspapers began to get "hundreds of phone calls", many of them from "people who apparently had seen such creatures in the past but wouldn't talk about it before." All this

Stan Gordon with a photo of a three-toed track found at his home town of Greensburg, Pennsylvania.

Ramona and Duane Hibner with Dennis Gates at the location of a Florida sighting report.

set off a rush of would-be monster hunters into the wooded areas behind Greengate Mall, where many of the sightings were said to have occurred, in daytime as well as at night. On August 14 two men reported seeing a huge gorilla-like creature only 40 feet away as it cut across the railway tracks behind the mall at 1:45 p.m. They said it had a very strange odor like something dead a long time. Meanwhile reports of similar creatures were coming in from many other places, particularly a heavily-wooded area between the towns of Latrobe and Derry. There some people claimed to have been approached by the creature, to distances of less than a yard.

At 2:30 a.m. on August 21 a woman living near Derry awoke to see a hideous face staring at her from an open window only three feet from her bed, but nine feet off the ground. A "rotten meat" odor hung around the house for two days. On August 24 a man at Herminie who had just finished mowing his lawn saw 30 feet away in his yard an apelike figure about seven feet tall, standing motionless. He smelled the strong odor of "rotten eggs." While he was in the house getting a gun the thing left. The same night a woman and her son near Derry saw outside the back door of their mobile home a large apelike creature with long arms. They had heard scratching sounds and what sounded like a crying baby and then their lights started to go on and off. After the creature left a sealed "power box" was found to have been ripped open. In the following weeks numerous other residents of the same trailer court reported seeing similar creatures by day and by night.

September 1, a woman visiting a grave in the cemetery at Youngstown, Pennsylvania, in the evening, smelled a strong stench of rotten eggs and looked up to see a large, hairy apelike creature walking out of nearby woods towards her baby, which had wandered about 30 feet from her. She grabbed the baby, ran to her car and fled to her father's house, five miles away. An hour after she arrived there several people reportedly saw the same or a similar creature looking around the corner of the house. At 4:30 a.m. on September 3, a couple called police to report a hairy manlike creature, over eight feet tall, standing by their mobile home, staring at the house next door where several children were sleeping.

From June, 1973, to February 18, 1974, the U.F.O. group documented 118 such sightings with 245 witnesses. Ninety-one reports were in Westmoreland County, seven in Allegheny, six in Indiana, five each in Fayette and Washington Counties, three in Beaver and one in Somerset. Only about 30 took place in daylight, while there were 24 at dusk, and 30 in the vicinity of 10 and 11 p.m. Forty-nine of the creatures were seen in wooded areas, but there were 19 near trailers and 18 near houses. Only 10 were seen on roads. Those figures were compiled after discounting a large number of suspect reports.

The majority of witnesses who were close to a creature reported a very strong smell. Most outstanding physical features reported were protruding fanglike teeth and eyes the diameter of golf balls that glowed even in total darkness, usually bright red but sometimes bright green or white. Most height estimates were in the seven to eight-foot range, and hair was usually dark. A number of witnesses reported creatures with dirty white, matted hair, and there were also some estimated at only five to six feet, as well as some smaller apelike things.

The creatures seemed curious about humans and not afraid of them. They would stand still in front of a car and stare at the driver.

During 1973, the group also investigated nearly 600 U.F.O. sighting reports. Explanations were found for the majority, but an unusually large number remained that appeared to involve "intelligently-controlled craft displaying flight characteristics far beyond our present technology." There were over a dozen cases where:

U.F.O.'s were observed at low level minutes before or minutes after a creature sighting had been made in the same immediate area. There are other instances where U.F.O.'s were observed at a certain spot, and within a few days a creature would be sighted at the exact same location.

In early September an anonymous male caller phoned to say that three women had seen a large, metallic, rectangular object resting on the ground as they were driving through a wooded area, and that they had stopped and watched as a door opened in the side of the "craft", a ramp came down and three apelike, hair-covered creatures about seven feet tall ran down the ramp and into the woods. The story could not be checked, as no names were given.

On September 27, two teenaged girls in Beaver County told of seeing a white, hair-covered creature with red eyes, "carrying a luminescent sphere in its hand." Several people told of seeing what looked like an airplane standing still in the sky over the woods into which this creature disappeared, and there were other strange goings-on.

On October 25, Stan Gordon was involved in the aftermath of the most startling U.F.O. incident of all. After receiving a call from a state trooper at Uniontown about a combination U.F.O. landing and creature sighting, he and five other members of the study group went to the scene. They were told by the main witness that about 9 that evening some 15 witnesses saw a large red ball slowly descend into a pasture on his father's farm. The witness, an adult, got a 30.06

261

rifle and with two neighbour boys drove towards the area, then approached on foot. The object, on or very near the ground, was now a bright white, and appeared to be about 100 feet in diameter. There were screaming sounds and a sound like a lawn mower motor.

At this point one of the boys yelled that something was moving along a fence line nearby. They all saw two very tall, apelike creatures with glowing green eyes. The creatures were making a crying sound and the smell of burning rubber filled the air. The man fired several shots over the creatures' heads without effect, then shot right at one of them three times. When hit it made a whining sound and raised its arms towards the other creature. At that exact moment the U.F.O. disappeared and the sound it had been making stopped. The creatures then turned slowly and moved back into the woods.

The state police were then called and a trooper arrived about 9:45 p.m. Where the craft was supposed to have landed they now found a glowing white illuminated area about 150 feet in diameter, light enough so that a newspaper could be read if it were held within a foot of the ground. Horses and cattle in the field were avoiding the lighted area. Over where the bipedal creatures had been they could hear something in the woods. When the U.F.O. team arrived about 1:30 a.m. the glow was gone, but the livestock were still avoiding the area where it was said to have been. About 2 a.m. a bull in a nearby field and a dog both seemed to sense something nearby, and then the main witness suddenly went berserk and afterwards collapsed. Another man had trouble breathing. Next, a strong smell of sulphur filled the air. Everyone decided to leave.

A week later more than 100 people watched a large disc-shaped object in the air near Midland, in Beaver County. Next morning a trail of three-toed prints measuring 11 inches long and five inches wide was found nearby, and the following day two hunters found a 42-foot ring impressed on a grassy area, with several triangular holes in the ground inside the circle, about 250 yards from the trail of footprints.

February 6, 1974, a woman near Uniontown reported that at 10 p.m. she went with a shotgun to investigate strange sounds on her porch and confronted a seven-foot, hair-covered, apelike creature just six feet away. It raised both arms over its head, but she shot it in the midsection nonetheless, and it "just disappeared in a flash of light." There was no sound and no smell, only a flash "just like someone taking a picture." Her son-in-law came over from his mobile home 100 feet away, and said that on the way over he saw four or five hairy people seven feet tall and apelike, with very long arms and "fire red eyes that glowed in total darkness." He also saw a bright red flashing light hovering over the woods 500 yards away. Police were called, but by then it was all over, except that dogs, cats and a horse were all acting strangely.

The son-in-law then said he had seen a similar creature in the woods in November, 1973, and had shot at it with a revolver, whereupon it disappeared into thin air, but he could still hear it running away. After this the investigators heard of a case in which a creature disappeared when hit by a car. There was also several instances of people being bothered by invisible intruders in their homes; of witnesses being visited by mystery interviewers, and other strange occurrences that are apparently encountered from time to time by U.F.O. investigators.

At the conclusion of his paper, Stan Gordon stressed that he did not believe that the creatures seen in central Pennsylvania were the same things as were involved in sasquatch/bigfoot reports.

Dennis Gates and I called on Stan Gordon on our trip east in the spring of 1976, and heard some of his taped interviews. The witnesses sounded entirely matter-of-fact and sincere. We also talked to two other people involved in the investigation in the Pittsburg area, Tom Shields and Joan Jeffers, and looked at a couple of footprint casts. There had not been much doing around Greensburg for the past two years, but Tom Shields said he had talked to more than 200 people who had reported sightings around their homes in a valley north of Pittsburg during that time, and he and Mrs. Jeffers had seen a hairy biped themselves while in that area at night. They confirmed that there were many reports of glowing red eyes when there was no apparent light source to reflect from them. They also confirmed that there were many U.F.O. reports in the same areas as the creature sightings.

I am satisfied that the investigations and the people making them are entirely genuine, and that they do have witnesses for the many strange reports. Some of the weirdest ones could just be the inventions of one or two people, but others involve a number of apparently credible witnesses. What is at the bottom of it all I am sure I don't know. I am rather thankful that there are the consistent differences in the description of the creatures, as well as their actions, sufficient to enable me to agree with Stan Gordon's conclusion that whatever may be involved it isn't sasquatches. That doesn't make it any less interesting a phenomenon. There is ample material to justify a major effort to establish just what is behind those hundreds of reports of unknown creatures, whether actual animals are involved or not. But as long as I am unable to stimulate any meaningful interest in a phenomenon considerably less incredible as well as far better documented, I have no inclination to shoulder the burden of explaining what is going on in that part of Pennsylvania.

During 1977 newspapers carried a couple of reports from places north and south of Pittsburgh. There was nothing in the stories to suggest a U.F.O. connection, but I note that Joan Jeffers was investigating one of them and Tom Shields the other, and they may take a different view. May 20, 1977, the Uniontown *Herald* reported that a Footdale, Pennsylvania couple and their three children got a back view of a large upright creature on a rural road in German Township near McClellandtown about 9:30 at night. They described it as reddish-brown, with shaggy hair, about six feet tall although walking in a slumped-over position. Their lights hit it at the edge of the narrow road as they rounded a bend and it dropped from sight down an embankment.

On July 18, 1977, the Pittsburgh *North Hills News Record* reported that John Tiskus, of Shaler, had seen an animal drinking out of his above-ground swimming pool at 2 a.m. the preceding Tuesday. He was taking a break from doing homework when he heard a gurgling sound together with a high-pitched squeal. Looking out the window he saw in the lighted yard a thing seven and a half feet tall with hair over most of its body. It scooped some water out of the pool with one hand and took a drink, then took off into an area of dense woods. The article concludes:

"News Record has talked to many residents of the Hampton-Shaler-Richland area of North Hills who are familiar with the squealing cries of Bigfoot."

Over in New Jersey the sasquatch situation is more normal. It may seem strange to suggest that there could be undiscovered giant animals in a tiny state that I am told has more human inhabitants to the square mile than any other, but half of New Jersey's 7,800 square miles is forest and the Pine Barrens in the southeast contain an area of 1,000 square miles of salt marshes and forests where virtually no one lives. They also contain the Jersey Devil, a local legend that comes in a number of different forms. About the first time I ever gave any serious thought to the possibility that there might be sasquatches on the east coast was in 1968 when a woman who had just seen the Patterson movie told me "that looks just like the Jersey Devil." I asked her what the Jersey Devil might be, and she said it was a thing that lived in the cranberry bogs in southern New Jersey where she came from, and that it looked like the animal in the movie. She seemed to consider it fairly commonplace.

I have learned since that other descriptions of the Jersey Devil give it cloven hooves and a tail, or four legs and feathers and a hoot like an owl, and are generally unsuitable for a self-respecting sasquatch. In any case it is not in the Pine Barrens that the bulk of New Jersey sightings have been reported, but in the moutainous area of the northwestern part of the state. A namesake of Roger Patterson's did find two men who reported seeing hairy bipeds over six feet tall while they were camping in the Pine Barrens back in the 1950's, according to a letter in the S.I.T.U. files. Both sightings were at night, but one was at a distance of only a few yards, and that observer also smelled an overpowering foul odor. The other sighting was at some distance.

There is also a curious story from the Vineland, New Jersey, *Times-Journal*, which carried several reports in late July, 1972, about an eight-foot creature covered with hair that was reported by some teenagers. They said they saw it about 2 a.m. "running near a sandwash on south Orchard Rd." Next day a farmer and his wife told police that they had seen the same thing that morning, walking on Elm Road, near the scene of the original report. The stories attracted crowds of people to the area and also brought out some interesting information, such as:

"Monster experts identified the beast as a Shushquan, a ball-footed, upright mammal native to Canada and highly intelligent."

After a few days, however, the police announced that the "thing" was only a 55-year-old, bearded swimmer who happened to be six feet, five inches tall, and who was in the habit of taking a bath in the sandwash at least three times a week.

"When he heard the voices of the four youths who reported the monster, he got scared and ran into the woods," the story went on. "As he fled he screamed with fright."

That explanation apparently satisfied the newspaper, but it leaves me a little pensive. I don't know what a "sandwash" is, but presumably some kind of sandy swimming hole is close enough. What is meant by "ball-footed" I cannot picture at all. I don't recall ever hearing of a "Shushquan" in Canada, intelligent or otherwise. I can't see how a naked human, no matter how scraggly his hair and beard, could account for the description "hair, brown in color, covered his body." I can't quite imagine a 55-year-old man, of near-giant size, running off screaming because he heard some voices while he was swimming. Finally, what about the sighting report from the farm couple?

It is an interesting thing about the sasquatch situation that the same "sceptics" who laugh off the most straightforward of reports will also swallow without a murmur the most idiotic of "explanations". I don't know if that is what went on here, but it has some of the earmarks.

There were also two reports recorded by the Morristown, New Jersey, police department in the 1960's, according to John Keel. In *Strange Creatures from Time and Space* he tells of four young people seeing a seven-foot monster covered with long black hair in

265

Morristown National Historical Park at dusk on May 21, 1966. It was ambling across a lawn. Thoroughly frightened, they drove to the entrance of the park and began stopping cars and warning people that a monster was loose. One hitched a ride to the municipal hall to report to police, and it turned out that the driver of the car had seen a similar thing in the same area a year earlier but had not reported it.

There are enough stories quoted already to rank New Jersey as a state that has produced a lot of reports for its small size, but that is hardly a beginning. Near Columbia, New Jersey, is Ivan Sanderson's former home and the headquarters of the Society for the Investigation of the Unexplained. Dennis Gates and I called there in late March, 1976, and were most hospitably received by Sabina Sanderson and Steve Mayne and Marty Wolf, cramming an extensive exchange of information into a two-day visit. In January, 1974, Allen V. Noe had written in the S.I.T.U. publication *Pursuit* that a man in northern New Jersey had seen a huge hairy creature standing close to his house in the mountains near High Point. The sighting took place as the man pulled into his driveway on a summer evening. He watched the thing walk around the corner of the house, then ran into the back yard with a flashlight and saw it walk across the yard and enter the woods. The following July, Robert E. Jones, who was then active in S.I.T.U., had an article in *Pursuit* that was entirely about such creatures in northwestern New Jersey. He had seen a story in a local newspaper about a man who claimed to have seen a large hairy biped on two recent occasions crossing the road in front of his car in an area of Sussex County known as Bear Swamp. When another story appeared with a different witness describing a similar encounter, he decided that the matter was worth investigating. By the time he wrote the article he was able to state that:

To date I have interviewed 18 witnesses who have seen something that, assuming this is not a huge hoax, can't be anything other than our old elusive friend, Bigfoot!

The oldest report he had come across went back 60 years, and there had been some in every decade since the 1940's. More than half the accounts described an animal crossing a road, but several were seen in the woods. Most sightings took place at night, dawn or dusk, but there were three incidents, involving five witnesses, in which the things were seen in the daytime.

One involved a contractor who, about five years ago, was driving his truck up a mountain road to do some work on a house. It was around 11 a.m. on a clear and sunny day. While he was rounding a curve on the road the animal stepped out onto the road directly in front of his truck. The driver immediately jammed on his brakes to avoid striking the animal, which was apparently just as startled as the driver, and it turned and stared at the truck, gave an extremely loud scream, then ran off into the woods "faster than any animal I have ever seen . . . At the time of this incident he had never heard of Bigfoot and knew of no unusual animals living in the area.

By far the most impressive story concerned two men, one of them a senior wildlife worker and deputy game warden of 20 years experience, who described seeing a "huge, hairy, apelike yet manlike creature standing in about three feet of water" battling with a large dog. Last spring I had the opportunity to talk to the wildlife man. He said that in the early summer of 1975, the two of them watched for at least half an hour as two dogs harassed a creature standing in the water in a swamp. When they first saw it, he said, all the animals were in the water and they thought two dogs were trying to drown a bear. But the creature soon "showed who was boss," and the dogs backed off, and the creature stood up on its hind legs. Then they could see that it was more like an ape, with a face like a chimpanzee, but shaped like a heavily-built man. It was about six feet tall and they estimated its weight at about 250 pounds. There was brown hair all over it except for a white spot on the throat.

There was a lot of skirmishing, during which it would cuff the dogs with its hands. It "had no trouble banging them around." However it was holding onto something, presumably a dead deer, that the dogs were trying to get. It never did get right out of the water. After watching for a long time they went to the state troopers' barracks, which was not far away, but when they came back with two troopers there was nothing to be found but the dead deer. The troopers concluded that the other men must have seen the dogs fighting with the deer.

I have not told the story in much detail, partly because Bob Jones indicated that he planned to write a book about what he had learned in New Jersey, and partly because the story we were told differed from the version we had previously read, particularly in that only one dog had been mentioned. We did not have time to determine the reason for the discrepancy, or to talk to the other witness. In a way the second reason stems from the first. Had there been no one doing anything about the situation in New Jersey we would have spent some more time there. Since there was already a group of people hard at work we limited ourselves to getting a look at the terrain and talking to enough people, both witnesses and investigators, to have some basis for an opinion on what we had previously heard and what we might be told in the future. From the far west, New Jersey does not seem like a logical place for sasquatch reports. To most New

Jersey residents it probably seems even less so. We saw enough to satisfy us that there was sufficient wild land in northwestern New Jersey for a few large animals to be hanging around there, and a great deal more across the Delaware in Pennsylvania that might support a breeding population from which the individuals could have come.

We also learned that, just as in every other place where we have checked, there are people who appear to be sane and responsible who tell of seeing hairy bipeds under circumstances that rule out any kind of mistaken identification as a reasonable explanation. Not all the witnesses leave that impression of themselves, and not all describe seeing that much, but there are some.

I don't know the latest count of witnesses in Sussex County. In October, 1975, it was 28. Some of them were known to have told their stories to other people before there was any publicity on the subject at all. Bob had half a dozen reports from the year 1975 alone. On February 21 of that year two park rangers, whom Bob interviewed, found two sets of humanlike tracks in the snow in High Point State Park. One track was 16 inches long, the other 18. They took strides from four to six feet long, and covered several hundred yards, not always walking the same route, before reaching an area where there was no snow. That summer two boys cycling at dusk on a trail around a lake near Rutherford, saw a dark brown, shaggy creature almost nine feet tall step across their path. It paid no attention to them.

After Dennis and I were there at the end of March, 1976, I received two more newspaper reports within a few months. On August 3, 1976, the Dover *Advance* had a story about four teenage boys seeing a 10-foot, black, hairy creature in the woods near White Meadow Lake. The story mentioned three-toed footprints 18 inches long, but Bob Jones wrote that there was nothing there that he considered a three-toed footprint. There was a partial print of a foot that he thought had more than three toes. Dover is just east of Sussex County. On August 12, there was a story in the Bernardsville *News*, which is a dozen miles farther south, of three children finding 11 three-toed footprints 17 inches long. It is easy to dismiss reports by children, and with reasons, because some of them do make up stories to get attention, but I have become rather cautious about doing so too readily, for two reasons. First, there is no doubt that at certain ages they tend to spend a lot of time playing in forested areas that other people don't normally go into. Second, over the years I have gone back and talked to a number of young adults who would now make impressive witnesses and who are still telling the same stories they did when they were young.

Another report recently received from New Jersey, although it happened some time ago, takes us back to the Pine Barrens, to the town of Lower Bank. Bob Jones, who now heads a new investigating group called Vestigia, wrote that a couple who lived there in the fall of 1966 claimed to have found 17-inch, five-toed tracks outside their house after they had seen a face looking in the window more than seven feet off the ground. Instead of panicking, they began leaving table scraps outside, mostly vegetables, which something ate. The only thing rejected was a peanut butter and jelly sandwich. Then came a night when they forgot to put out any scraps, and a loud banging was heard outside the house. The husband went out to find out what was going on, and saw a "Bigfoot-like" creature covered with grey hair throwing a garbage can against the side of the building. He fired a shot in the air, without scaring it off, then fired at it, whereupon it fled and did not return.

Most recent report I am aware of comes from the Newton, New Jersey, *Herald*, May 17, 1977 and later. The paper states that members of the Richard Sites family at Wantage, New Jersey, twice saw a seven foot, hair covered creature on their farm. Mrs. Sites and her 16-year-old daughter saw something standing by the road near the rabbit pen on the night of May 12, and her husband and several visitors ran out and saw "a big shadow—his head as high as the eaves. When my daughter screamed it took off." Next night a group of them waited for the thing, and when it arrived opened fire with two shotguns and .22 rifles. It ran into the chicken coop, emerged from a window at the opposite end and escaped through the apple orchard. Although it was always on its hind legs police said it was probably a bear.

A later report from Bob Jones confirmed the story with some changes in details, but the most interesting thing he learned came to light when he had a veterinarian examine a number of dead rabbits found where the thing had been. Four large ones had broken bones in the head and neck area and in the hind legs, and one had the head torn right off. Three smaller ones had shattered bones in the midsection only. No part of any of them had been eaten.

The stories from much of New Jersey don't add up to anything special, but Sussex County is another matter. If there continue to be even two or three substantial reports a year from so small an area, and if a substantial number of people continue working on the problem, it has to be considered one of the most likely areas anywhere for some kind of breakthrough.

Another area that seems even more promising is located in the central section of the west coast of Florida, the third eastern state that Dennis Gates and I wanted to look at. For about 10 years I kept my

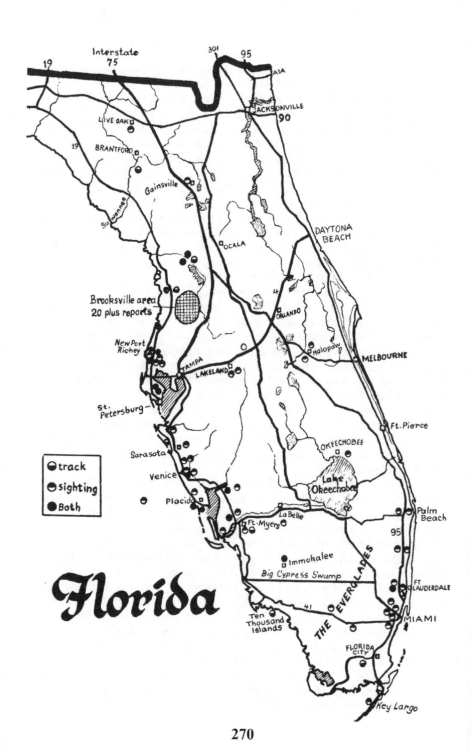

Florida

card file divided in five parts, a section each for British Columbia, Washington, Oregon and California, and another one for all the rest of North America. Most of the reports in the overflow section were from Alaska, Idaho and Montana, and the total number of reports in it was consistently less than in any other section except Oregon. That was still the situation in 1973. By June, 1976, I had divided that section into four, and reports from the eastern U.S. alone outnumbered those from Oregon and almost equalled those from British Columbia. By December 1, 1976, I had cards on more reports from the eastern U.S. than from British Columbia, but by then I had divided the file again to make a separate section for the Southern States, and another one for Florida alone. The Florida file by then had 87 reports in it, and only eight of them were more than 10 years old. Less than four years earlier I had a total of only three. For 1974 and 1975 I had more reports from Florida (32) than from any other state except Washington(34). There probably isn't a great deal from Washington that I don't learn of fairly quickly. I doubt that I am in touch with half of what happens in Florida.

Where have all those reports come from? There are several from the southeast coast near Miami, several more from central and northern areas inland, none at all from the western panhandle, and a mass of them all the way down the east coast from Crystal River to the Ten Thousand Islands, with the heaviest concentration a bit inland from that coast in the vicinity of Brooksville. Most of the information has come to me through the work of other investigators, but almost the oldest report I have, and they go back only to the 1940's, is in a letter to me from a resident of Lakeland, about 30 miles east of Tampa. The writer, when a child of four in 1947, got the fright of a lifetime from encountering at close range an upright animal with an almost human face standing under an orange tree behind their home.

The only other definite report I have earlier than the 1960's came via Loren Coleman from a resident of Deer Park, Texas, who wrote of an incident that happened while he and two other men were hunting in the Big Cypress Swamp in the spring of 1957. They were sleeping in hammocks slung between the cypress trees when they were awakened by heavy footfalls and splashes and breaking branches as something approached.

Only seconds passed when we were unzipped out of our hammocks and standing on the cold damp ground fumbling for our pistols. We crouched down with the fire at our backs to try to see what our visitor was . . . It wasn't a bear for sure. The Florida bear is small and didn't walk on two legs like this thing did . . . As the heavy footsteps came closer,

271

they stopped. Just about 30 feet away. Then we could make out a dark silhouette standing behind and in between the trees. The shape of a large man. First thought was that it was a big Seminole Indian . . . this creature was too big . . . We studied the form for about what seemed two minutes, then 'it turned and started away from us. As the thing turned, the head had a different shape, sort of a slump with what also seemed to be a heavy chin. Couldn't see any signs of clothing, just all one dark color. Only head to knees were visible. I wasn't sure if I saw glowing eyes, but Pug said he did from his position by the fire. Knowing that animals have reflecting eyes, was a thought that sent our blood racing."

Big Cypress Swamp is a huge area on the west side of the Everglades, crossed from east to west by only two main highways, the Tamiami Trail and Alligator Alley. Its name figures prominently in some later sighting and footprint reports that have received quite widespread publicity, but I am told that those events actually took place on an island off the west coast north of Fort Myers Apparently the people involved wanted the true location kept a secret.

John Keel in *Strange Creatures* had a report of a sighting on a ranch near Holopaw, on the east side of central Florida, in 1963. Then from the mid 1960's he had several reports from the area north of Tampa. One of the witnesses at that time was Mrs. Eulah Lewis, of Brooksville, whom Dennis and I called on in the spring of 1976. She said that she had gone outside her mobile home to water a little tree one evening, when she heard a sound like a crackling paper nearby. Looking up, she saw a thing about six feet tall and covered with dark hair standing looking at her. It had no neck and huge shoulders, and looked to her to weigh about 300 pounds. She ran for her front door and it ran there too, from a different angle, but she arrived first and shot the bolt behind her. During the night she and her husband heard noises outside as if two creatures were calling to each other, and next day her husband found big tracks in several places near the house. Three weeks later her daughter-in-law, who lived nearby, had a similar experience. There were other reports around Brooksville and a few miles to the southwest, on the Anclote River, in one of which the creature reportedly paid absolutely no attention to a dog that was biting at its ankles and feet.

If any of those reports were in the newspapers, I have no record of it. The oldest clipping I have is from 1971, when the "Skunk Ape" began finding its way into the headlines. It is from the November 7 *National Enquirer*, and I didn't take it too seriously at the time. The key incident was very similar to that reported from Big Cypress

Swamp in 1957. In February, 1971, a group of archaeologists camping on a huge, low Indian mound heard heavy footsteps approaching about 3 a.m. and looked out of their tent to see what appeared to be an eight-foot man standing just a few feet away. He was covered from head to foot with light brown hair, and he smelled awful. The next day five-toed footprints were found, 17½ inches long and 11 inches wide at the toes. They made a plaster cast of one. The only witness named was H. C. "Buz" Osbon, president of the Florida Peninsular Archaeological Society.

Next day the group was joined by L. Frank Hudson, vice-president of the society, who later became co-editor of the *Yeti Newsletter*. The following weekend and on later occasions more tracks were found, and Frank Hudson eventually saw one of the creatures himself. In December, 1971, he wrote to Ivan Sanderson about his experience. Those were the incidents that were said to have happened in Big Cypress Swamp but were actually elsewhere.

Other newspaper reports came from the Naples *Daily News*, in August, 1971. The incidents were in the Fort Lauderdale area, where residents of a trailer court formed a posse to hunt for two apelike creatures described by a pair of young children. There were several more reports of similar creatures in the vicinity.

On March 26, 1972, the St. Petersburg *Times-Floridian* carried a feature article on "The Abominable Florida Apeman" which stated that in Dade County the creatures were called "Squattam's Growlers," and that they were known as "buggers" to commercial fishermen and "The Sandmen" to Seminole Indians. The story noted that such creatures were reported more often from the Brooksville area than anywhere else. After recounting a number of stories, in which apemen (one at a time) yanked a sleeping truck driver out of his cab, ran beside a car at 80 miles an hour, and jumped on the running board of a car, almost turning it over; the authors gave a composite description of the creatures, as follows:

> He is a hair-covered creature eight to nine feet tall, except when he is smaller than that. He grunts, like a wild pig. He reeks, like a skunk. He has green eyes which shine in the dark like a panther's. His arms are long, like an ape's.

While all that activity did not attract much attention among the sasquatch fraternity on the west coast, it certainly did not pass unnoticed in Florida. Sometime about 1972 the Yeti Research Society was organized in St. Petersburg and began publishing *The Yeti Newsletter*, with L. Frank Hudson and Gordon R. Prescott as co-editors. They carried reports from all over, but the majority were from Florida. I had heard of the organization at the time, but assumed it was probably another of the many in-name-only operations with which the history of the sasquatch investigation is littered. Later, when I was shown a file of the newsletter by a friend I was not prepared to accept very many of the reports it contained. Most of them lacked specific detail, and I mistakenly considered it highly improbable that there could be so many stories from Florida. By the time I went to Florida in 1976 the newsletter was no longer being published.

One of the things members of the society used to do was to watch at night in places where "yetis" had been reported, waiting for them to show up, and several people claim to have been successful at this. I don't have any reason to doubt their word, although it would have been much more helpful if they had something to show for it. In 1974 I began corresponding with Ramona Hibner (then Ramona Clark), who had just resigned from the society and was not impressed with it. However she seemed to rattle off sighting reports in quantities equal to those in the newsletter, and she said that she had seen the things herself on more than one occasion, and had found footprints in several sizes up to 19 inches, the toes of which dug in very deeply, always deeper than the heels.

She reported that the creatures never wander very far from water, that they apparently eat small animals and various types of vegetation like cactus pulps, water lily bulbs and oranges, and that despite the name "skunk ape" they sometimes have no noticeable smell. She suggested that the smell comes not from the animal itself, but from rolling on rotting carcasses: "It will actually gag you, it is so disgusting." Her dog sometimes did the same thing, she said, and the smell was the same. On another occasion a yeti she was close to had just an earthy smell, not at all unpleasant. Plotting the reports around Brooksville on maps she had found that a lot of them were concentrated in an oblong area of about 4,000 acres. Along with the general information and the old reports were also statements like the following:

This past week has been an exciting one. A lady who lives about four miles from me saw one of these creatures sitting in the middle of a dirt lane, she said it was so wide it filled up the road and was black, completely covered with hair. It had its back to her, just sitting there! She turned the car around and went to a gas station to calm her nerves. Her companion also saw it. They smelled a rotten odor, very strong. Her home is about one mile from where they saw the creature.

You would almost have to be a Pacific coast sasquatch hunter to appreciate how unbelievable that sort of thing sounds. In this part of the world only the luckiest or most persevering can expect to see tracks that someone else didn't find first, and no one has any realistic expectation of ever seeing the creature that makes them. I don't know how many reports there were in a year in that 4,000-acre area near Brooksville, but it would certainly be more than in any 4,000-square-mile area in the far west. Combined with the stories that there were hundreds of reports from western Pennsylvania, even though we weren't receiving any details of those, the Florida information began throwing the whole picture badly out of balance. If it were true it indicated that the logical places to settle the mystery were in the east, not the west. But with so much channelled through a very few sources, it might be some kind of a put-on. And if there was anything doing in Florida at all, why did Bob Morgan, one of the few people who was ever able to raise money to put a substantial number of people in the field, always take his group 3,000 miles across the country to Mount St. Helens to look for sasquatches?

If they were true, the eastern reports certainly could not be ignored, but if they weren't true anyone who took them seriously was going to end up feeling pretty stupid. A tour of the east was obviously necessary for anyone who hoped to have a reasonably accurate grasp of the overall situation. By the time Dennis and I made the trip, Ramona had married her yeti-hunting partner, so it was Mr. and Mrs. Duane Hibner that we called on at Brooksville. They had quite literally made their home there to be near the yetis. We soon found that whether the information was accurate or not it wasn't anything the Hibners and a few others were making up. Local people didn't necessarily believe in the yetis, but they had certainly heard about them.

Other questions that were answered immediately were whether there were large enough wild areas to support a population of such creatures and why there were so many reports now when there had been virtually none earlier. The same answer covered both. The wild areas were there alright, both swamp and dry forest, but they were disappearing. Locations where encounters like Mrs. Lewis's had taken place in the 1960's were in the middle of town in the 1970's. Huge subdivisions, while by no means threatening to drive out all the wild-life, would plainly be putting pressure on any type of wild creature that had to have its own territory away from others of its kind, and might also be affecting the food supply of any large animals. The pattern of the yeti reports fitted well with that line of speculation. A lot of sightings had taken place at specific locations where the creatures were said to be seeking food in dumped garbage or going after small livestock like rabbits and chickens. Those activities were sporadic, and unfortunately nothing of the sort was known to be going on while we were there. Under those circumstances the idea of people being able to sit up nights and wait for the creatures seemed

275

much more acceptable. It also appeared that none of the people who had been doing the observing so much as owned a heavy rifle, or had any other equipment that could be expected to produce some sort of evidence that the sightings actually took place.

It turned out that there had been some reports the year before 100 miles down the coast near Venice, that the Hibners hadn't found time to look into. When you have just driven from British Columbia to Florida by way of Los Angeles, 100 miles seems just around the corner, so off we went. There we talked to a 12-year-old boy who, the preceding summer, had seen a six-foot creature covered with dark hair trying to get at his pet ducks right behind his house. It had damaged a heavy chain-link fence. His family hadn't seen it, but they confirmed that he had run in the house screaming that there was an ape outside.

A clipping that Ramona had stated that a man named Mike Corrodino, who operated a monkey sanctuary, had tried to track the night raider and had dismissed out of hand any suggestion that there was such a thing as an upright ape in Florida. We found Mr. Corrodino living just down the road from the boy's house. His property contained several cages full of monkeys, some gibbons and even a large siamang. It turned out that he was in the business of capturing monkeys and apes that had escaped and were living wild. He also edited a newspaper in Venice. We learned that not long after he started his investigation of the neighbour boy's story he had completely reversed his opinion and was now entirely convinced of the reality of the skunk ape. Not only that, he had seen one himself, crossing the road near his home. Here is how his paper, *The Sun Coast Times*, told the story of the boy's experience in a feature story published January 17, 1976, under the heading:

WILL THE SKUNK APE STRIKE THIS YEAR?

At about 10:30 p.m. on Saturday, June 7, 1975, 12-year-old Ronnie Steves lay in his bed, just drifting off to sleep, at his home on Jackson Road, east of Venice.

Suddenly he was awakened by the sounds of his ducks thrashing about in their pen just outside his bedroom window. With sleep-fogged eyes young Steves looked out the window into the enclosure to see a vague, misty gray figure moving about in the cage, which was surrounded by a six-foot-high heavy chain-link fence.

By the time Ronnie got out of the house and around to the cage, the figure was gone, so he ran to another, smaller duck pen south of the house to check those animals.

There it stood: Just three feet away, leaning almost

casually on one of the cage posts and staring down at the frantic ducks, was a six-foot-tall, dark, hairy, apelike creature.

"The next thing we knew, Ronnie ran in the house like a shot, yelling there was an ape out there," the boy's father, Ed Steves, recalled.

The creature also took off, heading east toward the Lazy T Ranch.

Representatives of the Saratoga County Sheriff's Department and the Florida Monkey Sanctuary (FMS) were notified and they began a search for the "ape-man", along with a number of neighbours.

Scanning the area with flashlights, the investigators found several fresh tracks beside the ditch and others following a fence line. The tracks measured six-to-eight inches and had a discernible arch.

FMS Executive Director Michael Corradino—now editor of the Sun Coast Times—said the prints resembled those of a chimpanzee but did not have the chimp's characteristic thumb-like big toe.

Plans were made to return and make plaster casts of the tracks, but heavy rains ruined them during the night.

The elusive creature vanished with no other traces of its presence, but both Deputy Dennis Bosze and Corporal William Gordon of the sheriff's department agreed that something had been in the area.

Young Ronnie Steves said later that the creature was definitely "like an ape" and was "about as tall as my father (six feet)." He didn't really get a good look at the creature's face, but he did notice that it had shaggy-haired arms that appeared to be brown.

A check of the fence surrounding the large duck pen showed evidence of a visit by a large and apparently powerful creature. At one place the heavy-gauge metal chain-link was bowed out, and two other small areas were bent out of alignment as if from the pressure of a strong hand-grasp.

Two other accounts of a similar creature in the same general area were reported the previous week.

With that incident, young Ronnie Steves joined an ever-growing number of south Florida residents who have claimed seeing what has come to be called a Skunk Ape or Swamp Ape, which some think is the Florida-grown variety of a creature that is also known as the Abominable Snowman, Yeti, Sasquatch or Bigfoot . . .

In recent years, however, more such encounters have come from residents in all walks of life in more populated areas of south Florida.

As of a year ago, Venice seems to be one of those areas.

A flurry of sightings and strange occurences took place in the Venice area between January and June of last year. It was a period of such activity that coincided with similar happenings throughout southwest Florida . . .

We found that in Venice as in Brooksville the sightings would go on for a while in the same area and then stop. A reasonable explanation seemed to be that because of the reduction in forested area there might be periods during the year when the animals were short of food. There was also a report in the summer of 1975 farther down the coast at North Fort Myers, and in the summer of 1976 there were again reports at North Fort Myers and at Brooksville, while in the Venice area, Ramona reported, Mike Corrodino had been on an open-line radio program and had received hundreds of calls. There were also reports north of Brooksville, near Dunellon, and to the east, near Orlando. At North Fort Myers the Hibners interviewed two boys who said they had been chased, in the summer of 1975, by a skunk ape that smelled terrible and ran on all fours in a crablike fashion as well as running upright. When first seen it was eating a bird.

Among so many Florida reports there are quite a few of special interest. One from the *Yeti Newsletter* which Ramona confirmed, took place a few miles north of Brooksville, where a five-year-old boy who had been playing with a tricycle ran and got his older brothers and sisters because two animals had come out of the woods and stood watching him. He had not run until the small animal climbed the fence towards him. The older children reportedly saw a baby skunk ape spinning a wheel of the overturned tricycle. When they approached it went back over the fence and the mother hid it in a bush. She was about seven feet tall and covered with dark hair. There was no noticeable odor.

On January 10, 1974, the Miami *Herald* reported that Richard Lee Smith, who lived on State Road 27 near Hollywood, had told police that just after midnight he had run into something on the road "huge, seven or eight feet tall, dark-colored and human-like." He said the thing, resembling "a gorilla, a strange creature of some sort" had suddenly jumped in front of his car. After it was hit it ran limping into the Everglades. There was another report of one hit by a car near Gainsville the following February.

Gordon Strasenburgh turned up perhaps the best story of all, from the files of the Dade County Public Safety Department:

SKUNK APE

Dade County Public Safety Dept., Miscellaneous Report #72168-1.

Reported by: Bennett, Ronald, 46 w/m, 2820 SW 106 Ave. 226-3619.

"Suspicious Incident at Black Point-Goulds Canal, 12:00 a.m. 3/24/75".

Police dispatched: 2:26 a.m.; Arr: 2:31 a.m.; In service: 4:22 a.m.

Remarks:

The reporter stated that at the above time, date and location of his son Michael Bennett and a friend Lawrence Groom w/m 54 yrs 223-0108, while driving down a dirt road toward Black Point near the water dyke they observed what appeared to be a giant ape-like man, approximately eight (8) to nine (9) ft. tall and very heavy set, black in color with no clothes, standing next to a blue chevy and rocking the car back and forth with great force. The witness further stated they observed a man getting out of the vehicle in a hysterical manner and yelling for help. When the lights of the vehicle that the witnesses were in lit up the ape-like man he turned and ran into the mangroves. The witnesses stated they could hear the thing running through the mangroves. At this time the witnesses' vehicle was turned around and leaving the area. Upon departing they did not see where the man in the blue chevy ran to.

South District Sta #4 was notified and responded to the area. A search of the area for the blue Chevy and the ape-like man produced negative results.

Note: The location of the incident can be variously described as the eastern end of 248th Street SW, 87th Avenue SW and Biscayne Bay, and Snapper Point.

Comments: To get from the location of the sighting to the Bennett home would take no more than an hour, and probably thirty minutes. So it is clear that the incident was discussed at least an hour and a half before the elder Bennett decided to call the police. Mrs. Bennett told me that her son began his story with the predictable: "Mom, you're not going to believe this, but . . ."

With ample publicity in recent years, the fact that people constantly report seeing apes in Florida must by now be quite well known, but there is no sign that it is yet being taken seriously. An attempt by Representative Paul Nuckolls from Fort Myers to have the state legislature protect skunk apes from molestation with a one-year jail term and a $1,000 fine was defeated in June, 1977—"shouted down" according to the St. Petersburg *Times*. But if things keep happening at the present rate it seems likely that something decisive will have to happen before long, either through a change of official attitudes or through someone killing one of the creatures or finding the remains of one.

Throughout 1977 reports have continued to come in. The Miami *News*, February 18, 1977, had been told of several received within the past week by Delray Beach police lieutenant Lorenzo Brooks. Most interesting was the story told by the superintendent of a golf course west of the town who said he had seen a black, hairy seven-footer down drinking from a lake near the second tee. About the same time Duane and Ramona Hibner were investigating reports from people at Moon Lake near Port Richey, where there had been several sightings of gorilla-like creatures. Three members of one family saw three of them in the yard of their home, the largest 10 feet tall.

On March 31 the Boynton Beach *News Journal* had a report of yet another sighting at Delray Beach, by two teenagers who roared past one on a motorcycle at night on an isolated road. David Smith and Bob Sauer said they thought at first there was a tree on the road. By the time Smith realized it was an animal it was too late to do anything but accelerate past it, which he did, while the thing just walked calmly on down the road. They estimated its height at 10 feet. It had strong, thick legs but otherwise was not heavily built. Its hair looked slick, as if wet.

In April the Hibners reported two more sightings at Moon Lake. In one, a nine-foot black creature was seen in the light of a full moon, leaping a dirt road in one stride in front of a truck with three men in it. In May they talked to two men at Nobleton who claimed to have seen a big, hairy upright creature in a back yard fighting with some dogs. On May 26, the Belle Glade *Herald* quoted four witnesses who saw a black seven or eight-foot skunk ape cross the road on the morning of May 23 as they were starting to walk to civilization from where their van had broken down on State Route 29, 10 miles south of LaBelle.

In July the action shifted, for the first time that I know of, to the Florida Keys. Here, in part, is how the Marathon *Florida Keys Keynoter* told the story on July 28,

> KEY LARGO—For Charlie Stoeckman, a skunk ape is no laughing matter. He thinks one has a personal vendetta against him.

The reason, he figures, is that on July 14, he and his son
Charlie Jr. were out bottle hunting in a desolate stretch of
woods adjacent to his oceanside home between mile marker
94 and 95, and removed a lignum vitae tree stump.

That was the first time they saw the skunk ape, then just a
blur through a small opening in a thicket of mangroves.

The same night, the skunk ape approached their home,
Mr. Stoeckman says. Charlie Jr. saw it first, looming about
eight feet tall, a silhouette against the darkening sky,
approaching the rabbit hutch outside the 13-year-old youth's
bedroom window.

Charlie Sr. saw it then, too. Before he could get a good
look at it, it fled.

It didn't return until a week later, on July 21. It was about
9:30 p.m., as he and a neighbour, Sawn Tubbs prepared to
go out lobstering. They saw it running across a field as the
lights of Mr. Stoeckman's pick-up truck shone on it . . .

Saturday, a posse that included Sergeant R. Chinn and
Deputy John Fay of the Plantation Key sheriff sub-station
and Capt. Jack Gillen of Pennekamp State Park, combed the
woods for 90 minutes looking for the animals. They found no
evidence of it.

But Sgt. Chinn said Monday, he would not discount the
existence of the skunk ape, or some other exotic animal.

Citing the fact that a neighbor, Mr. Tubbs, as well as Mr.
Stoeckman and his family had seen the skunk ape, Sgt. Chinn
said he "would not disbelieve the case, because the people are
realistically scared. Something has them scared to the point of
thinking of moving out.

281

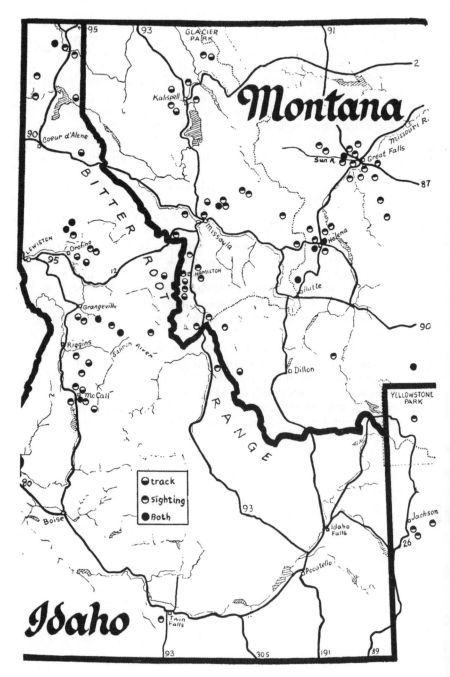

Track
Sighting
Both

Montana

Idaho

-15-

Along the Great Divide

Eastern Montana is part of the Great Plains, so you would not expect sasquatch reports to come from there, and they don't. Southern Idaho is very mountainous, but like the adjoining Rocky Mountain States to the south and east it too has produced virtually no information. Northern Idaho and western Montana are just the opposite. North from the middle of the Bitterroot Range, where it separates Montana from Idaho and the Columbia watershed from the Missouri—just about the spot where Teddy Roosevelt's mountain men had the fatal encounter with the monster that walked on two legs— there are many reports from both sides of the continental divide.

Besides the Roosevelt story, I know of only one other old written report from that part of the United States and it is a vague one. From Tim Church I received a copy of a story from the New Haven, Connecticut, *Evening Register*, November 11, 1892, quoting the Anaconda, Montana, *Standard*, about a man who claimed to have seen a creature covered with hair, but in form not unlike a man, in the mountains on the Wyoming line. The story referred to "the creature's habit of rising on its haunches and walking on its hind legs after the manner of a gorilla," but there is no more specific information. In addition to that, on an open-line radio show Rene and I once received a phone call from a woman who said that in the Snowy Mountains in Montana when she was a child, in the early 1900's, everyone knew that the cattle had to be kept at lower elevations in late fall and early spring or the hairy men that lived in the mountains would kill them and eat them. She said that her father and another man had seen a tall, hairy creature walking upright, and that one her uncle saw ran away when he reached for his gun. There are Big Snowy Mountains just about in the center of Montana and a Snowy Mountain in the northwest, so I am not sure which she meant, and I have never heard a similar story, from anywhere.

From Idaho there is one second-hand report dating in the 1930's. A woman wrote to me that members of her former husband's family told of returning to their ranch to find an animal kneeling on a bench looking in a window. At their approach it got down from the bench and walked away upright. That was near Priest Lake, at the northern tip of the state, and she said that when she lived there herself in later years she had often heard screams that were different from cougar screams.

The next report from Idaho that I have is not until the middle of the 1960's, but from then on there are many solid ones. June 27, 1968, the Grangeville *Idaho County Free Press* carried the story of Frank Bond, from Council, Idaho, who claimed to have seen a pair of humanoids, seven and eight feet tall, covered with silver-grey hair, while he was fishing alone near French Creek on June 16. The owner of the property, game guide and former government trapper Wayne Twitchell, checked the location and found half a dozen light-colored hairs, which the newspaper sent to Ray Pinker in Los Angeles, a veteran of 36 years in police crime-lab work who was then professor of police science at California State College in Los Angeles.

I read at the time a newspaper story attributing to Professor Pinker the statement that the hairs showed both animal and human characteristics, so I got in touch with him. He said that the hairs did not match any samples that he had, that they resembled animal hairs in showing a variation of color and thickness from the root to the tip, whereas human hair is uniform in color and thickness, but that the scale pattern on the outside was similar to that of human hair, and that there was no continuous medulla, or core, visible in the center as would be the case with almost all animal hairs. There were both coarse outer hairs and fine hairs from an undercoat. All were light in color, so he asked me if I could send him samples of white hairs from the known animals of the area, and with the help of Charles Guiget, the curator of mammals at the British Columbia Museum, I was able to do so. Later I went to Los Angeles, and on the way south I picked up samples of all the suspected sasquatch hairs I knew of, hoping that they might all prove to be the same as the Idaho sample.

A few hours spent in the lab with Professor Pinker pretty-well wiped out my interest in hair samples. I was already aware that the method used to identify hair was to match it in a comparison microscope with a known sample, but I had not realized what a time-consuming process that was. Most of the hairs I had brought with me had first to be mounted on slides, and then we found that none of them matched the Idaho hairs and no two of them matched each other. The prospect was that after many hours of painstaking comparison all of them would be identified as belonging to different known animals, leaving the Idaho hairs the only interesting ones. The alternative was that one or more of the other samples would prove unidentifiable, and then we would have two or more types of hair that were unidentified but different. It seemed to me that neither outcome would begin to justify the amount of work involved in reaching it—and I wasn't even going to have to do the work. Professor Pinker agreed with me. Another thing that made me question the whole operation was his fairly diffident suggestion that one sample of hair on a slide already labelled "Coon" looked to him more like bear hair. Perhaps it was just politeness, but he didn't make it sound as if he was certain a mistake had been made. The hair in question was one that was already mounted when I got it, and the slide bore the surname of its owner, Ken Coon.

There are a couple of other points regarding hair identification that have to be kept in mind. One is that nobody is likely to have a sample plucked from the animal itself. The suspected hairs are normally found by searching around in a place where a sasquatch has supposedly been. There is nothing to establish that some other animal could not have left hairs there, whether a sasquatch was actually there or not. Hairs will last a very long time and they are often very hard to see. The other thing to keep in mind is that the incident may be a carefully-staged hoax, so there is no use comparing only hairs from animals that could normally be expected to be in the area. Any exotic variety of hair from anywhere in the world might be available to the hoaxer. Whether there is any place in the world where samples of every type of hair from every part of every animal are available for comparative study I very much doubt, and if such a collection does exist there is no prospect that any qualified person is going to be available to work his way through it every time a sasquatch hunter shows up with a prize clamped between thumb and forefinger.

It would be a different thing entirely if there were standards by which specialists could establish just by study of the hair in question that it had to come from an animal in a limited group—so that they could say, for instance, whether a hair was or was not from a primate. If it were possible to find hairs that were definitely from a primate but not from any known primate, that would be real progress. Apparently that cannot be done, so what is really needed—as Professor Pinker told me— is a sample of known, proven, sasquatch hair. If anyone is ever able to provide that, it will be possible to establish in short order whether suspected hairs come from a sasquatch or not—but there will not be the same need to know. In the meantime, if anyone wants to collect suspected hairs, more power to them; and if anyone who is in a position to try to identify them is willing to do so, that's just fine. I see no real prospect, however, that any results will take matters farther than the study of the hairs from Idaho—unidentified but with similarities to both human and non-human samples.

Many of the Idaho reports I know about have come to me from Russell Gebhart, of Lewiston, who has been an active investigator for quite a few years. He notes that there has been a whole series of reports from a wide area both north and south of the lower Salmon River in recent years, the earliest being in the fall of 1966, when two elk hunters found tracks in the snow about 6,000 feet up on a mountain near Burgdorf. There were two sets of tracks, one estimated at 16 inches long, the other only eight to ten inches. They encountered them about a mile apart, but both were headed downhill and before they could get down out of the snow they had meandered across each other several times with no indication of joining up. One of the men, from Riggins, told Russ that after his own experience he had made enquiries and found that there were other people who had seen similar tracks but said nothing about them.

In the summer of 1968, besides the French Creek incident, there was a report from farther north on the Clearwater River near Golden, where something entered the camp of three prospectors, left tooth marks on a coffee can indicating a tremendous spread of jaw, and made off with a five-pound sack of pancake flour, apparently eating it as it travelled. They found tracks, not too distinct, which were twice as wide at the toes as at the heel. One of them in a letter to me estimated the length at 14 inches. Another, in a news story, is quoted as saying "about a foot." Both likened the shape to a swim fin. There were no claw marks. Two of the men at another time got a look at a cinnamon-colored creature as big as a bear that did not move like a bear as it took off. That fall a logger driving a skidder saw a sasquatch just north of his home at McCall, according to a letter he wrote to Roger Patterson.

In the summer of 1970, according to what they told Grover Krantz, a father and son from McCall, with about three other men, saw a black, heavy-set biped about a half mile away and studied it through field glasses until it noticed them and moved off. They estimated its height at seven feet. Grover talked to them in 1973, and the father said that in the fall of 1972 he had driven past another of the creatures, about six feet tall, heavy-set and light brown in color, on a forest road. Both sightings were in daylight. Grover was told that a number of other people around McCall had had similar experiences, but did not talk about them.

Russ Gebhart talked to two men in Peyette, Idaho, who had reported seeing two sasquatches while hunting near Dixie in September, 1972. They had spotted them first in their headlights while driving back to camp, and backed up to get a good look. A short time later they saw them again by flashlight, in the brush, less than 100 feet from their camp. Both were completely covered with reddish-brown hair and were around eight feet tall. They were eating something, the men could not tell what, and shots fired in the air did not frighten them although they slowly moved away. Their eyes reflected bluish-green. The next day the men found two tracks near a stream, flat, about 18 inches long and 11 inches wide, with only four toes.

In October, 1974, six or seven miles south of Grangeville, two youths hunting deer found a half dozen 15-inch, five-toed tracks in deep dust on the edge of an old skid road, and made a cast of one of them but got poor results as the plaster was too thick. That December, back at McCall, a woman told Russ, her son and another boy playing near a pond about a quarter mile from her home saw a heavy animal seven or eight feet tall that was "standing upright flopping its arms and screeching" at them from about 50 feet away. They ran back to the house. That story and others he investigated turned Deputy Sheriff Barry Brightwell from a sceptic to a believer, according to the Boise *Idaho Statesman*, March 13, 1977. The following September, in the mountains between the Salmon and Clearwater Rivers south of Golden, several youths watched a dark figure below them near Kelly Lake, about three quarters of a mile away. One of them told Russ that it seemed to be much larger than a human, but it kept moving around on its hind legs. Although it did not go anywhere, it was seldom still. They watched it for more than a hour, until nightfall.

In July, 1976, there was another sighting somewhat farther east, about 18 miles southeast of Elk City. A woman from Weippe, Idaho, told Russ that during a family camping trip she stayed with her granddaughters while her daughter went on farther up an old road by Seigel Creek. Her daughter returned hurriedly and said that from a distance of about 70 yards she had seen a creature about seven feet tall, covered with short, dark brown hair. It was standing with its arms around a tree. As she hurried away she looked back several times and saw it paralleling her path higher up the hill. It had a flat face covered with shorter, lighter hair, and it walked completely upright. That December, Mrs. Virgil Donica and her 15-year-old son

287

saw an eight or nine-foot black creature walking up a hill near their home just two miles from McCall, about 8 a.m. Russ talked to her, and this story was also publicized in the Hamilton, Montana *Bitterroot Journal*, May, 1977.

Perhaps the most detailed description Russ had recorded was given by a Lewiston man who told of seeing a sasquatch at Poet Creek campground about 25 miles east of the Red River Ranger Station in July, 1975. The man had slept in his car and woke up about 4 a.m. He was headed for the restroom when he saw the animal walk out in front of him. It never looked his way, but apparently became aware of his presence and walked rapidly away with a very smooth, fluid gait. Russ wrote:

He described the animal as approximately eight feet tall and weighing around 800 pounds. It was covered with long, heavy, shaggy dark grey (not black) hair. Its arms were long and its legs slightly short. It had hands similar to human hands but he didn't notice exact features although they were covered with short hair. Its neck was short and as wide as the head. Its face was flat and the forehead was very sloping. He could not see ears but there were lumpy-looking places where they should have been under the hair. It was built very massively but not fat. It appeared to be extremely strong. He detected no sounds or smells. He did not hear brush or twigs snapping but could hear the ground crunch under each step.

He described its posture as very peculiar and this feature seemed to catch his attention most. Its legs were always bent, as if in a slouch and its buttocks stuck out far to the rear. Its back was in an extremely pronounced swayback position when it walked. When it came to attention upon sensing his presence its back became almost straight but its legs remained in a slouched position. When it walked its arms swung similar to humans' and did not come to the ground.

All those reports came from a fairly limited area near the Salmon River, and there have been several sets of large tracks seen there in recent years as well. Another, smaller, area where Russ Gebhart has investigated several reports is around Orofino, about 40 miles east of Lewiston. The first of those was in June, 1969, and he wrote a report on it early in 1971 for the *Bigfoot Bulletin*, the newsletter published by George Haas at Oakland, California, from January, 1969 until June, 1971. Russ had talked to a man at Orofino who was the night watchman at a small sawmill in the forest 20 miles north of the town.

The animal was described as being about six feet tall, completely covered with shiny dark hair except for its face, hands and the nipple area of its very large breasts. All visible

288

skin was pink. He thinks it may have been nursing a baby. The eyes were large and fiery red. It was built like a very heavy, stocky person except that its arms were very long. It walked upright, just like a person, and moved gracefully. It appeared to have great speed when wanted. It had a very disagreeable odor which he could not describe but would never forget or mistake for anything else. He observed the animal for about five minutes from around 35 to 40 feet.

He and several other employees had seen many tracks in and around O Mill all during the summer of 1969. There were at least three different sets of tracks, one very large set, one human size and one child size. The tracks indicated that they had explored in and around and under the buildings. They played in the sawdust pile and ate sandwiches he had put near the carriage. He could hear them jabbering among themselves and throwing boards which got in their way. They were in and around the mill most of the summer and he saw a very large, enormous dog once which he believes was running with them.

Another man whom Russ, Grover and I have all talked to, tells of seeing sasquatches twice on his farm near Orofino. The first time was in the fall of 1972, early in the morning. The creature walked across an open field that slopes uphill behind the house. It was in sight for about a minute at a distance of 50 to 75 yards, and he watched it step over a four-foot fence without breaking stride. Although it seemed to be walking casually, its long arms were swinging widely and it covered the ground very quickly. He estimated that it was over eight feet tall and extremely stocky, with shoulders four to five feet wide and a body very deep from front to back. It was covered with long greyish white hair that seemed very thick.

The second sighting was in the afternoon, in June, 1973, and the thing was farther up the hillside. This creature had the same heavy build, but he thought it was only about six feet tall. Russ and Grover talked to him only two days after the second sighting. The ground was too hard to show tracks.

There was another sighting in August, 1973, on a mountain east of Bovil, Idaho, which is about 30 miles north of Orofino. A man from Lewiston phoned Russ to tell him that at about 10 in the morning on the second day of a camping honeymoon he and his bride saw something in the thick brush that they estimated was about 10 feet tall. They did not get a good look at it, but it was dark in color and appeared to be very heavy. The following May, Russ heard from a cougar hunter who had found huge tracks of an apparently bipedal animal in January, 1974, in deep snow near Weippe, a few miles

289

southeast of Orofino. He did not follow them far because they were going through thick brush.

Another sighting, about 18 miles north of Orofino in October, 1975, is accompanied by a super-8 movie. The film shows so little, however, that it depends for credibility on the story that goes with it, rather than the other way around. An Orofino man says that on October 24, 1975, he was hunting elk early in the morning and had parked his pickup on a logging road while he used an elk call. He heard something coming and waited outside the truck, rifle in hand. Considerable snow was falling. Instead of an elk, a sasquatch appeared and stood by a bush about 60 yards away. The hunter put his rifle away, got out his zoom movie camera and took a few seconds of pictures, moving around to try to get different angles of the motionless figure. At a time when he was moving and the camera was not trained on it, the creature stepped out of sight. Russ, Grover and I saw the movie the following January. There was no apparent reason to doubt its authenticity, but a combination of distance, poor light and large snowflakes made it impossible to get a good look at the dark figure. If the man is telling the truth, then it is undoubtedly a picture of a sasquatch, but it doesn't show enough to be a good still picture and the figure never moves.

All of the information quoted comes from the narrow northern portion of Idaho, and there is a fair amount more, including two sighting reports and two track reports from up near the Canadian border. From the southern part of the state, which covers a much greater area and includes a substantial proportion of the mountain forests, I have only one report, going back to the fall of 1969. Robert Walls, an anthropology student interested in sasquatches, located a man who told of seeing a strange creature at night while hunting rabbits from a pickup truck on an old army rifle range about nine miles south of Twin Falls. Seen at the edge of a spotlight beam, it appeared to be eight to 10 feet tall and three to four feet across the shoulders, and covered with short, dark hair. Apparently no one was in a position to move the spotlight to follow it and it was seen to make only one stride, upright, before it was out of the light.

It seems to me that the geographic pattern of the Idaho reports is worth some consideration. Russell Gebhart's efforts, including advertising for information, presumably account for a good part of the uneven distribution, but a fair proportion of the reports did not come through him, and several were known before he became active. To have sighting reports you have to have people to do the seeing as well as animals to be seen, and there are several times as many people in the southern part of the state as in the north. Why would mankind's supposed need to imagine hairy wild men become

290

atrophied in southern Idaho, even though there is plenty of country there of the type the wild men are supposed to inhabit? If one considers, on the other hand, that the frequency of reports of animals of that type is related to whether or not such animals are actually there, then a reasonable explanation is apparent. Climate, rainfall and vegetation maps all indicate changes that conform roughly to the distribution of reports, not only in Idaho but in the adjacent areas of surrounding states. Just as was noted for the United States as a whole, the areas where there are few sasquatch reports are the drier areas.

Moving over into Montana there is another interesting pattern. South of Missoula for about one hundred miles, sighting reports dot the Montana side of the Bitterroot Range like a string of pearls. On the Idaho side there aren't any. There is often a considerable difference in the weather conditions on opposite sides of a mountain range, but the eastern side would be the dry side. Along the base of the mountains in Montana, however, there is a highway and a string of towns. On the Idaho side there is a wilderness area. To have sighting reports you have to have people.

For a long time I didn't have much information from Montana, but in recent years there has been a flood of stories, one going back as far as 1916. There is even one by a newspaper writer who claims to have seen the thing himself. Tom Tiede, who was a reporter for the Kalispell *Inter Lake* sometime in the 1960's, wrote a feature story long afterwards for the NEA syndicate in which he told of going to interview a hermit who lived in the woods near Columbia Falls. He failed to find the place, and was sitting down to rest before starting back when he heard something:

I swear the noise was like wailing, only not mournful wailing. Happy wailing? I can't help it, so help me, it sounded rhythmic, patterned—I shall say singsong.

I remember thinking it must be a hunter. And, knowing about hunters as I did, I remember wishing I had worn a red cap to distinguish me from a deer. I got up, to be as conspicuous as possible, and looked around. I had intentions of calling out, but thought better of it. And as the moments passed, and the singsong continued, my hours of viewing TV Westerns overcame me—I reached for my gun.

Suddenly, and I swear by the Abominable Snowman, I saw it. About 50 yards away. Coming down off one of the interconnected hills, passing at a moderate speed through the woods, disappearing and reappearing in the trees. I don't remember feeling anything. I could see plainly that it was not like anything I had ever seen before. It had swinging arms, like a B-grade gorilla movie, a gray coat of hair, and a small head which I could not make out. And it was moving parallel to me.

Now to be honest, I don't know if it saw me. But it stopped. And seemed to look in my direction. As it stopped, so did its song. I raised my rifle, forgot to take the safety off, but did nothing anyway. The thing paused for just a moment, then moved, silently now, off in a direction my shaky compass said was north . . .

I beat it out of the forest then, and to hell with the old hermit, I never went back. I never wrote the story before either.

A story that did receive considerable publicity, about a year after it happened, was told by R. W. Rye, of Billings, Montana. He wrote to *Saga* magazine late in 1960 to describe an experience he had while bear hunting in the foothills of the Mission Mountains near Seeley Lake in 1959. He said that he was following large tracks in the snow, and while crossing a small clearing got a feeling that he was not alone:

He looked around to his left and then to his right. There it was, twenty yards away, looking straight at him was this thing. Its head and arms were resting on a fallen tree that was five or six feet above the ground. Seeing only the head and arms, Rye thought it was a large bear, a very large bear. He began to raise his rifle to his shoulder. The thing, still looking at him, still leaning on the fallen tree, grinned at him—or so Rye thought. Then the thing let go an eerie, half-human scream. Its head began to rock from side to side and Rye could hear a rumbling sound. His rifle at ready, Rye back-pedaled. The thing moved a moment and Rye could see it from the waist up. It had a large flat head, stubby ears, a short neck and sloping shoulders with long arms. It was all covered with brownish grey hair.

That is how the story was told in *Saga*. Its publication in December, 1960, resulted in the local news media getting in touch with Mr. Rye for confirmation, which he gave. The Billings *Gazette* story on December 4 noted that he had been a licenced game guide and had shot plenty of bears. The *Daily Missoulian*, December 2, had checked another aspect of Mr. Rye's story, that three men had gone missing in the same area, and found it to be true. The Missoula sheriff's office confirmed that a student from Montana State University had never returned from a trip to Crystal Lake; a Kalispell man had gone missing on a hunting trip to Swamp Creek, and while a Missoula man who went hunting in the area was not exactly missing, only parts of his body and his broken rifle had been found. It was assumed that he had been killed by a grizzly bear.

The *Montana Sports Outdoors* magazine also took up the story in its December, 1960, issue:

When we read the Saga account, we immediately suspected the whole thing had been dreamed up by an imaginative editor shy a couple of pages of copy at press time. We were certain that Saga had picked the name "Rye" from a bottle tucked in a desk drawer and that Montanans were the victims of some "Rye" humor. Then someone checked the Billings phone book, and what do you know, the name "R. W. Rye" was listed . . .

When we first asked if he could have been mistaken and had actually seen a bear, he quickly agreed. Too quickly, we thought. So we talked about hunting in the Seeley Lake area. Pretty soon we could see that here was a perfectly rational and experienced woodsman who knew the Seeley Lake area quite well, a man with an unpleasant experience that even he could not fully explain.

Then we again asked him if he thought he had seen a bear.

"That was no bear," he replied firmly, "I've seen lots of bears and have killed a few, and I've never seen a bear that looked anything like that."

Rye thought the creature looked more like a huge ape . . .

And what had Mr. Rye done after confronting this apparition? *Montana Sports Outdoors* told that story too:

At the time, Rye was armed with a .270 rifle and a shoulder-holstered .357 magnum pistol. What did he do next? Why he did what any other red-blooded, well-armed hunter would have done under the circumstances. He broke all records for cross country running in getting the heck out of there.

In 1965 I received two reports from Dr. Joseph Feathers, a professor at Western Montana College of Education at Dillon. He said that in the fall of 1960 a senior student named Dean Staton had seen huge footprints in the snow on a mountainside 15 miles southwest of Jackson in southwest Montana. They looked like bare human tracks but were nine inches wide at the ball, five inches wide at the heel and 17 inches long. They went over a windfall four feet off the ground without breaking stride. After following the tracks about 300 yards across the open mountainside, the student lost the trail on a rocky slope.

In May, 1964, two students, Gary Simons and Sid Richardson, had taken Simons' Boy Scout troop from Butte camping about 15 miles from the town up Brown's Gulch. At 4 a.m. one of the scouts was awakened by a form in front of his tent. Dr. Feathers wrote:

He got up and met face-to-face a Sasquatch man who was hairy all over, brown hair with silver tips. The man also had a heavy beard. The boy screamed and the man ran away. Gary came to investigate and they heard the creature splashing in the creek below. He made giggling sounds like a human.

Later in the morning, they found many barefoot, human tracks, twenty inches long and six inches wide by measurement. The stride was seven feet. They could track him for one hundred and fifty yards before the trail also disappeared in the rocky strata.

There were other Montana reports as well, but not many. Then in 1974 I began to get letters from Greg Mastel, in Missoula. At first they seemed to be the sort of thing a young boy would write, but soon there began to be references to sasquatch reports that I had never heard of. It turned out that Greg was indeed a young boy, at that time he would have been 10 years old, but that didn't stop him from being a highly-productive researcher. I operate an information exchange for a widespread group of people active in sasquatch research, and for the past two years Greg has been one of the top three in digging up new information. Only Ramona Hibner in Florida is able to come up with more reports from her home area than Greg does from Montana. They both get information from other areas as well.

A few months after I first heard from Greg, Tim Church, who comes from Rapid City, South Dakota, but attends university at Missoula, also began to write to me. He passes on stories from all over the United States and is the other member of the top three in contributing reports to the exchange. With both Greg and Tim active there, although unfortunately in the same town, it is not surprising that a lot of Montana reports are accumulating, but there are two other reasons as well. Montana has the only sheriff's department I know of that is actively investigating sasquatch reports, and it has a publication, the *Bitterroot Journal*, at Hamilton, that devotes a lot of attention to the subject. In 1975 Greg named the Bitterroots, the Kalispell area and the Mission Mountains as the places where there was the most information, and since early 1976 Great Falls would have to be added to the list.

The first reference I have to the Bitterroots is contained in a letter from Mrs. Callie Lund, at Oakville, Washington. She wrote:

When my husband was alive we had a summer cabin in the Bitter Root Mountains in Montana and once in a while

Denis Gates and Greg Mastel, spring, 1976.

Captain Keith Wolverton with his map of hairy monster sighting reports at the Cascade County sheriff's office in Great Falls, Montana.

some old miner would start to tell about seeing a strange ape-like creature in the area.

The *Bitterroot Journal* published several reports in November, 1974 and later. The earliest dated back to November, 1962, when Reed Christenson, of Hamilton, and his wife and daughter saw a six-to-seven-foot, brown biped run up an embankment at the side of the road near the top of Lost Trail Pass at two in the morning. In August, 1964, also Lou Bigley, of Hamilton, while hauling a load of logs out of upper Grid Creek, saw a brown biped about five feet tall, with broad shoulders, standing in the middle of the road about 100 feet ahead of his truck. He slammed on the brakes and it turned and ran up a rocky draw. About the same year, Gary Hall, hunting north of Bass Creek, followed tracks along the crest of a mountain in a few inches of snow that he thought were made by a bear walking on two feet.

In the fall of 1965, Bob Shook, of Hamilton, and several friends were hunting up Piquett Creek south of Darby, Montana, when they found a five-toed track in some soft dirt so big a man's arm from the elbow to the fingertip would fit inside. They had smelled an overpowering odor and had been bothered by some whistling animal at their camp the night before, and ended up spending the night locked in their camper truck. There was a similar story from the same area a few years later told by some out-of-state hunters who stopped in Hamilton to get a tire fixed. They said that they had not only smelled something horrible but had seen an ugly black round face looking in the window of their camper and after the face disappeared something rocked the heavy vehicle until they thought it was going to tip over.

In September, 1969, a man cutting lodge pole just east of Lost Trail Pass sensed that someone was walking behind him. Shutting off his power saw and turning around, he saw about 25 feet away a seven-foot, 300-pound "apeman" covered with black hair. The *Bitterroot Journal* story continues:

> Thinking to defend himself, Ted started the saw. The creature didn't move, but the hair stood up on its neck. Ted turned the saw off, the hair on the animal's neck settled back down again. Without taking his eyes off of Ted, the beast turned sideways and began walking away with huge graceful steps . . . The incident occurred about 3 p.m. in broad daylight.

In May, 1977 the *Bitterroot Journal* carried a story about two women who watched a creature in Pfeiling Gulch for half an hour in July, 1976. They had driven up a remote logging road and were picking berries when the daughter noticed a huge hairy creature several hundred yards away watching her mother. She guessed it was about eight feet tall. It had straw-colored hair on its head and shoulders, while the rest was dark brown. It moved its head from side to side as it watched them, then was still again. After a while it sat down, but they could still see it. Then it stood up again. The young woman started walking slowly towards it for a better look, but she ran out of courage and returned to the car. After watching it for about half an hour the women finally left.

There are several other recent reports from the Bitterroots, and a lot more from other areas—more than a dozen in 1976 alone—but that is a fair sample. Most interesting development in Montana in recent years has been the involvement of the Cascade County sheriff's department in the sasquatch investigation.

Bipedal animals, no matter how big or how hairy, are not really a matter for the police if they do no harm. Neither are unidentified flying objects. Mutilated cattle are. To date I have not seen a great deal written about it as a nation-wide phenomenon, but almost everywhere Dennis Gates and I went in the United States in the spring of 1976 we encountered local interest in who or what was removing bits and pieces from dead livestock, mainly beef cattle. It is a mystery that I don't propose to get involved with. I understand that the number of animals found mutilated, while substantial, is only a small proportion of the number that die natural deaths, and I do not know if it has ever been established that animals from which blood, or ears, or organs, or whatever, had been removed were killed for that purpose. But it is certainly reasonable to assume that animals that have been mutilated may first have been killed, and that certainly makes the mutilations a matter for police investigation.

Like law enforcement agencies in many other areas of the United States (I have no knowledge of whether it also happens in Canada) the Cascade County sheriff's department at Great Falls has been involved in recent years in investigating mutilations. I don't know when they began elsewhere, but the first reports around Great Falls were in the summer of 1974. Predators might be to blame in a few cases, but certainly not in those where hides were precisely cut with sharp instruments or where hollow needles were used to remove the animal's blood. In most instances there were no tracks around the dead animal, of animal, human or vehicle. Usually the victims were in open pastures, but some mutilations were near homes and in enclosed yards.

Perhaps the whole matter will have been solved before this book appears in print, but at the time of writing I have not heard of any solution. "Satanic cults" are frequently blamed, since it is possible to imagine uses that they might make of the materials removed, but I have not heard of any direct evidence implicating any such groups, if indeed there are any in the areas where the mutilations take place. There have been reports of helicopters flying at night, sometimes without making any noise, near the scene of some mutilations. This has led to suspicions that members of the armed forces are involved, although there is no suggestion as to why. Flying saucer occupants have come in for a share of the blame, of course. So have sasquatches.

Captain Keith Wolverton of the Cascade County department has co-authored a book about the mutilations, *Mystery Stalks the Prairie*, Lacking any conventional explanation, the department has been prepared to investigate unconventional ones, and since there have been mutilations reported from almost every part of the county, almost any other unexplained phenomenon reported anywhere has to be looked into. The helicopter and U.F.O. reports in Cascade County became numerous about a year after the mutilations started. Hairy giants didn't enter the picture until late December, 1975, although there had been one report in December, 1974, that had been laughed off.

The first report that officers investigated was received on December 26, 1975, from two badly frightened teen-aged girls at Vaughn, a few miles west of Great Falls. They said that in the late afternoon they noticed some horses acting very strangely, pawing the ground and rearing, so went out to see what was the matter. They then saw a strange creature about 200 yards away which they estimated to be over seven feet tall and twice as wide as a man. One girl looked at it through a telescopic sight on a .22 rifle. She described its face as "dark and awful looking and not like a human's." Trying to frighten it, she fired in the air once, but it did nothing. When she fired again it dropped to the ground and pulled itself along with its arms a short distance before getting up on two legs again. The girls ran away, but one of them said she looked back and saw three or four of the creatures apparently helping the first one into a nearby thicket. A most unlikely story, but they agreed to take polygraph (lie-detector) tests, and both passed.

Aside from several reports of strange screams, there was nothing further until February 21, when two boys at Ulm, a few miles southwest of Great Falls, reported seeing a hairy creature just south of the bridge over the Missouri River. One saw only a hair-covered arm extending from some brush. The other, who was ahead, described a very tall creature covered with dark brown hair and with glowing whitish-yellow eyes. That was not reported to the sheriff's office until more than two weeks after it was supposed to have happened, so investigation at the site was fruitless. Both boys passed polygraph tests.

In the meantime the sheriff's officers had talked to a man who claimed to have seen a nine-foot, hairy creature close up and two similar creatures farther away, a couple of miles east of Ulm on the morning of February 22. He had stopped his car and run towards one of the creatures, which was walking in a field about a quarter mile away from the highway, but when it turned in his direction he thought better of it and ran back to the car. The man also reported seeing a grey object hovering in the sky in the distance. Another report was relayed by city police on the same day. A man claimed to have seen a large brown hairy creature on a back road 15 or 20 miles east of Sun River that morning. He gave a name, but could not be located later.

Shortly after midnight on March 7 a woman phoned the sheriff's office to report seeing a reddish-brown, hairy creature standing in the ditch, one arm forward as if it were about to cross the road, on the highway north of Vaughn. She refused to identify herself. Later that month a boy reported that he saw a hairy creature standing in the middle of Dempsey Road, in Great Falls, as he was riding his bicycle about 9 p.m. It took off through a hedge into the yard of a house. That boy also passed a polygraph test.

Some of that was going on while Dennis and I were travelling, and we added Great Falls to our itinerary on the way home. After years of careful sparring with polite but sceptical policemen on the subject of hairy creatures, I found the reception at the Cascade County sheriff's office all but overwhelming. Captain Wolverton devoted a day to showing us around and Undersheriff Glenn Osborne was equally hospitable. What was really mind-blowing was being taken into a special room in the county building and shown on the wall a large map with reported sightings of hairy monsters (several more than I have mentioned) carefully plotted on it. The strange thing is that Great Falls is not properly situated to have sasquatches. It sits out on the open prairie, with mountains about 50 miles away to the west and somewhat closer to the south and east. However all but one of the sightings were very close to one of the two rivers, the Sun and the Missouri, which have brushy valleys cut below the level of the plain. I couldn't imagine sasquatches living in the vicinity. If for any reason they wanted to move through the area via the river valleys that would be sensible enough, but what the reason would be is not apparent. Still, there must have been big hairy bipeds around Great Falls that spring, or the sheriff's department ought to go out of the polygraph business. I am quite sure there is no connection between

299

the sasquatch sightings and the cattle mutilations, but I hope that conclusion isn't reached too quickly at Great Falls.

Dennis and I were in Montana April 7 and 8, stopping in Kalispell and Missoula as well as Great Falls, but neither Keith Wolverton nor Greg Mastel had heard that the sheriff's office in Lewis and Clark County, at Helena, had been called out on April 4 to investigate a report of two hairy creatures seen that morning. Sixteen-year-old Bob Lea told the investigating officer that he looked out his bedroom window at about 5 a.m. and saw a large humanlike thing walking in the field about 600 feet away. It met a smaller creature of the same sort, paused to handle some large object for a few moments, then started walking directly towards the house. When it began to get close he ran downstairs and wakened his parents, but when they got back upstairs there was nothing to be seen. In a signed statement he described the larger creature as 10 to 11 feet tall, with a large head and bulky muscles. The smaller one was up to the large one's shoulder. The officer took the call at 6:30 a.m., and he and the youth searched the field for nearly an hour, but found no tracks.

During the day, however, Bob's sister Debbie found three huge tracks, 17½ inches long and seven inches wide. They had a stride of six feet and had only three toes, the longest in the middle. One of them was cast the next day. Captain Wolverton went to see Bob Lea a couple of weeks later and found his story convincing. The footprint cast, in a photo, does not look like anything I am familiar with.

Greg Mastel later wrote that he had talked to another person who saw a dark creature about nine feet tall near Helena on April 23, and that he had found three-toed tracks at the scene of the report. Later he passed on two other reports from the Helena area, in February and June, and eight from near Great Falls, in July, August, October and November.

I was back in Great Falls in June, 1977. There had been more reports in February, March and May that year, one of them by a brother of a deputy sheriff. He and two friends had seen an animal in a plowed field as they were driving near Gerber, and headed into the field after it. It started to run away as their pickup approached, but then turned and stood its ground. At that the men became frightened and drove off. They said it was about six and a half feet tall, all black, with hair about four inches long all over its body. They had been drinking, but Captain Wolverton found their story convincing.

In late August, 1977, the Great Falls *Tribune* carried a story out of Missoula about a group of campers shooting at a 15-foot creature on a ridge in Belt Creek Canyon, 20 miles southeast of Great Falls, at 2 a.m. August 20. Staff Sergeant Fred Wilson from Malmstrom Air Force Base was the only one who agreed to be identified, but two others confirmed his account anonymously. On August 31, the three came to the sheriff's office, and Captain Wolverton went to the scene with them, but could find nothing. Three men and two boys had been on the camping trip. They had decided to break camp because of rain and were walking back to their cars when the men heard noises and then saw by flashlight a large hairy creature standing near a clump of bushes. The boys had gone on ahead and did not see it. Fred Wilson described the creature as being covered with long tan hair, having no neck, standing on two legs. He estimated its height at 15½ feet. The three men ran to the car and got a shotgun, firing two shots in the direction of the creature. It then started running towards them, but they got in the cars and drove away. Fred Wilson was tested on the polygraph and passed.

I have no doubt at all that it is the encouraging attitude of the sheriff's department that has resulted in so many reports coming to light in Great Falls in such a short time, and that if it ever becomes general practice for police departments to solicit reports and record them with some central agency they will mount up at a rate of thousands a year.

The only state I have not dealt with, aside from the Pacific Coast trio, is Alaska. Several of the stories from there are mentioned in other chapters, but they do not give any idea of the overall picture. At the present time there are two widely-separated areas in the state from which there are concentrations of reports, and by widely-separated I mean a thousand miles. One is at the southern end of the panhandle, around Ketchikan and Wrangell. The other is on the Yukon River in the western part of the state, from Ruby to Nulato.

The Yukon River reports are all from Indian sources, and were obtained in the course of a single two-week visit to the area in 1970 by a team of four men led by Bob Betts, an anthropologist who worked with the Bureau of Land Management fire-fighting division in Alaska for several summers while a student. He first learned of reports of such creatures in the summer of that year, as he outlined in a letter to me:

> I've been getting out into the remote villages more than ever this year and have come up with some interesting information. The legend of the "Bushman" is widespread throughout interior Alaska. I've talked to natives from Huslia who claim people from their village have seen the "bushman" recently, they describe it as larger than a man and completely covered with hair and very strong. I've also talked to natives from Allakaket and Tanana who also have a real belief in the

"Bushman." The natives will not usually talk to outsiders about this.

That September he and Jim McClarin and two other men spent September 10 to 25 near the village of Ruby, with a rotating team of three men at a former homestead 18 miles down river keeping a 24-hour vigil for one of the creatures that was reported to return there for a few days each year. The fourth man stayed in Ruby to take advantage of any opportunity to obtain more information from the residents. No Bushman showed up. The Indians said it was already too late in the year. Besides a great deal of background information and legend the group compiled 15 reports of what were considered to be encounters with the creatures, although only eight involved sightings. The incidents had taken place at various towns and fish camps along the river, and only one actual eye witness was available in Ruby. Paul Peters said that he had seen a Bushman coming along the beach near his fish camp 10 miles west of Ruby in August, 1960. When about 75 yards away it turned and climbed a steep hill, disappearing in the brush. It was covered with black hair, was tall and broad and walked upright like a man. The other stories spanned a period of more than 60 years, but included three sightings in 1968 and 1969 near Galena.

The creatures, whose Indian name was "Nakentlia", were said to frequent the area only in late summer, when it was still warm but was already getting dark at night. They were mainly nocturnal, and were credited with stealing dried salmon from smokehouses, throwing sticks and rocks at people, and making a characteristic high-pitched whistle which one informant likened to the whine of a jet engine starting up. The creatures were considered to have various combinations of animal, human and supernatural characteristics.

That so much information was available in a single community of course raises a question as to whether the situation there is special or typical. Bob Betts also learned, from people he worked with at Fairbanks, of two recent reports from the vicinity of Fort Yukon, about 300 miles farther east on the Yukon River, and one near Kotzebue Sound, on the Kabuk River. There are also two old stories of the capture of hairy bipeds by groups of Indians, one at Holy Cross, almost 200 miles farther down the Yukon than Ruby, one at Kulukak on Bristol Bay. With the stories quoted elsewhere from Portlock and Nelchina, that provides a fair distribution of reports around the state, suggesting that there may indeed by a great deal more to be learned. It contrasts sharply with the lack of reports from the same latitudes in the Yukon and Northwest Territories of Canada, but that may result partly from the lack of potential witnesses.

Although Alaska is only about two-fifths as large as northern Canada it has several times as many people.

The reports from the southern panhandle come from a wide variety of sources. There is an area east of Thomas Bay where several men told of seeing hairy "devils" back around 1900. One of them, the late Harry D. Colp, of Petersburg, wrote a small book about it, *The Strangest Story Ever Told.* The creatures were described as "neither men nor monkeys," but they were much smaller than men. In 1942, from the deck of ship, Bob Titmus had a good look at an upright ape on the beach, but refused to believe his eyes. He is not positive where the ship was at the time, but thinks it was in Wrangell Narrows.

Bob Everett, of Santa Barbara, California, tells of a similar sighting from a Fisheries boat as it was being anchored for the night near the southern tip of the panhandle on the Inside Passage. That was in August, 1956, a few years before a reported sighting of such a creature that swam away under water, also in the Ketchikan area.

Late in August, 1968, two hunters saw what they thought was a bear as they drove along a mountain road, but when they stopped to shoot it, it walked off up the hill on its hind legs, leaving them gaping. I have heard the story from one of the men involved and have read what is obviously the same report in a letter by a friend of the two men. One places the occurrence above Stewart, B.C., the other north of Hyder, Alaska. It doesn't make much difference, since the two towns are side by side.

Finally, the *Bigfoot Bulletin* in November, 1969, carried a report by J. W. Huff, of Ward Cove, Alaska, telling of an experience he had in July of that year while prospecting in the mountains above Bradfield Canal. He and a partner flew in by helicopter, and while they were setting up camp saw what they took to be a man observing them from a ridge about 300 feet above them and 500 yards away They watched him for 10 to 15 minutes, expecting him to come down to their camp, but he just stood there. He seemed extremely large and was very dark, but they could make out no hat or clothing. When one of them yelled and waved at him he "took off with a lumbering gait rather rapidly and soon disappeared off the ridge . . ." Next day they went over to where they had seen him, but found nothing.

-16-

So Close to Hollywood

To many people, California means Los Angeles or San Francisco. They know it has the most people of any state in the U.S., and it sometimes takes a bit of persuading to convince them that in the northern part of the state there are vast wild areas that are all but empty of people, where a large unknown animal could easily be hidden. Their impression is not unreasonable either. Wild areas notwithstanding, California said goodby to its last grizzly bear a long time ago. After seeing the country for myself I had no trouble accepting the idea that if there were such things as sasquatches there was plenty of suitable habitat for them in northern California, and also in the Sierra Nevada. Around Los Angeles, forget it.

But enormous as it is, greater Los Angeles occupies only about one quarter of Los Angeles County, and a lot of what remains is mountainous. In fact the area of national forest within the county seems to be greater than the area occupied by the metropolis, and in the San Gabriel Range, directly north of the city, there are half a dozen peaks ranging from 5,000 to 10,000 feet in height.

Most of the mountain ranges in California tend north and south. The San Gabriels, and some connecting ranges, run more east and west, forming a barrier between Los Angeles and the greater part of the state. A few reports of hairy giants from various places in that high country have been known for a long time, but no one paid much attention to them. There was an active group of Los Angeles area policemen, including Ken Coon, who used to go "Bigfoot" hunting in northern California, and they were also interested in an area in Nevada, but no one was home minding the store.

When the computer survey was first underway in 1970, Ken Coon had the job of trying to find the people involved in a most unlikely report that had been carried by United Press International in September, 1966. It told of a girl who claimed to have been grabbed by a seven-foot beast covered with moss and slime, as she was driving near Lytle Creek wash somewhere north of Fontana, which is just west of San Bernardino. With some difficulty he succeeded in tracing the girl, and was surprised to discover that the story, while not exactly the same as the newspaper version, was entirely serious. Along with the computer booklet he sent the comment, "Either I am becoming gullible or Bigfoot is alive and well in the mountains of Southern California."

Having risen to command a bureau of 64 detectives in the Los Angeles County sheriff's department, Ken Coon did not seem a likely candidate to win any prizes for gullibility. Some time later he sent along a report on the whole Southern California situation, including in his introductory remarks the following:

> We are all aware of the difficulties in attempting to con-
> vince the world of the mere existence of Bigfoot/Sasquatch
> type creatures. I can well imagine the howls of laughter that
> would meet any attempt to sell the idea of such creatures
> bounding about on the very outskirts of one of the largest
> metropolitan areas in the world.
>
> This report will concern itself with certain mountain and
> desert areas of Los Angeles, Ventura, San Bernardino, River-
> side, San Diego and Imperial counties.
>
> The mountains in question are commonly called the Coast
> Range although certain portions of them are known by other
> names. Actually they run in a fairly well connnected mass
> from near the Mexican border to a point well beyond the
> limits of the area herein discussed.
>
> Throughout most of this region the mountains are bordered
> on the south and west by heavily populated plains and on the
> north and east by desert.
>
> Those who reject the idea that Sasquatch may inhabit this
> area give the following reasons:
>
> Lack of cover. At all but the highest elevations these tend
> to be brush-covered rather than forested mountains.
>
> Shortage of food and water. These mountains become
> exceedingly dry by late summer when most streams and rivers
> have dried up.
>
> The entire region is fully explored and heavily travelled.
> On some weekends it appears that most of the seven million
> residents of the Los Angeles basin have taken to the

mountains. Skiers, hunters, motorcyclists, equestrians and hikers flock into them during appropriate seasons.

All of those statements are true to a degree, but the opposite view can be taken when considering the following:

As heavily-travelled as most parts of these mountains may be, there are some fairly remote locales. For instance, there are five regions set aside by the federal government as wild areas or primitive areas.

Recent reports from the Pacific Northwest suggest that the creatures are omnivorous to a greater degree than previously suspected. Wild game of many kinds including deer and mountain sheep inhabit these mountains. Fruit orchards of various kinds are found in or near the foothills over much of the range.

The mountain range in question is nearly two hundred miles long by twenty five to fifty miles wide, with many peaks of six thousand to ten thousand feet. Private planes disappear and are not found for days or even weeks. Hikers become lost and some of the nation's finest search and rescue teams are required to locate and rescue them. Some are never found.

Any mountain range that is inhabited by cougar, black bear, Bighorn sheep and the rare condor can hardly be classified as a city park.

The report noted that the first mention of such creatures in the area was apparently in the writings of early California mission priests. The fathers commented upon the Indians' fear of the "hairy giants who supposedly live up certain dry arroyos." It also referred to a number of vague stories, lacking detail. The oldest definite story was the newspaper account of the sighting of a manlike beast near Warner's Ranch, east of San Diego, in 1876, which has been quoted in full in an earlier chapter.

The next story is dated more than 50 years later, in 1939. A Los Angeles store owner told of being approached by several giant hairy bipeds while he was camped alone in the desert in the Borrego area on a prospecting trip. He told Ken Coon that they were covered with a light colored hair and had glowing red eyes. They came quite close and he was of the opinion that one of them was eager to attack either him or his burro, but was held at bay by the fire. The man's story, told more than 30 years after the event, sounded impressive.

On November 9, 1958, the United Press carried a story that a man named Charles Wetzel, from Bloomington, California, had encountered a monster while crossing the Santa Ana riverbed on North Main Street near Riverside. The thing jumped in front of his car, reached back to the windshield and started clawing at him. He drove on and it fell back from the car. He described it as having a round "scarecrowish" head, very long arms, no ears, a protruding mouth, and eyes shining as if they were flourescent. It appeared scaly "like leaves." I remember reading the story at the time, which was shortly after my first trip to Bluff Creek, and feeling annoyed that people were making a farce out of what I had come to consider a serious subject.

In November, 1964, the Los Angeles papers carried an item about a number of youngsters claiming that they had encountered a hairy "animal-man" while playing around the ruins of an abandoned dairy near Fillmore in Ventura County. The next report was recorded in the July, 1970, edition of George Haas' *Bigfoot Bulletin*, as follows:

Donald K. Allen, San Bernardino, California, investigated and sent us the following report on July 20, 1970:

In the summer of 1965, Jim and Jan Gorrell, both employees of the Continental Telephone Co., Victorville, Calif., were picnicking near a place called "Bowen's Ranch", adjacent to Deep Creek at the foot of the San Gorgonio Mountains. The area is south, south-east of Hesperia, Calif.

They were cooking hot dogs over a fire after dark when Mr. Gorrell heard a noise just outside the clearing they were in. He looked up to see two eyes reflecting light in the darkness. At first he thought it to be an owl, but the eyes were too large and too high off the ground, for there were no trees to support an owl in that spot.

Soon his eyes became adjusted to looking into the darkness, and he saw a "big black mass" just watching them, about 9-10 feet in height.

He didn't tell his wife what he saw right away, but said, "Let's go," packed up and headed for their car, parked a short distance away on a ridge. He didn't want to alarm his wife, but when he left the fire going and the hot dogs at the site, she asked and he told her what he saw.

The thing followed them to their car, gaining on them and almost reaching the car by the time they drove away.

I called both of them at different times, and their descriptions of the thing were identical.

It walked totally upright like a man, had dark brown or black hair completely covering its body, had arms longer in proportion than a man's, with a large head and short, thick neck or no neck at all. It appeared to be merely curious, for it followed them instead of staying near the smell of the food. Mr. Gorrell stated that they were cooking the hot dogs on

307

coat hangers, and they left hangers and food right there by the fire when they left. The thing made no noise or sound itself during the encounter. The couple tried to ignore whatever it was and get out of there fast. Needless to say, they were very scared.

Ken Coon interviewed the Gorrells for the computer survey. The place where they were picnicking is about 3,000 feet up on the north slope of the San Bernardino Mountains. San Gorgonio is not the name of the range, but the name of an 11,000-foot peak a bit farther east. The next incidents led up to the one involving the girl and the creature "covered with moss and slime." Ken reported on them as follows:

In July, 1966, some teenage boys reported that they had seen "an ape in a tree" just north of the city on Fontana in San Bernardino County. This is on the opposite side of the mountains from the Deep Creek incident reported above.

One of the boys claimed that he had been grabbed by the beast. He exhibited scratches and torn clothing as proof of his story. Apparently no one paid much attention to the boys' report except the other teenagers. On August 27, two teenage girls spent the evening driving around the dirt roads north of Fontana in hopes of spotting the monster. Later that evening they rushed into the local sheriff's station screaming that they had been attacked by a seven-foot monster. One of the girls, Jerri Mendenhall, 16, had a badly scratched arm which she stated was grabbed by the beast as he reached in the car after her.

The sheriff's department took all of this none too seriously, as might be expected, but one officer supposedly admitted that he found a 17 by 6-inch footprint at the scene. According to the press report the print was human appearing except that it had only two toes.

Bill Early and I recently interviewed Jerri Mendenhall. She tells what, to us at least, is a very convincing story emphasizing particularly the terrible odor of the creature, "like a dead animal." She also pointed out that the hair was very matted and slimy or muddy.

She stated that she was backing slowly down the dirt road when the monster stepped out of the bushes and grabbed her through the driver's window. She says that she screamed and stepped on the accelerator. The creature then left at "a fast walk," disappearing again in the bush.

The girl involved, now married, agreed to relive her experience under hypnosis for the filming of the motion picture *Mysterious*

Monsters in 1975. The result apparently supported the story but she became so terrified that the hypnotist hastily brought her back to the present before she could be questioned in detail. Robert and Frances Guenette state in their book *Bigfoot—The Mysterious Monster*:

> For those of us in the room during Geri Lou Welchel's hypnosis, there is no doubt that she encountered what she said she encountered.

There were other sighting reports during the 1960's that were not followed up. Sometime in 1966 two young men reported to the Los Angeles County sheriff's department that they had seen the outline of a sasquatch-type giant on a hilltop above the highway on the outskirts of Quartz Hill in the high desert north of the San Gabriels, just southwest of Lancaster. Deputies went to the scene but found nothing. A woman who wrote to Roger Patterson said that some boys hunting in a bushy area near San Diego saw a large hairy creature poking around in the car when they returned to it. It left when they yelled at it.

The Charles Wetzel report and the two at Fontana are all beside the Santa Ana river system which I presume is often dry. It would be interesting to know whether there was water in it or not when the various sightings took place. In his report Ken Coon made the following comments:

> The Charles Wetzel sighting near San Bernardino occurred at the Santa Ana River. This river has its headwaters in the San Bernardino National Forest, near the San Gorgonio Wilderness area. The Deep Creek sighting occured at the mouth of a canyon which originates in the same mountain area, although it runs to the opposite or desert side.
>
> The Fontana incidents occurred in the Lytle Creek wash. This arroyo originates in the Cucamonga Wilderness Area.
>
> The Fillmore report comes from a location near where Sespe Creek winds out of the Los Padres National Forest. Sespe Creek has its headwaters in the wild and rugged Sespe Wildlife area, set aside to protect the giant condor.
>
> Of the nine reports of recent years it is interesting to note that six of them occurred in areas with fruit orchards. As an example, the Lytle Creek wash from the point of the sightings of the mountains is bordered by orange and olive groves . . .
>
> The only case wherin body covering was described as other than hair was the Wetzel incident. According to the press report the creature was supposedly scaley or slimy. The description of the slimy, matted fur as related to me by the Mendenhall girl sounds like a condition that might easily be mistaken for scales by a startled observer . . .

309

It has been pointed out that if Bigfoot does inhabit the mountains there should be a greater number of reports from this area than any place else on earth. The reason being that the mountains are so heavily travelled and the cover so sparse. It is true that surprisingly few reports have come to our attention, but I feel there could be several reasons for this.

By remaining almost completely nocturnal in habits and by living primarily in the remote primitive or wild areas, few encounters with man would take place. In addition, some reports are doubtless written off by law enforcement officials and others as "impossible". As a matter of fact all but one of the sightings of modern times took place after sundown. The single exception being the Fillmore incident which is supposed to have occurred in late afternoon.

It seems important to point out that all but one of the primitive or wild areas in these mountains has had at least one report originate nearby. Seven of the reports in fact originate from locations connected with such wild areas by canyons and arroyos. This brings to mind the reports of the mission fathers.

The only remote area without such reports is the Devil Canyon, Bear Canyon Wild area, which is the one closest to Los Angeles. I do not wish to indicate that I feel there is a connection, but it is a fact that several children have disappeared in canyons that wind out of that area.

The local sceptics have pointed out that not only are there very few reports from Southern California but many of them are made by teenagers. Of course, the subject of teenager reliability is food for thought. Actually only half of the cases involved non-adult informants and I have interviewed several of them since they have reached maturity.

An example is Jerri Mendenhall, who is now twenty, and married. It is important to note that she was willing to relate her story to me in detail, four years after it happened. Her mother and brother also commented that she has never wavered from her story despite suffering considerable ridicule . . .

It appears to me that we are faced with much the same situation here that we have had in the Northwest all of these years. We have reports, some going back many years, which describe encounters with Sasquatch type creatures. These reports have been made by people of all walks of life, and under many different circumstances. We have in the past gone all through the "explanations" that they are lies, hallucinations, or mis-identifications. We know that such arguments eventually break down and turn out to be no more than attempts to explain away something difficult to accept. I believe the incidents herein reported ring true from a Sasquatch report standpoint. I am convinced the terrain is capable of supporting such creatures . . .

There was one set of tracks reported during the 1960's, near the Superstition Hills on the east side of the Anza-Borrego Desert. Major Victor Stoyanow wrote an article in *Desert* Magazine, which I have read but do not have, describing strange humanoid tracks he found in the sand, 14 cm long and nine cm wide. He made plaster casts and photographs. When I saw the article I did not consider the tracks to have anything to do with the sasquatch, but they were probably no stranger than some of those that have turned up in California since. A more conventional type of track report came from a young lady who visited at Ken Coon's home in Palmdale during the early 1970's and commented that she and some friends had seen prints just like Ken's "Bigfoot" cast. The prints they saw were going along the edge of a dirt road in the foothills near the Fairmont Reservoir, just north of Portal Ridge in northeast Los Angeles County.

Ken Coon and his friends also participated with Joel Hurd, a researcher now living in Sells, Arizona, in investigating some reports about 20 miles east of San Diego in 1971, near the town of Alpine. Here is what he had to say about that in an article in *Saga* in July, 1975:

> The strangest looking footprints I have ever seen were found near San Diego . . . The story behind the discovery of those prints, and the apparent reliability of our informant, make this one of the most interesting cases.
>
> Our informant, a respected San Diego physician, had purchased a ranch home in the mountains a few miles east of San Diego. One night, shortly after buying the property, as he, his wife and their daughter drove into their driveway, the headlights illuminated a large, upright, hairy creature, apparently eating apples from the orchard in front of the house. Upon the car's approach, the "big monkey" as the doctor refers to it, disappeared into the darkness . . .
>
> Upon searching the area in daylight he found a number of strange, four-toed tracks, that were very broad toward the front and tapered to a very pointed heel.
>
> The doctor also noticed other strange circumstances. Many of the trees is his orchard had been stripped of apples to a

311

height well beyond a man's normal reach and there was no evidence that anyone had been there with a ladder. He also observed the same strange footprints near his chicken house, and, even more odd, the small structure itself appeared to have been moved slightly off its foundation.

The doctor did some checking with his neighbors and found that they, too, had experienced some strange occurrences in past weeks.

By the time news of this sighting reached me, the doctor and his family had been badly frightened by further encounters with "the big monkeys" and he had taken to the habit of carrying a .44 magnum when ever working around the ranch. He described the creatures to me as heavily built, upright walking, dark colored "things", looking more like apes than men. The doctor said he and his family had seen as many as three of these creatures in the orchard at one time, the largest about six feet tall. He also pointed out the large floodlights he had installed and was pleased to report that the creatures had not returned since he started using the lights.

In the course of our investigation, we were contacted by a nearby rancher who showed us a strange footprint which he had found on his property. Sure enough, this lone print in powder dust clearly showed an odd vee-shaped foot with four toes. The print was 14 inches long and 10 inches wide at its broadest point. The following day, while searching the bank of a reservoir a short distance away, we discovered two sets of similar, but smaller footprints.

Subsequent investigations in that area have failed to turn up any more prints. However, we have spoken with several ranchers who have told strange stories of "the monsters" frightening their livestock and occasionally carrying off great numbers of chickens or ducks, leaving only their strange-looking four-toed footprints behind.

In July, 1971, Ken Coon also learned of a report that hikers had seen "an upright walking bear" in the mountains near Newhall, at the northern tip of greater Los Angeles.

A new phase of activity opened in Southern California in February, 1972, when some 17-inch tracks with a seven-foot stride were found on the hillside above the road up Nine-Mile Canyon west of China Lake, and were followed for over two miles. Those tracks don't altogether fit in with the reports from around Los Angeles and San Diego, since the location is on the east side of the Sierra Nevada, bordering on fairly normal sasquatch country. Their significance lies in the fact that they are item one in the search records of the California Bigfoot Organization.

The C.B.F.O. was the creation of Louis R. (Rich) Grumley, then a resident of Palmdale. Most "organizations" involved with the sasquatch search are very ad hoc affairs, if they have any existence at all beyond a name on a letterhead. Rich Grumley works on the opposite principle. The organization he founded was structured in every detail, its meetings were formal and its operations were conducted with all-but-military discipline. It began simply enough, back about 1969, when he and Floyd Smith got a few people together and began going to the Kennedy Meadows area in the mountains east of China Lake to spend weekends in the bush, and also holding regular meetings in Palmdale. They concentrated on Kennedy Meadows because a man had told them that he saw something reddish-brown and about seven feet tall run off on its hind legs while he was working on a truck there one afternoon. They would stay awake all night and then spent most of the day moving around in the area talking to people.

Thousands of man hours were spent without very much result. In the meantime the organization became increasingly elaborate. There were committees for such things as communications, research and development and capture and confinement, as well as for more mundane functions like publicity and ways and means. They had their own radio codes and a full set of operating rules and regulations. News coverage attracted both additional members and information. Among the people they talked to were a youth who claimed to have shot a sasquatch four times with a 30-30 behind the Little Rock Dam, just south of Palmdale, about 1971, and four young people who told of watching such a creature for a long time by moonlight as it moved aimlessly around in a campground where they were staying at Lake Isabella in February, 1972. Then in March, 1973, things started happening right in town.

Ken Coon was still living in Palmdale for part of 1973, but he and his police associates by that time were all working in Los Angeles and he was also busy winding things up for his retirement from the sheriff's department and a permanent move to Colville, Washington. He did not hear about most of the incidents until several days after they happened, and while he did interview some of the witnesses and spend some time looking for footprints, it was the C.B.F.O. members who were on top of the situation.

The first sighting was reported at the sheriff's office at Lancaster shortly after 1 a.m. on March 14, 1973, by three marines. They said that just a few minutes earlier a huge hairy thing the shape of a man had suddenly jumped in front of their car as they were driving on

Avenue J a quarter mile east of 110th Street East. The driver slammed on the brakes and the monster ran off into the desert. Deputies found the ground too hard-packed to show prints. They didn't get at all excited about the story, and this reaction is reported to have upset the marines. The *Ledger-Gazette* at Lancaster didn't mention the report until March 24, and then they noted that *The Legend of Boggy Creek* was playing at a local drive-in.

It was only 10 days later that the sheriff's office received another report, from a location within two miles of the marines' sighting. Nineteen-year-old Kim McDonald, returning to her home on East 115th Street after 2 a.m. from babysitting at her sister's, heard one of her dogs whining when she got out of the car and went looking for it. Here is part of the story as told in the *Ledger-Gazette* on March 30.

The girl walked behind the trailer, expecting to find the dogs, calling "Shad", thinking the cries were coming from a grassy area 100 feet away near a telegraph pole.

She walked up to the pole—on a night, remember, when the moon was so bright her step-dad could read his wristwatch clearly—with a warm smile on her face expecting the dogs to leap out with their customary friendly overture.

Not that night. Suddenly from the rustling grass a huge figure rose as if awakened from sleep and stood straight up like a man. She described it as being about seven feet tall, almost as high as the trailer in which she lives, which is eight feet tall.

The girl said that the huge hairy monster stood straight up and was completely covered with hair—except for the face—and ran away on two legs . . . but not at a high rate of speed . . .

"She saw something," Hawkins (her stepfather) said, "and she has no fear of the dark or coming home alone. She has lived here all her life and she is not afraid of the fields and is not the scary type."

That's why Hawkins got his rifle and patrolled the area for a quarter mile radius until daylight without finding the hairy thing or his dogs, which had run off and did not come home until daybreak.

C.B.F.O. members later questioned the girl in detail and recorded the interview. She told them the creature was far wider than Rich Grumley, who stands six foot six and weighs over 250 pounds, and that it never stood entirely straight, but was about two feet taller than she was. She was not afraid at first when it started to get up, but it seemed to keep getting taller forever. Then she found she could neither breathe nor scream. When it turned and loped off she ran and pounded on the trailer door, forgetting that she had a key.

Searching the area, Rich found footprints 13 inches long and far wider than a man's foot, in an old reservoir near 110th Street. They had the first and fifth toes widely splayed out.

The next sighting was at 10 in the morning on April 14, at Avenue J and 180th Street. The C.B.F.O. had become widely known by that time, and were summoned within an hour. They found some footprints nearby in the desert with a stride between five and six feet. A 16-year-old boy described the creature as a dark, hairy biped about eight feet tall with a very heavy build but a small head and no neck.

A month after that an ex-marine named Ron Bailey, arriving at his home in Palmdale at 4:15 a.m., saw an eight-foot, dark figure standing near a telephone pole. It was heavily built and hairy. Once it sneezed, like a person, but there was no other sound except a thumping as it walked away. He noticed a foul smell, like rotten apples. A search in that area turned up prints only 13 inches by 4½ inches, but indicating very great weight. The toes were long and splayed, with the longest in the middle.

In June two young children reported seeing a thing like an ape looking at them from behind a boulder, on a butte near their home on the edge of Lancaster, at about 10 a.m. That summer also, Mike Pense, from Lancaster, reported seeing a black hairy figure standing upright among the rocks on a butte in the Mojave Desert east of the city. He said it threw a rock in his direction, almost hitting his motorcycle. On September 17, two young people delivering papers in Quartz Hill by car at about 10 p.m. were turning around in a driveway when the car lights hit a huge, heavily-built creature about 50 yards away, covered with dark matted hair. Its eyes had a golden glow. C.B.F.O. members found footprints the following day, showing strides of four to five feet and great weight. Finally, on December 27, a young woman returning to her home in Palmdale at about 5 p.m. saw walking towards her in the desert something that looked like "a big person, all one color." She couldn't tell what it was, so went in the house and got her brother. It then took off running, "as fast as if it was on a motorcycle." They estimated its height as twice that of a man.

The use of street addresses makes it seem as if some of those reports were in places in the middle of town, but actually all had open areas close by. Lancaster, Quartz Hill and Palmdale are all laid out within a 10-mile square on the desert and there is a pattern of paved streets at one-mile intervals throughout the area between them, so there are street addresses right out in the desert.

Ken Coon was extrememly sceptical about the whole thing at first. Ron Bailey lived only a few blocks from his home. In June, 1973, however, he distributed a report in which he said that after investigating he was convinced that at least some of the reports were genuine. He was particularly impressed with Kim McDonald's story. Most of the footprint photos and casts collected by the C.B.F.O. could be attributed, he thought, to a single individual. He pointed out that loose sand blows over the ground all the time and there are high winds every day, so tracks soon lose shape, but any track found is certain to be fairly fresh. The tracks found near Ron Bailey's home, although they had an unusual toe pattern, were very convincing. They were not much larger than a human foot but "Grumley's 260 pounds did not make nearly as much impression at the same spot."

The terrain in the Antelope Valley is flat farmland, with alfalfa the main crop, but about half the area is still desert. A very large tract of desert northeast of Lancaster is occupied by Edwards Air Force Base, and while I am not aware of any more specific reports in the valley away from the mountains after 1973, I received a second-hand story in September, 1974, that Air Police had seen some such thing half a dozen times within the past year while patrolling the far reaches of the base. My informant said that the armed patrolmen did not shoot at the creatures, nor were they encouraged to make any official report about them. I have not been able to obtain any confirmation of that information, but I can't help being intrigued by the idea that all one has to do to wander freely around a closed military installation is to wear a fur suit. Considering the normal attitude of any sort of officialdom to reports of hairy bipeds, it wouldn't really surprise me.

In speculating about the possibility of wild creatures coming down into the desert in the spring of 1973, Ken Coon noted that the previous winter had been severe, with heavy snow in the high country. Following three nights of patrolling the roads with a spotlight he also noted that he saw no rabbits at all, whereas there would normally have been many of them in the brush and dead on the roads. When a rancher's wife told him she had lost 50 chickens he suggested that coyotes might be responsible, and she replied that no coyotes had been seen or heard all spring, an unprecedented situation. In May, he also looked over the area by helicopter:

> From the air I was able to observe some things that make sense relative to the sightings. To begin with, the most logical route out of the San Gabriel Mountains which border the valley on the south is down Bigrock Canyon. It is a major canyon with year around water and good cover. The mouth of the canyon opens up into the desert directly opposite the buttes which are about fifteen miles due north. It was in the

316

farmlands around and just west of the buttes that the reports and tracks originated.

Big Rock Canyon also leads up towards the north boundary of the large Devil Canyon, Bear Canyon wild area about which Ken Coon had commented in 1971 that there were no sasquatch reports from there or from the canyons leading from it. The canyon itself had been the scene of several reports in 1973 and activity continued there after it had stopped in the desert.

On April 22, 1973, William Roemermann, Brian Goldojarb and Richard Engels, three youths from the Van Nuys-Sherman Oaks area of Los Angeles, reported to the sheriff's office in Lancaster that they had seen a huge biped as they were driving downhill past Sycamore Flat camp ground in the Big Rock Canyon. Richard and Brian were sitting in the back of their pickup truck as they drove down the canyon in the dark, about 10 p.m. Suddenly they saw the creature run out on the road behind them and follow for about 15 or 20 seconds. They could see it clearly, running on two legs, with its arms swinging back and forth in front of its chest. On the following day the boys went back and found hundreds of three-toed prints along the road and some deeper ones off in the bush, one of which they cast.

In October members of the C.B.F.O found 21-inch tracks with a 12-foot stride at South Fork campground, farther up the mountain than Sycamore Flats. The following month Ron Bailey's wife, Margaret, reported seeing a huge figure move by the moonlight while she waited in the car at Sycamore Flats during a C.B.F.O. exercise. Later the men found more of the enormous tracks at the base of the mountain there.

It seems that 1973 was the big year for the California Bigfoot Organization. Besides the activities already mentioned, members gave lectures in schools, a newsletter was started, a second chapter was in operation at Oroville, and there were plans drawn up for establishing affiliated organizations all over the country. But the activity in the Antelope Valley brought in a lot of new people, some of whom operated on their own from the beginning while others, like Ron Bailey, joined the C.B.F.O. but later split off. As I mentioned before, sasquatch hunters are not naturally inclined to come under one umbrella. When Rich Grumley had to drop out for personal reasons the organization became largely inactive, although I would not say that it may not be revived.

I have studied their tape-recorded interviews with witnesses, which were very thorough, but when Dennis Gates and I were in the area in the spring of 1976 we were not able to see many of their casts and those we did see were not of impressive quality, although certainly of impressive size. We spent two days with Rich Grumley and met Stan

Lusack, Dave Frisbee and Bill Darch from the organization, plus Mike Pense, and Ron and Margaret Bailey among the witnesses. We also sneaked in an after-midnight hunt in Big Rock Canyon in which we were much impressed with Ron Bailey's attempt to call a sasquatch, but none came. One thing that we found strange was that they considered the creatures to be dangerous and went into the bush cautiously and well-armed, although they had no intention of killing anything. Of course they may have considered it equally strange that Dennis and I, whose announced intention was to kill a sasquatch if we had a chance, had no guns.

From the Antelope Valley, Dennis and I went to Los Angeles, where we spent the next three days trying to sort out some of the personnel involved in recent sasquatch activities and to get some grasp of the resulting information. I am still not sure how many groups were represented or which ones normally work together. We were expecting to be told of incidents linking sasquatches with U.F.O.'s, since two of the people we met, Barbara Ann Slate and Peter Guttilla, had written a magazine article on that subject. It turned out that they were not prepared to contend that there was any direct evidence of such a link in Southern California. They did insist that things had happened that seemed to require supernatural explanations.

There was a story in the Newhall *Signal* and Saugus *Enterprise*, October 9, 1974, mentioning a boy having seen a nine-foot hairy man in a pig pen wearing a blue belt, but that did not stand up well to direct inquiry. There had been a sighting report alright in June, 1974, at Lost Creek Canyon, northeast of Saugus, which is in the same general area as the "upright walking bear" of 1971 and several recent reports, and there had been a boy who claimed to have seen a creature in a blue belt, though not in a pig pen. He had been living with a foster family and was no longer available, but was thought to have been making a bid for attention.

Barbara Ann Slate works with the "Angeles Sasquatch Association", a group started by the three youths who reported the sighting from the pickup truck in Big Rock Canyon. She calls them "trackers" and they have certainly assembled a remarkable collection of casts of tracks. As recounted in her book *Bigfoot* they have also recorded several sightings, all on the north slope of the San Gabriels.

In the spring of 1974 an Indian couple living at Little Rock, about a dozen miles west of Big Rock, told of several encounters with hairy giants. The wife said she had seen an immense white one with pendulous breasts and a flattened nose drinking out of the hose at the rear of their cabin, and they both told of seeing a black hairy form about 11 feet tall standing in a stream as they drove home one afternoon. The husband also reported a hairy face seen at a window, and there were huge three-toed tracks found in the vicinity of their home on several occasions.

Terry Allbright, one of the A.S.A. members, told of encountering a seven-foot, black, hair-covered form that was just clearing the road as he rounded a curve. Three other members told of seeing a shaggy white, crouched figure, six feet tall when bent over, for a few seconds at the Big Rock campground in August, 1974. In September two others reported spotting a black seven-footer by flashlight in a rocky area a quarter-mile beyond the end of Avenue J. Later they found narrow three-toed tracks 15 inches long in the vicinity. One of the witnesses was Richard Engles whom we met at Barbara Slate's home. That fall he and Willie Roemermann had found six fresh three-toed tracks that somehow vanished after they went to get someone else to look at them. In January, 1976, Willie Roemermann reported seeing an eight-foot creature by flashlight in the Sycamore Flats campground. Five-toed prints 16½ inches long were cast the next day.

Two things are disturbing about all the reports from the north slope of the San Gabriels. One is the apparently high success rate on the part of the hunters, but that would have to be weighed against the number of man hours spent in the hunt. The other is the footprint casts, several of which totally dwarf a normal "Bigfoot" print, yet show only three toe marks, about the size of baseballs. Along with their other peculiarites, many of the casts are not at all flat on the bottom, but rounded or even somewhat V-shaped from side to side; hardly a practical shape to walk on. The narrow three-toed prints bear some resemblance to the considerably-smaller ones from Fouke, Arkansas. The wide ones are unique. None of the casts were made in material allowing for really clear detail, but they are certainly casts of tracks, whether genuine or otherwise, not of holes in the ground of any other origin. I have not seen any good photographs of the three-toed tracks either.

As if the three-toed casts of the Angeles Sasquatch Association were not enough of a problem, Dennis and I were later taken on tour by another Los Angeles group, including Pat Macey, Dan Shepard and Susan Shaw, who have been investigating reports from the Newhall-Saugus area. They produced a very presentable four-toed print found all by itself in the vicinity of a sighting in that area. They had located the print themselves, early in January, 1975, in the hills near Canyon High School, while looking around after a sighting report December 31 by five children aged five to 14. The youngsters said they saw at a distance a nine-foot, apelike creature that raised its

Peter Guttilla and Barbara Ann Slate with casts of three-toed tracks from Los Angeles County.

At left is a narrow three-toed cast from. Los Angeles County.

Smokey Crabtree, below, holding a cast of a Fouke, Arkansas, three-toed track and a poster advertising his book.

arms at them, then jumped down off a rock and disappeared in a side canyon. It was dark in color and not heavily built.

Another sighting report they know of was received second hand in December, 1975, from south of Saugus, where a man was supposed to have seen the top half of a large creature looking like a gorilla that was holding a ground squirrel or rabbit. He saw it illuminated by a flashing advertising light and it disappeared between two of the flashes. The only problem was that the spot referred to is open and flat, providing no place for the creature to disappear to, and no hole in which its bottom half could have been hidden.

To complete the record, there are three more reports involving locations near the Santa Ana River. The Riverside *Enterprise*, August 19, 1975, carried a story about two youths who told police they saw a half-human, half-ape monster in an orange grove at Main Street and Chase Drive in Corona, which is southwest of Riverside. In July, 1976, two young boys claimed to have seen a tall hairy "big ape" in the Santa Ana river bottoms near their home in the vicinity of Riverside. Peter Guttilla talked to them and found them convincing. The following month a man reported finding 21-inch, five-toed tracks about five miles farther east in the Santa Ana River bottoms. I have a picture of one of those, and it is an excellent, normal sort of track. If it was faked, someone knew what he was doing. Peter Guttilla failed to find those tracks in the area indicated, but did find five-toed tracks some miles farther east, which he cast.

For me, the present situation in the Los Angeles area falls somewhere in between. It is not nearly as strange as that in Pennsylvania, which I will happily leave to the U.F.O. people or anyone else who is interested, but it is nevertheless a long way from the relatively consistent and comfortable type of thing I am accustomed to dealing with in the Pacific Northwest. If the footprints were consistent and reasonable adaptions of the basic primate pattern they would provide solid corroboration for the sighting reports, even if they were different from prints found elsewhere, but being as diverse as they are, and of such illogical shapes, they have the opposite effect, at least for me. Of all the non-conforming footprint types so far recorded, half of them must be from Southern California.

It would be easy to conclude that all the strange footprints have to be fakes. Perhaps they are. But why would anyone spoil a good hoax by manufacturing obviously unacceptable footprints? It can be argued that the C.B.F.O. and A.S.A. members fabricated their reports, but I have no reason to think that of them and no such explanation will cover the reports from other Antelope valley residents. If those were hoaxes, someone was wearing a fur suit and taking a very real risk of being shot full of holes.

321

Grover Krantz is of the opinion that a biped would never lose any of its toes, because they are too useful for balancing—that it could only show up with less than five if the others had been lost at some stage of its evolution before it started to walk upright on two feet. I can find no fault with that argument, but neither can I disagree with what Ken Coon had to say in a paper he wrote about the Southern California footprints in 1974:

If you are thinking that a few eye-witness accounts and a few dozen footprints are very slim evidence for the existence of large three and four-toed unknown creatures, let me remind you of one significant fact. When most of us started in this Sasquatch business that is exactly what we had to go on . . . a few eye-witness accounts and a few dozen footprints.

-17-

A Mystery in Stone

Chuck Edmonds, at Ashland, Oregon, was the first sasquatch researcher to collect information about footprints in stone, and it was from him that I learned about the "giant human tracks" at Glen Rose, Texas, as well as humanlike tracks in stone at other locations. The idea of finding fossilized sasquatch tracks had a lot of appeal, since it would be a little difficult to dismiss them as modern fakes (so I thought), but none of the locations was anywhere I could readily get to, and I didn't feel the prospects were good enough for a special effort.

Several years later I received a letter from Bill Dowson, an Alberta newsman who had been looking for sasquatches in the Rockies for a decade but who by then (1972) was working for a television station in Dallas. He wanted to know if I was aware of the work of a Christian film group who were excavating giant footprints in a limestone layer near Glen Rose. When I wrote that I was not, he provided me with several articles on their work, as well as going to the site to see for himself. He reported that the footprints were there alright. However they were all mixed in with dinosaur tracks, which removed any prospect that they could be sasquatch footprints. But what were they? I was sufficiently intrigued that when I learned in 1974 that the film the church people had made was available in Vancouver, B.C., I arranged to rent it. I still didn't expect much, and on the Richter scale for shock value the jolt that movie gave me exceeded the impact of the Patterson film about 10 to 1.

Reports of fossil tracks resembling human footprints have been

323

known in North America for a long time. Senator Thomas Hart Benton wrote about them in *The American Journal of Science* back in 1822. An article in *Scientific American*, January, 1940, told of such tracks in Virginia, Kentucky, Illinois, Missouri and other places. Dr. Wilbur Burroughs, a geology professor at Berea College, Kentucky, devoted years to studying a series of 12 prints in Coal Age sandstone near Berea that looked like those of human feet with widely-splayed toes, and wrote about them in the *Berea College Bulletin* in October, 1938. Those and other prints in sandstone drew the interest of Charles Gilmore, curator of palaeontology at the Smithsonian Institution, according to the October 29, 1938, issue of *Science News Letter*. The following May, Dr. Roland T. Bird, from the American Museum of Natural History, wrote an article in the museum's publication, *Natural History*, telling how he found the first brontosaurus tracks known to science in the limestone bed of the Paluxy River near Glen Rose. He also referred to stories told by local residents of "giant man tracks" in the same formation, and said that he himself had seen a print "about 15 inches long with a curious elongated heel". He expressed a wish that he had been able to see more of them.

In spite of the expressions of interest by Dr. Gilmore and Dr. Bird, it does not appear that any specialist in fossil footprints ever followed up any of those reports. The only people to do so, to my knowledge, have been creationists who do not accept the theory of evolution and are seeking proof of simultaneous creation of creatures that are supposed to be millions of years apart on the evolutionary calendar, in this case men and dinosaurs.

In 1969 the Films for Christ Association, headed by the late Dr. Stanley E. Taylor, went to work at the Glen Rose site. They had seen various depressions in the limestone that local residents considered to be human footprints at several locations within a short stretch of the Paluxy River. None of these are very far from the spot where 30 years before Dr. Bird had removed several tons of limestone containing dinosaur tracks for the American Museum of Natural History and the University of Texas. In 1970 Dr. Taylor returned to try to find more prints by digging in the riverbank. That year and in 1971 he and his crew did considerable excavating, diking the river around their work site and pumping the area dry. They even went to the extent of jackhammering through an unbroken limestone layer in the bank and removing the slabs with a front-end loader in order to continue following a line of tracks on a lower layer. Dr. Bird had done the same thing, with hand labor, in his quest for good dinosaur tracks.

The film Dr. Taylor and his people produced, titled *Footprints in Stone*, showed all these operations and included footage showing two lines of tracks that they had uncovered under the riverbank, one apparently of human size, the other much larger, crossing a line of dinosaur tracks. Whatever made the smaller set of prints had actually stepped right in the toe of a dinosaur track. It showed people fitting their feet into the tracks, including some child-sized tracks, and also stepping from track to track in a typically human pattern, although the strides were extremely long. There was a lot in the film giving interpretations of what had been found which I did not consider convincing, and they had moistened many of the tracks to make them photograph better, which I think was a mistake. Nevertheless they presented very convincing evidence that there were indeed humanlike prints in the same limestone as dinosaur tracks in the bed of the Paluxy.

Thinking that there had to be a catch to it somewhere, and that experts in such matters must know all about it, I wrote to a number of scientific institutions, including the American Museum of Natural History, the Texas Memorial Museum at the University of Texas, the Hall of State Museum at Dallas, the U.S. National Museum, the Dinosaur State Park at Glen Rose and the National Geographic Society. I also made inquiries through university contacts.

The National Geographic Society referred me to an article by Roland Bird in their May, 1954 magazine about the Glen Rose dinosaur tracks, in which he neglected to make any mention of the "giant man tracks."

The assistant curator of fossil reptiles and amphibians at the American Museum of Natural History—the people who should know all about the Glen Rose site—wrote as follows:

Yes, we are aware of the creationist movement and the efforts on their part to "disprove" evolution. I am familiar with the Texas track situation, and in fact, there are "human" footprints in the same rocks as dinosaur prints. However, the "human" prints appear to have been formed by local people initially as a joke (as early as the 1930's they were selling human prints cast in concrete as dinosaur tracks) and the proximity of a bible camp insured their discovery. There is no way to prove that this is being continued as a hoax, although it is quite obvious to me. The creationists want to believe in the human prints and logical arguments are usually pointless.

I couldn't be very impressed by that reply, since it seemed obvious that science must have advanced at least to the point where it would be possible to tell the difference between a shape carved in a piece of limestone and one impressed in the material while it was still mud. However it did provide an explanation, convincing or not. Scientists have determined that the tracks are faked. Then I received a message, via Grover Krantz, from some unnamed authority in the palaeontology department at Southern Methodist University, as follows:

> The Glen Rose area dinosaur is early cretaceous (circa 120 million years) and the "footprints" are merely "fluted river products." He says the religious groups are out of Chicago and are a bunch of "nuts", He is aware of the sasquatch prints and said these are NOT even similar.

Science had spoken, but it couldn't keep its story straight. Not about the tracks, anyway. There seemed to be a consistent attitude regarding the creationists. I received no other replies containing any information whatever, and I didn't find the two I did get very helpful. For one thing, the pictures made clear that some of the Glen Rose tracks were extremely similar to sasquatch prints. For another, the large numbers of people involved in doing the excavating and just hanging around, as shown in the movie, would have made it very difficult to fake anything.

Further, although I am not religious myself I know a lot of people who are. I may frequently disagree with the interpretation they put on things, but they are the last people I would expect to fake anything to support their arguments. Finally, if they were faking tracks why would they have faked any giant tracks? According to the film they found the human-sized tracks first, and they showed one stepping right in a dinosaur track. That made their case with tremendous clarity and precision. They had what they were looking for. Why did they keep on digging and uncover tracks 18 inches long? That just left them with a problem they couldn't really explain and they had to spend much of the latter part of the movie demonstrating that there is the occasional freak human (they found two among 200 million) with a 16-inch or 18-inch foot.

It is apparently true that carved representations of both dinosaur and human footprints were sold at Glen Rose during the depression, and that may indeed have had a lot to do with frightening off the palaeontologists, although according to Dr. Bird's article it was seeing just such a carving in a store in Gallup, New Mexico, that led him to investigate the Glen Rose area in the first place. The Films for Christ crew did not ignore that aspect of the matter. When I called at their headquarters early in 1976 I found that they had investigated such reports and learned that there were few carvings of human prints made, and none of them were in stone still in the riverbed. As a matter of fact they showed a refreshing scepticism about the opinions of some of their fellow creationists, compared to the unquestioning way that scientists seemed prepared to swallow conflicting explanations. One track that they considered to be hand-made had been cut in half and showed a compression in the material underneath the track, but they argued that the person carving the track had simply started with a piece of stone that already had a depression in it—probably a genuine track alright but of no clear shape—and had turned it into a giant human track. They were prepared to take it seriously only if the owner could cut it across the toes and show indications of compression under each separate toe.

My own inquiries having produced no indication that any non-creationist scientist had ever so much as looked at the prints that the film crew had uncovered, I decided to visit the site myself, and I was able to do so in March, 1976, and again in July. On the first visit I learned that all the prints excavated for the film had been buried again by the river the same year, and were now under several inches of silt as well as under water. I did not find anyone who could tell exactly where they were, and didn't have time to do a lot of digging looking for them.

Dr. Cecil Dougherty, a chiropractor who has been studying the Glen Rose prints for a decade, kindly showed me several track-like marks that were exposed, but none of them showed any clear detail that would identify it beyond doubt as the kind of track I was interested in. One set, right in the Dinosaur State Park, did show a sequence like a set of human tracks. I did not learn anything to suggest that the movie contained any fakery, and I was assured that there had been near-perfect tracks exposed in the past, only to be eroded into their present condition or washed away entirely. The river is reported to be very violent in flood, and it undermines and breaks away slabs of the limestone layers, exposing fresh layers underneath. This process exposes new tracks, which then begin to weather and wear away. Dr. Dougherty told me that some of the exposed tracks he was able to show me had been much clearer in former years, and that he had never seen the river make any such tracks, it only eroded them.

On the same trip I went to the Films for Christ headquarters at Elmwood, Illinois. Stan Taylor was too ill to see me, but his son Paul showed me casts of many of the tracks shown in the film, and Marvin Herrmann loaned me a copy of a cast of an 18-inch track. Some of their casts were a good deal better than anything I had seen at Glen Rose, although none approached the quality I was accustomed to in casts of sasquatch tracks. Several were human-sized and showed heel or toe impressions that looked very human indeed.

In July I returned to Glen Rose to take part in a dig organized by Fred Beierle, of Commerce, Texas, and Dick Caster, of Seattle, Washington, who are members of the Bible Science Association. I had never met either of them before, and of course I do not share their point of view, but they proved to be eminently reasonable men, well educated, and able to hold their own with anyone in logical argument. I had always been under the impression that the debate over evolution had ended decisively a long time ago, and I think most of the general public have the same view, but having been exposed to people on both sides of the question I would now suggest that the evolutionists are seriously underestimating their opposition, and overestimating the completeness of their own case. The object of our efforts in July was to expose new tracks by digging down in the bank several yards back from the river, close to the area that had been exposed when the film was made. That effort failed, as it turned out that we had neither the time nor the equipment to uncover a large area. The limestone layers are not level, and we were not even able to be sure whether we had reached the right layer. However the pump we had rented to remove water from the hole proved very suitable for dredging away the layer of silt covering the tracks out in the river, and when Fred and Dick had left I kept the pump for three more days and recruited the assistance of Dr. Jack Walper, a professor of geology at Texas Christian University in Fort Worth, to clean out, dike off and pump dry perhaps half the area that the film crew had originally exposed. We uncovered most of the dinosaur trail, plus three of the five "giant" tracks crossing it and nine of two dozen "human" tracks.

It then became clear that the human-sized casts I had seen could not have come from this spot, since all the tracks we uncovered were 18 to 19 inches long, with a stride of more than four feet. To that extent, the movie had given me an incorrect impression, but it may have been a mistake on my part. I would have to see the movie again to be sure. In any event, we uncovered two sets of tracks, apparently made by two different individuals, one with a foot about seven inches wide, the other only five inches, but all 18 or 19 inches in length. I am not suggesting that the movie showed any tracks that did not exist, but the small ones certainly were not from the area we uncovered.

The prints had apparently been made in very sloppy material and did not show any detail at all. Some of them were very irregular at the bottom, as was also true of the dinosaur tracks, which were about four times as deep. Of the nine narrow tracks only six were in an unbroken sequence, but it appeared that they probably all belonged to a single trail, in which two prints had failed to sink in the mud. No reason for that was apparent. None of the elongated prints showed any suggestions of claws, which in the dinosaur prints were quite distinct. In some of the prints there was little to indicate which end

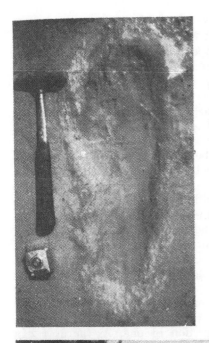

Dr. Cecil Dougherty, above, with a huge track of human shape

The ridge pushed up by whatever stepped in the mud more than 100 million years ago shows clearly in the picture on the left

Below is a 17-inch track in the limestone partly filled with water

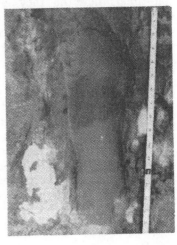

Dr. Jack Walper, left, and Fred Beirle examining humanlike tracks in the limestone bed of the Paluxy River

was which, but whenever one end was wider than the other or showed any marks suggesting toes it was always the same end, while the opposite end of all the tracks appeared rounded. The prints were more than four feet apart in the direction of travel, and less than a foot out of line from left to right. In every way they were consistent with the tracks that would have been left by a human walking in sloppy mud, except that both the tracks and the stride were longer. A human could certainly match the stride without excessive effort, but the prints were definitely too large to be human.

As to any possibility of the tracks being faked, it simply couldn't be done. Some of them had very distinct ridges around them where the original mud had been pushed upward as the foot sank in. For the river to erode a shape like that would be impossible, and for it to be done by human hand would require cutting down the level of the surrounding riverbed over a considerable area. To suggest that anyone had enjoyed the undisturbed opportunity to do that successfully during the making of the film would be ridiculous. Dr. Walper, who is not a creationist, confirmed that the tracks were unquestionably those of a creature with a long, narrow foot walking upright in the same manner as a man. He did not know of any dinosaur that could have left such a trail, and inquiries I have made since have not produced any suggestion of one. As far as I can determine there is no creature known to have existed that could make a similar trail except man or one of his extinct relatives. But no mammal bigger than a cat is known to have lived that long ago, and primates were not due on the stage for another 100 million years.

Having established that tracks of this type do exist in the Glen Rose limestone, there seems to be no reason to doubt the statements of old-timers there that they have seen much better ones which showed five clear toes of human pattern. Tracks such as the ones that can be seen now they call "moccasin prints" because no toes can be distinguished. There appears to be every indication that they are the tracks of a creature unknown to science, and at the very least they indicate the evolution of a foot and stride very similar to that of Homo sapiens well over 100 million years before he got around to it. The only alternative would involve the simultaneous existence in Texas of several types of dinosaurs and at least one bipedal higher primate. You would think that ichnologists (specialists in fossil footprints) would be elbowing each other aside in the rush to study them, but this is one itch they apparently have no urge to scratch.

I am certainly not contending that any of these are sasquatch tracks. It is a separate mystery altogether, which I stumbled on because of my interest in modern giant tracks, and there appears to be no direct connection. There is an indirect connection, however, in

330

that it is another matter of fascinating potential that the scientists who should be interested will not so much as glance at, because it conflicts with their beliefs. I have tried to get some specialist in fossil prints to go and examine the ones I uncovered for them (Dr. Walper's specialty is continental drift) but at time of writing not even those who are only a few hours drive away have bothered to do so, as far as I know, and the river presumably has buried them again.

Where this differs entirely from the sasquatch situation is that there is nothing left to find. The evidence is all there, it only requires someone to look at it.

-18-

Gentle Giants

The gorilla enjoyed a century or so of terrorizing human beings by reputation and appearance. Then people started to study him in his native haunts and discovered that he was a placid fellow most of the time and that when he did throw a tantrum it was mostly bluff. In his book *The Gentle Giants*, Dr. Geoffrey Bourne has this to say about the observations of Dian Fossey, who has spent more time among gorillas in the wild than anyone else:

> Dian also stresses the gorilla's lack of aggressiveness and says that in 2,000 hours of observation she saw only about four minutes of behavior which could be called aggressive. Most of this was bluffing behavior. The closest call she had was when five large roaring males charged at her. The leader was only three feet away when she spread her arms wide and shouted "Whoa". And the whole group stopped. She described mountain gorillas as nothing but introverted, peaceful vegetarians.

It is the unknown that terrifies, and the sasquatch, never having been much bothered by prying zoologists, remains unknown and terrifying. Not everyone considers them fearsome, and not everyone panics when confronted by one, but a lot of people do. Even some of the people who are most outspoken against shooting a sasquatch go out looking for them armed to the teeth, apprehensive and on the alert at all times. I would not deny that the first few times I was alone at night in the Bluff Creek area after seeing the giant footprints, it didn't take much to make the hair stand up on the back of my neck. What's more, if I should be awakened while spending the night in my minibus to find it being rocked back and forth by a nine-foot monster, or even to find a mere seven-footer peering in the window at me, I would be unlikely to remain calm and detached. That would be an emotional reaction, however, and would not make sense, because the sasquatch is not really unknown and should not really be terrifying.

The same point applies here as to a great many other matters. We may not have provided a positive answer to the question of whether sasquatches exist at all, but if they do exist we know a great deal about them, and what we know does not suggest that they are dangerous. Each year there are reports of people being mauled by bears, and sometimes people are killed by them. Each year there are no reports of people being mauled or killed by sasquatches. It cannot be reasonable, therefore, to consider a sasquatch as dangerous as a bear. Yet very few people fear bears to the point where they would not walk in woods where bears are known to be, and the average person, on seeing a bear, will just stay and watch it unless it acts in a threatening manner.

Man is a very poorly-muscled and poorly-armed animal. If he is not equipped with a weapon, there is no animal of equal weight that he would be likely to win a fight with, and probably very few of half his weight. The reason that not just men, but women and children too, can safely go into the wild areas of North America, and not only wander around during the day, but sleep in the open at night, is that there is no animal there that will normally attack a human. There are certainly plenty that could. Cougars on the average are a little larger than leopards and have the same equipment for killing, but I know no record of a man-eating cougar. Of course the very few that have killed a human have quickly been hunted down. Wolves could kill humans easily. As far as I am aware, in North America, they simply don't do it. Bears are more unpredictable, with grizzlies representing a definite danger, although statistically a small one, but only polar bears have the reputation of being actual man hunters. There are many other animals—wolverines, lynx, moose, elk, buffalo, even deer and antelope—that would be capable of killing people, they just don't act that way. Surely the sasquatch should be judged as the others are, by his actual record rather than by his looks.

In the early 1950's, Clarence Fox, who was then a school teacher at Wenatchee, Washington, set out with a friend to hike through the valley of the Napeequa River high up in the Cascades. With precipitous sides, and extremely dense undergrowth, the valley was considered to be all but impassable, and they had a hard time to get through. Right in the middle, however, they woke up one morning to find bipedal tracks, fresh, and larger than grizzly tracks, crossing the gravel bar where they had slept. Clarence Fox saw tracks high in the Cascades on several other occasions, and also collected reports by other people, so he goes back the farthest of any of the current crop of sasquatch investigators, as far as I know.

Another researcher, Keith Soesbe, in Oregon, learned of two occasions when people watched sasquatch approach sleeping men. One report was by three fishermen who were out on a lake in the mountains southeast of Portland, when a seven-foot creature "like a bear on its hind legs, hairy but almost human," circled a member of their party who was asleep on the shore. Another man, making coffee early in the morning, saw a sasquatch come out of the trees and look in at his companion, who was still asleep in their station wagon. The man yelled, and it left.

I woke up once on Mount St. Helens with a herd of elk browsing just a few feet away from me, and they did not move off until someone else woke up and disturbed them. It was a fascinating experience. Why should it be any different to wake up and see sasquatch nearby? They aren't any bigger and they have not shown any greater disposition to be dangerous.

I am not making these statements off the top of my head. It is my practice to do a thorough analysis of the information on my file cards every two or three years, and from more than 1600 reports every instance of behavior even suggesting aggression can be listed on a couple of sheets of writing paper. It is true that many Indian traditions describe sasquatch-like creatures as man eaters, and I don't contend that they never were. They easily could have been. But we do not live in those times, and we have facts to deal with, not legends.

Looking over the list, I find not a single account of a sasquatch killing a human anywhere in Canada or anywhere in the United States east of the Rockies. In Washington there is only the Indian who told Paul Kane that the "Skookums" had eaten his companion on Mt. St. Helens. In Oregon, Ivan Sanderson had a story of two men being killed near the Chetko River in 1890. I don't know where he heard it, and I have never run across it from any other source. From California there was the legend of Deadman's Hole in San Diego County, where two people actually were found dead in the 1880's. Those killings were traditionally blamed on a sasquatch-like creature, but Patrick Householder, in San Diego, did us all the favor of digging up the reports of the inquests. He found that one victim was knifed and the other shot.

The only substantial account of a human being killed is the story from the Idaho-Montana border country well over 100 years ago that was told to Theodore Roosevelt. While it is impossible to be certain, I am inclined to believe that one. But one vicious individual in an animal population in a century is surely not anything to make a case out of.

The exception to this almost unbroken record of peaceful co-existence is found in Alaska, where there are two direct accounts of killings and two indirect ones, all from Indian sources. Bob Betts, who made inquiries in several areas in Alaska a few years ago, was told two stories by Indians living along the Yukon River in the western part of the state. They said that about 1920 at Nulato a man named Albert Petka was attacked by a "bushman" on his boat, where he lived, and later died of injuries, although his dogs had driven the intruder away. The second fatality was at DeWilde's camp, near Ruby in 1943, where a man named John Mire, alias The Dutchman, fought with a hair-covered man and later died of internal injuries. Again, the man's dogs drove the creature away. Alaska in the 1940's or even the 1920's is not exactly ancient times or unknown territory, and I find it hard to believe that such events could have taken place unknown to the rest of the world. Also, if such a creature decided to attack a man I find it strangely coincidental that in each case the man should have survived temporarily with fatal injuries, or that the dogs should have conveniently driven it off. However there is certainly nothing impossible in the incidents described.

The other two stories are even less specific, although much more elaborate. The Anchorage *Daily News*, April 15, 1973, carried a feature article on the abandoned cannery town of Portlock on the Kenai Peninsula. The writer had learned the story during an evening spent with the school teacher and his wife at English Bay while on a boat trip. It went, in part, as follows:

Portlock began its existence sometime after the turn of the century, as a cannery town. In 1921 a post office was established there, and for a time the residents, mostly natives of Russian-Aleut extraction, lived in peace with their picturesque mountain-and-sea setting.

Then, sometime in the beginning years of World War II, rumors began to seep along the Kenai Peninsula that things were not right in Portlock. Men from the cannery town would go up into the hills to hunt the Dall sheep and bear, and never return. Worse yet, the stories ran, sometimes their mutilated bodies would be swept down into the lagoon, torn and dismembered in a way that bears could not, or would not, do.

Tales were told of villagers tracking moose over soft ground. They would find giant, man-like tracks over 18 inches in length closing upon those of the moose, the signs of a short struggle where the grass had been matted down, then only the deeper tracks of the manlike animal departing toward the high, fog-shrouded mountains with their deep valleys and hidden glaciers.

Finally, the villagers could take no more. In 1949 they left their houses and tanks, their wharfs and pilings, and bundled their families off to English Bay and Port Graham, where no rumors of giant, manlike creatures would trouble their sleep. Even today, they refuse to stay overnight at the ruined site of the village, smaller because of 1964 and its waves, and are hesitant to speak of those troubled times.

Perhaps Alaska does have a Sasquatch, perhaps not. We talked to numerous people about the story, and no one would admit having heard it. It is possible that it's just another long winter fabrication of active imaginations.

However, two things tend to lend support: The village of Portlock was abandoned over a very brief period of time in 1949 for no apparent reason, and the villagers of today at English Bay and Port Graham (some of whom are ex-residents), will not spend the night there. They are heavy weights on the scales of believability.

Finally, there is a story published in *Sports Afield* in 1956, in which the writer, Russell Annabel, tells of an Indian being carried off, presumably for dinner, by "Gilyuk, the shaggy cannibal giant sometimes called The-Big-Man-With-The-Little-Hat." The Indians knew that Gilyuk was around because they had seen his sign, a birch sapling about four inches through that had been twisted into shreds as a man might twist a match stick. The scene is set on the Nelchina Plateau, south of Tyone Lake, sometime about the 1940's. According to the Indians, their chief, named Stickman, went down to a lake where they were camped, at night, and nothing was ever found of him but his red flannel underwear. Oddly, "stick man" is a name sometimes used for the sasquatch.

Fatalities in seven of 1,600 reports, even if all were true, would hardly be the record of a dangerous animal. Reports of injuries are even rarer than reports of killings, in fact there are virtually none. One black eye and a few bruises and scratches make up the entire list. There is not even one broken bone. That could indicate either that there is no real animal involved, or that if there is, then it exercises remarkable self-restraint. Anyone finding themselves in close proximity to such an animal might well draw considerable reassurance from the second explanation.

Towards domestic animals, sasquatches have reportedly been

somewhat more violent. One in Oregon is said to have torn apart two hounds, and there are several similar reports in northern California. Dogs have been reported killed on two occasions in Missouri, and in Michigan, Indiana, and New Jersey. I have only one such story first hand, from a hound hunter in Willow Creek, California, who says two of his dogs were ripped in two when they chased a shadowy thing that left huge footprints near their campfire. There are other stories of the creatures being seen fighting with dogs, without casualties.

Other domestic animals have come away unscathed in the entire western half of the continent, but in Missouri, calves, sheep, cows, horses and goats, as well as dogs, have been reported killed. In fact, out of 10 reports from Missouri, there are five in which the sasquatch killed domestic animals and one in which it tried to make off with a little boy. Four of the stories are very old and vague, but still it is interesting that there should be such a concentration of stories of that type from one state. There are also reports indicating that sasquatches killed a cow, a calf and two horses in Florida, a cow in Pennsylvania and a calf in Tennessee, and a few others. In most cases the reports are vague or the evidence is circumstancial. The same can be said of stories of the killing of chickens, rabbits and ducks. It may be happening, but apparently not often.

With regards to dogs, it should be kept in mind that the sasquatch can hardly be considered the aggressor if he turns on animals that are pursuing and harassing him. There are other stories, that I suspect can be found almost anywhere, of dogs being mysteriously killed by something that struck them a heavy blow without leaving claw marks. Of course such injuries could be caused by cars, but I have talked to at least two people who told me they had heard something slam against the house at night and later found their dog dead of such an injury. Probably I shouldn't go into this, since there is nothing to connect those stories with sasquatches other than that a huge beast with a hand could easily cause such an injury. The stories belong in the broad category of strange incidents that might most easily be explained by blaming the sasquatch if it were established that such a creature existed, but that cannot be used as evidence to prove that it does. As I keep no files of that sort of information I can't discuss it with any authority.

Another thing that interests me about dogs is that they appear to have an instinctive fear of these creatures. I can't cite chapter and verse, I just have the impression from listening to many and various stories that there is a pattern indicating that dogs are instinctively afraid of them. There would be nothing unexpected about dogs learning to be afraid of them, and perhaps that is the actual situation, but in some cases it is hard to see how the dog would have had any previous opportunity to encounter one. The interesting thing about an instinctive fear would be that dogs do not originate in North America and would have to have acquired the instinct on some other continent.

Another form of activity that could be considered as aggression is chasing people, whether on foot, horseback, bicycle or in a vehicle. People on foot have reported being chased on 17 occasions that I have heard of, mainly in British Columbia and Washington. There are also five reports in which the person was rushed at but did not flee, and in each case the animal called the whole thing off, so it may well be that nothing more than a bluff was involved in any of the incidents. That seems even more likely when one considers that by no means all of the 17 chasees had any secure place to run to, yet none was overtaken. The two people chased on bicycles presumably could also have been caught, and also the one on horseback, if the witnesses who have estimated sasquatch sprinting speed have been anywhere near correct. People in vehicles, or should it be vehicles containing people, were chased on eight occasions. Cars have also been shaken or hit six times and jumped on five times, and there have been three apparent attacks on occupied buildings, although nothing else like the Ape Canyon attack. There are also ten reports of sasquatches throwing things at people, rocks on eight occasions and other objects twice. That doesn't cover all the accounts of things being thrown, only things being thrown apparently at a human target. In the only instance when anyone reported being hit by a rock (but not hurt), the sasquatch was not reported seen until afterwards.

The idea of an animal shaking or rocking a car with a person in it takes a bit of getting used to, but it seems that sometimes a sasquatch will feel the urge to do just that. They sometimes try to shake houses too, but without much understanding of construction, since they always seem to try to do it from a corner. Here is a fairly typical story, from The Blue Lake *Advocate*, September 24, 1964. Blue Lake is in northwestern California, about 40 miles south of Bluff Creek and near the coast:

> An abnormal brown bear (some call it "Bigfoot" or the Abominable Snowman) was reported seen at 1:00 a.m., Sunday, September 13th by Benjamin Wilder who told of being awakened while sleeping in his car at his camp on the water pipeline which is at the four thousand foot elevation.
>
> Wilder told how he was awakened by his car being shook up two different times. He said at first he thought it was an earthquake, but thought, with such a hard shake, there should be rocks rolling. Then after the second time, Wilder

338

put on his flashlight and to his surprise there was a large animal standing beside the driver's door of the car.

The animal had its two arms on top of the small car. He said it had long shaggy hair, about three inches long, on the chest. Wilder said he tried to scare it by shouting, but the bear did not move. It only made noises like a hog. Then Wilder honked the car horn. This scared the big animal and it took off, walking on its hind feet, and disappeared over the hill. Wilder said the animal never got down on its four feet, and he never got to see the face of the animal.

Something else that is worth considering regarding stories of aggressive behavior is that the percentage of such reports for which there is a witness now available is very low. Very few of the reports of violence to people or the killing of animals are what I would class as direct reports, where a witness was interviewed by me or someone that I know to be a serious investigator, whereas on the average close to a third of all reports involve witnesses who have been interviewed. That is not to say that all reports involving interviews are correct, but I think the percentage is likely to be substantially higher than for indirect reports. On the other hand, a story involving violence is likely to spread farther than one lacking that interesting element, so it would be logical that reports of that sort would be more likely to reach us from indirect sources than would more mundane ones.

There is another category of reports I have not touched on, that of people being picked up or carried off by sasquatches. For actual kidnappings the only direct story is Albert Ostman's and all the others I know of are old Indian tales. There was one hunter carried some distance by a sasquatch, but it was running away from another hunter at the time and appears to have collided with the second man by accident. At Kinlock, in Missouri (where else?), there was a story on KXOK radio in August, 1968, of a creature picking up a four-year-old boy, and dropping him only when his aunt screamed. The woman said the animal was a bear, while the boy when shown a model of a gorilla said that was what it looked like. The story was included in an article by Loren Coleman about manlike monsters, so I assume that the thing stood upright and picked up the boy in its forelimbs, but that is not specified.

In eastern Florida there are three stories of truckers being grabbed in the cabs of their trucks while they were sleeping, but they suffered nothing worse than being pulled out on the ground, and there is one similar report from Texas involving a man sleeping in the back of a pickup truck. There are also a couple of people who tell of being knocked down in sudden encounters, but when someone is in physical contact with a thing as powerful as a sasquatch and ends up

339

horizontal but unhurt, I would consider that to be evidence not of aggressive intent but of deliberate restraint.

Perhaps the most interesting story of that sort was told to me by a young woman at Wilsonville, Oregon, who wasn't just picked up, but was thrown, landing on the seat of her jeans in a patch of thistles. It happened so fast that she was left with only impressions of the thing that grabbed her, but I think they were sufficient for identification. It happened on August 29, 1970, about 5 p.m., and thanks to a call from one of her neighbours I was there to talk to her the following day. She said that she had heard some shooting over beyond a patch of woods on their farm and had gone down from the house across the field with the intention of ordering whoever was shooting off the property, taking her own shotgun with her. The woods were separated from the field by a barbed wire fence. She told me:

I put my gun up in the corner and started to go through the fence. I went through the fence and it just grabbed me and threw me back over the fence. I was still kind of crouched. I'm sure I didn't even have time enough to get straightened up from going through the fence.

How did it pick you up?

I'm not sure, I just know I was flying through the air.

And what did you see?

Well, it had hairy . . . I don't think it had hair on its palms . . . it was just hairy, and then afterwards I noticed an odor, a terrible odor.

When did you see the hand?

When it grabbed me.

Do you remember where on you the hands were?

I saw it on my arm.

Did you see the face of this thing at all?

I don't think so. It was just big and hairy and threw me, and by the time I got through rolling and back up, I don't remember seeing it again. I don't know if it went right back in the bush or what happened.

Did you see its eyes?

I thought I did. I thought they were beady, but I'm not sure if it was its eyes or what it was. They were reddish, what it was I saw.

Did you see anything to indicate how big it was?

I thought when it threw me it stepped over the fence. It straddled the fence.

At the same time it was throwing you?

Yes, it straddled the fence, and by the time I got through rolling and stuff it was gone.

340

Could you see that it was something upright?

Yes.

You never really saw the whole outline of the thing?

No. I know it had to be big though, to throw me.

Between the distance she was thrown and the scrambling she did before regaining her feet, she was perhaps 10 or 15 yards from the fence, and she walked back to get the shotgun. When I talked to her she could not understand how she could have done that. She wasn't very keen on going back down to the end of the field even with several armed men with her. As she was starting back to the house with the shotgun her husband came running towards her. He said he had heard her scream. She remembered trying to scream but thought she had been unable to make any noise. Aside from being stuck by some thistles she was not hurt.

The small patch of woods certainly was not the permanent abode of any large animal, but it was strategically placed for a daytime stop for a wild creature crossing between the Coast Range and the Cascades. We thought it possible that the animal had moved over to the edge of the wood to get as far as it could from the shooting, and considered itself threatened when someone else approached from the opposite side. What it did certainly got across the message that it did not want company, but to have done it so swiftly, yet without causing the slightest injury, it must have been deliberately very gentle. No one ever did find out what the shooting was about.

The neighbour who phoned me said that the creature's smell was still obvious on the sweater the lady was wearing. However it was half a day before I could get there, and by that time there was no smell that I could discern.

That's all there is regarding violence on the part of the sasquatches, but there is another element to consider. Far more often the animals have been on the receiving end of far more deadly forms of violence on the part of humans, and the only tale of attempted retaliation is that at Ape Canyon. In short, there is no evidence that even wounded sasquatches are dangerous to people.

Lumping all those stories together may be a mistake, creating the impression that there actually has been a lot of aggressive activity and that sasquatches are indeed something to be afraid of. I hope not, because the truth is the other way around. Everyone tends to begin with the idea that such a creature would be very dangerous, but the more one learns the more obvious it becomes that if there is such an animal, then like its cousin the gorilla it must be basically a gentle giant.

The gentle gorilla, however, does put on a good bluff at times, and the sasquatch apparently does the same. Take for instance the

341

experience of two young couples from Yakima who went on a shopping trip to Seattle in June, 1970, and didn't want to spend the money for motel rooms. They drove back up the road towards Yakima until they got away from civilization, then parked their car on a side road in the mountain forest and got out their sleeping bags. It was about 11:30 p.m. Here are excerpts from a tape recording as one of the young men talked to Roger Patterson:

My friend started talking about different calls, you know. He said he was from Missouri, said he could make a black panther sound, like a woman's scream, he said. So he did that, and I did a wolf call. About that time, after we completed our calls we first noticed a big form on top of the hill . . . To me it looked a good nine foot tall, very, very broad. I didn't get that good a look at it . . .

We thought we got a glimpse of it coming down to the left of us and that was the last time we seen it for a while. About half an hour later it was coming down the road. N-- shone the light on it. He started screaming, "Look, look, there it is!" and it was coming down the road and I turned around and I saw it.

He said he was going to stop it by blinding it in the eyes and it stopped cold and jumped behind the embankment, behind some bushes, and it looked like it was going to stalk us. We could see its hair and the outline of it and the eyes. You know like you'll shine another animal in the eyes, in the light they'll glow.

I jumped in the car and accidentally locked the door on N-- and the thing got up and rushed us and poor N-- couldn't get in because the door was locked. He started screaming and everything else so I got the door unlocked and he jumped in the car and all of a sudden that thing just hit the car and really made a ruckus.

It seemed to want to get in for some reason, or maybe it was trying to be friendly or curious. Whatever it was, we didn't like it. So then it went away and we decided, well, we're going to split, take off, but he couldn't find his keys. He figured he left them in the trunk. So we got out and sneak around there and no keys.

Then we heard . . . saw the thing so we think we better get back in the car, it would be safer'n out there, so we kept hearing the thing, kept seeing it and we decided well we've got to get out of here. We decided the keys have got to be around there, so we both got out and I took a pop bottle and I busted it as a weapon, but I busted it all to pieces so we stood there and here the thing came and we never got nothing . . . had no chance of leaving cause we didn't have the keys and couldn't find them.

So the thing started rushing around . . . just walking around us out of curiosity . . . well it didn't do any more damage after that, didn't attack the car and we just stayed there all night.

Next morning, it was about eight o'clock, we got out of the car and walked around to make sure the thing wasn't there. Actually we looked around first, we figured it wouldn't show in the daytime. We found the tracks, down in the cut we noticed these big tracks of this here particular thing, and they was good sized, they looked like a human's footprint.

We followed the tracks down the road for quite a ways and we lost them going into the brush. They went down towards a ravine. We could see some of the tracks and then we lost them. We searched around again for more tracks but they just all seemed to lead down that way.

The face was sort of a human-type face. It was pointed like on the top, it had a large pointed forehead on the top. A human's is sort of rounded off, his wasn't, it was dome-shaped. It wasn't really that it had a lot of hair on its face, except the side features, it had hair on that. And he had a nose, it resembled a human nose except it was flatter and wider than a human's nose would be.

His mouth was somewhat like a human's, except it was bare, but a larger mouth. And he had something like carnivorous teeth, like a dog's teeth, and his eyes were good sized, and like I said, they shone in the dark. Real bright, like whitish-yellow, very very glowing. He had a bull neck, didn't have an extremely long neck, in fact there wasn't hardly a neck there, it was a bull's.

His skin had a yellow tinge to it like it had been bleached in the sun, like leather would be, buckskin would be, sort of tanned like you get in a taxidermist's and stuff, that's what it looked like. His torso was very muscular. I didn't get a good look at his hands because we didn't have the light actually on his body that much, you know, when we were trying to blind him in the eyes is when we actually shone the light on him outside. And he was bent over, stooped over. He had a pretty long stride to him. His legs were long and muscular and his waist tapered. He did have actually a large waist. I didn't really get a good view of his back. He had brown hair, real good shape, wasn't ragged or nothing.

He was about nine feet. I'm six-two, and he stood a good three feet taller than I was. Weight I'd say a thousand two hundred, something like that.

The witness said he had hunted Kodiak bears and this looked about the same size, but didn't resemble a bear, except for its fur. He thought about it being someone dressed up, but that wasn't possible because of the size and the way it ran, it was "lightning quick." He said that the dog they had with them, "he screamed too" and the hair on its neck stood up. "He just about went berserk."

The giant creature sometimes made a sound like a moaning dog. None of the four in the car got any sleep, they could see the thing once in a while and hear it walking up and down. It never peered into the car, except one time it looked in the back window. The thing looked angry, but they thought that it couldn't actually be angry or it would have ripped the car to pieces.

Roger and the witness went back to the spot a week later to spend the night, and tried to attract it by imitating animal calls, but without success. I have never talked to any of the four people involved, and I do not have addresses for them. As a rule if someone has already been interviewed by someone active in the investigation those who are in contact with the first interviewer do not bother them further. I did talk to another man who had written to Roger in 1966, because his story was a most unusual one, and I wanted to size him up for myself. Here is part of what he wrote:

I had been reading a story written by Ivan T. Sanderson about a Bigfoot monster that people had seen in Northern California. As I was reading the story I suddenly had the eerie realization that I, too, had had a similar experience. As the story came snaking its way out of my subconscious, I began to remember more and more of what happened. I had me a bad case of the jitters as the memory uncoiled.

The first part of the story took me back to 1952, when I had gone down to Orleans to start preliminary work on a logging operation with two men by the names of Lee Vlery and Josh Russel.

One evening Josh told me Lee had gone up to Happy Camp, but not having transportation back, wanted me to take the Mercury and go up and get him. I had driven the extremely crooked and dangerous road up there, but not being able to find him started back alone to Orleans.

It had been raining very heavily and after going back a few miles I found there had been a slide across the road. There was a man with a flashlight there who told me I could still get back to Orleans by way of a detour across the river.

He said it was a dirt road that went through Bear Valley and would come out at the mouth of Bluff Creek a few miles below Orleans.

I had been driving slowly down this road for about twenty miles, I guess, sort of daydreaming, when I saw it . . . dimly in the headlights and the rain was the shaggy oranutang-like apparition of a human. For an instant I had the impression the shaggy hair of the creature was a hoary blue grey in the headlights. An ogre! I remember thinking, but the thing swiftly back-pedalled off the road and behind a tree. I automatically passed it off as imagination and drove on by the spot.

Suddenly, without warning, the car went into a violent and unreasonable skid. I brought the car back under control, but for some reason glanced into the rear view mirror. In the dim light of the tail lights and licence-display bulb, I thought I could see a savage looking face looking through the rear glass. I continued on, and when I looked again there was no face, so again concluded it was imagination.

I had gone another quarter mile I guess, when across the road was a small six-inch sapling—I stopped the car and got out, intending to drag it aside if possible. Suddenly I heard the swift thud of flying feet of something coming down the road. Reality was upon me and I remember cursing myself for not paying attention to what I had previously seen. It was the shaggy human-like monster I had seen in the headlights.

It at once started circling around me, snarling and acting very menacing. It kept this circling up for some time and once came up quite close, and I could see its face reflected by the headlights much better. The eyes were round, and rather luminous, the hair on top of its rather low and rounded head pretty short. Its eye teeth were far longer than a human's, also the chest and upper part of the torso was rather bare of hair, and also leathery looking. It wasn't too tall—not much more than my own 5 feet 9 inches, although it had a stooped, long armed posture.

Then it suddenly changed tactics—it would stalk off down the road but would come charging back, like a bat out of hell, when I started toward the car. The hour was late, the thing was becoming more and more menacing, and I was almost paralyzed by this time, paralyzed by fear.

Suddenly a plan of escape, born out of desperation, popped into my mind. Since the monster seemed to think I couldn't get away, why not, when it went down the road again, playing cat and mouse, try and get in the car and smash through the sapling. This I did, and sprang for the door of the car a dozen feet away. No sooner was I inside when there it was, trying to claw through the window. I jerked the car into gear, floored the accelerator, and can vividly remember the wet sapling glistening whitely in the headlights as the car slashed it aside.

I remember then the scream of rage and frustration it then gave. It was a curious trumpeting sound like the scream of a stallion and the roar of a mad grizzly. The car then felt as though it were being held back by something half riding and attempting to stop it, but the powerful Mercury proved too much for it, and after a couple of hundred yards I felt no more resistance.

To top this unbelievable experience off, believe it or not, I promptly forgot the whole experience. Then and there it went out of my mind. Not even the next day when Lee asked me if I had seen anything unusual on that road last night did I remember. (He had come later from Happy Camp with another man he hired to take him to Orleans.)

A few days later an incident happened that should have brought the experience back but didn't. Lee noticed a big dent in the grill of the car and asked me how it got there. I told him I didn't know. Incidentally, Lee told me that something had tried to push them off the road, when they came through on the detour. He said there's something strange going on around here and let the matter drop.

Talking to him several years after he wrote the letter I found this man completely serious about the story, yet still unable to be certain that it had actually taken place. He had a normal recollection of the trip to Happy Camp, and of the car's owner asking about the dent he had put in the grill, and he was fairly sure that the experience with the creature was real, but had been wiped out by some kind of shock reaction until he saw Ivan's article, more than seven years later.

He wrote to Roger two years after George Schaller's book *Year of the Gorilla* tipped off the general public to the fact that a gorilla's ferocious charge is generally a bluff, but before that information was by any means common knowledge. He told me that when the thing was circling him at first it was five or six feet away but once it stepped up very close to him and looked right in his face. He raised his fist as if to hit it in the face and it jumped back. He felt that it was getting more and more angry and would attack him sooner or later, but it never actually touched him.

There is no doubt that if a sasquatch attacked a lone man in the woods it could easily kill him, and if his body were never found he would be just one more of the many people who have disappeared in the forests and mountains all over North America over the years. That being so, there is no way to be sure that no one has been killed by a sasquatch in recent times. There are many other ways a person alone can die, however. If there is no one who will come looking for him within a few days, any illness or injury that leaves him unable to walk is likely to prove fatal. Getting lost can be fatal too. In spite of the many things that can and do happen, only a very tiny proportion of those who take the chance of going alone into the woods fail to come out again, so even if attack by a sasquatch were to be added to the list of potential perils it would still be a negligible danger.

Assuming that there is a real animal out there, the chances of any particular person ever seeing one are almost infinitely small. If they do see one, the chances that it will take a run at them are much fewer yet, and the chances that it will touch them much fewer than that. As to the chances that it will injure them, up to the present I have encountered no clear cut case of that ever happening.

-19-

Footprints, But Whose?

Sighting reports are generally far more interesting than footprints, but they are not of equal quality as evidence. A sighting report can be entirely imaginary, in fact all of them could be. Footprints are real. They have physical existence and can be studied by people other than those who first reported them, either directly or in photographs and casts. They also require a physical cause. Where there are footprints there has previously been some living creature there to make them. And whether animal or human that creature is subject to being caught in the act.

A thousand reports of hairy bipeds can be explained away as a thousand peole claiming to see something that wasn't there, but 600 footprint reports can not be dismissed so easily. A fair number of filed reports undoubtedly immortalize miscellaneous depressions in the ground that were never footprints at all or else were originally made by unassuming animals that wouldn't dream of claiming to be sasquatches. It is instructive to see how excited people can get sometimes about a couple of overlapping cow tracks. Then there are a lot of reports that stand on the same footing as the sighting reports, since there is only some person's word for it that anything ever existed. But when all those are deducted there are still hundreds of sets of good tracks that independent witnesses were able to go and see, and more than 100 sets that have been photographed, or cast, or both.

Those tracks present an apparently simple problem, but one that has never been solved. It should, on the face of it, be easy to prove what makes them, but it has not been done.

The steady accumulation of sighting reports without the creatures

348

described ever being caught, killed or found dead is probably weakening the case for the existence of any such animals, but the steady accumulation of verified reports of clear, manlike, giant tracks surely has the opposite effect. If they are all made by humans, then humans should sometimes be caught making them, and that isn't happening.

There have been some faked tracks, of course. I know of several instances in which people claimed to have made tracks that were generally considered to be genuine, and then produced the equipment they did it with and demonstrated their techniques. There have also been a number of other attempts in which the fakery was obvious. Only one of those incidents involved manufacturing really impressive-looking tracks, but one is enough to establish that it can be done. It is not, however, all that easy.

Some people think all it takes to make false sasquatch footprints is a pair of giant rubber feet from a joke shop. I have seen newspaper stories quoting "the sheriff" as giving that comforting explanation in more than one instance. Of course there is no strength in such contrivances to make the toes and heels dig in at all. Others think it is just a matter of nailing carved wooden feet on an old pair of shoes. I have tried carved feet and also fibreglass copies of actual footprints. In some kinds of soft material they work fairly well, but it is not possible to create any illusion of individual toe movement. It is quite obvious that all the prints are made by the same rigid objects. No doubt someone who spent a lot of time practicing could do a better job of trying to imitate toe movement than I have done, but I don't think they would fool anyone who made a serious study of a long string of tracks.

Another problem is the depth of the tracks in materials like wet sand that can support a lot of weight. The best pair of wooden feet I have carved is only 14½ inches long, six inches wide and four inches across the heel. A typical Bluff Creek "Bigfoot" track has half again as much area as that. The carved feet are curved to roll the weight in, rather than spread it over the whole area of the foot at once. I weigh close to 200 pounds and I have used those feet while carrying a man who weighs more than 250 pounds, for a total weight of a little over 450 pounds. With that weight we did not make a single track in sand that would be worth casting. In loose, dry sand they had no shape. Where the sand was wet and packed, they didn't sink in far enough.

I am satisfied that the weight problem can be solved by a hoaxer with enough ingenuity, but I don't know any way to combine weight with the degree of individual toe movement that was evident, for instance, in the prints at the scene of the Patterson movie. I have casts of nine different tracks from that set, and the distance between the big toe and the little toe varies by almost an inch. It would obviously require a very ingenious device or technique to duplicate that, and to do it with the impression of great weight, while taking six-foot strides, and adding such requirements as walking up and down cutbanks and going under logs only three feet off the ground. If there is anyone who can do it, I wish he would come forward and give a demonstration. He would certainly save a lot of people a great deal of time and trouble. As a matter of fact it would save me so much trouble that I'll make an offer right now of $5,000 for the first person to show me how it is done.

There are a couple of things that give away tracks made by someone wearing carved feet. One is that the middle of the print doesn't sink in very far. It is possible to put all the weight on the heel and on the toes to make them dig in, but not on the center of the foot. The other is that if an attempt is made to produce long strides and deep prints by bounding along, the toes dig up a lot of loose dirt or sand and leave it heaped behind the toe prints. I expect that a sasquatch could leave the same effects if it wanted, but I don't know of any animal that walks that way.

Aside from fakery, the big problem with tracks in recent years is the variety of strange-shaped ones that have been turning up, particularly three-toed tracks. It would be easy and convenient to maintain that only one standard type of five-toed track is genuine and dismiss everything else. That would avoid a serious complication which I am sure must be totally unacceptable to many people with scientific training who might otherwise be open-minded regarding the possible existence of a giant primate. It would also be very much in accord with my own feelings.

I recognize that footprints with not enough toes are just as much of a physical fact as those with the right number, and just as much in need of an explanation. I also recognize that it would be silly for someone faking tracks to advertise the fact by anything as obvious as making the wrong number of toes. He might as well fake three legs. Someone might do such an odd thing once, but surely not over and over in places hundreds or thousands of miles apart. Four-toed tracks are not a problem if it's just a case of the little toe not showing. There are tracks, including the ones Roger Patterson and Bob Gimlin cast after getting the movie, with a little toe small enough that it would hardly be missed if it failed to mark. It is easy to imagine a mutation in which that toe did not develop at all. But four-toed tracks, and worse yet, three-toed ones, in which all the toes are about the same size, have no explanation.

The closest thing to a four-toed primate is an orangutan, with just a stub of a big toe more than half way back on its foot. Superficially its foot has some similarities to the thin three-toed tracks at Fouke, Arkansas and some thin but much larger three-toed tracks in southern California, but the orang's foot isn't adapted for walking flat on the ground at all, so it could hardly represent a stage in the development of a foot for a ground-living biped. Looking at photos of gorillas and oranutan feet, it is possible to imagine the evolution from the gorilla of the typical "Bigfoot" and from the orang of the Fouke foot, but whereas the Bigfoot type is reported all over the place, at least out west, the Fouke foot has no near relatives and no resemblance to most of the other three-toed prints.

Returning to the more conventional types of tracks, there are a number of reports well worth some special study. I have referred to the tracks at Ruby Creek, British Columbia, in 1941, reported to have crushed potatoes in the ground and stepped over a railway fence without breaking stride. How well are those claims documented? I have spoken to six people who saw those tracks when they were fresh, and have listened to two couples reminiscing about them 16 years after the event, at a time when they had not seen each other for many years. I also have statutory declarations made by those four. I don't recall whether any of them but Esse Tyfting mentioned the crushed potatoes, but all four were certain that the prints indicated very clearly that whatever made them had casually stepped over a wire fence four to five feet high, with no indication that it had jumped or had stepped on or pulled down the wire. One thing that made quite an impression on me during the conversation was the fact that while the two ladies remembered the tracks probably as well as the men did, they were quite casual about them. To me, the idea of such tracks being real had tremendous impact, and that had certainly been the reaction of the two men. Apparently their wives had not found it so even when they were looking at them.

I usually tend to downgrade reports of tracks in snow, but there are exceptions. One of the most notable of them concerns a set of tracks seen by several elk hunters on Coleman Ridge, near Ellensburg, Washington, on November 6, 1970. Those tracks were studied by more than one hunting party, and reports and photos of them have come from more than one source—in fact I think that Roger Patterson was notified in time to see the tracks himself, although I did not hear of that until after his death. I have a letter from one of the hunters, Oscar Hickerson Jr., of Renton, Washington, giving a brief description of the tracks, but there is a better account, a letter which Nick Carter, an old friend of Mr. Hickerson's as well as a sasquatch researcher, wrote to George Haas. The following is an excerpt from it:

Oscar is about 55 and has spent most of his recreation time in the Washington, Idaho mountains and wild areas all his life. He had read about the California Bigfoot sightings in *True* magazine but had never seen anything like that line of tracks in the snow before. At first he, and his hunting partner, Mr. Jess Helton, thought it was some sort of hoax, but a little investigation convinced them otherwise. Both are experienced hunters and trackers.

The two men had gone to a campsite, previously selected, in Hunting Unit 4 E on a Washington Game Dept. map. The spot is about 30 miles N.E. of Ellensburg, not very far east of a game preserve. They followed Coleman Creek up to their camp, which was on Coleman Ridge. They pitched their tent on Nov. 5 and intended to spend the 6th cutting wood in preparation for elk hunting which opened Nov. 7 at dawn. During the night Nov. 5-6 some six inches of fresh snow fell, the first of the year. It stopped about 5 a.m. Nov. 6.

They woke up, knocked the heavy wet snow off their tent where it was sagging in, made breakfast and were getting out the chain saw when a man, name not given, came up from his camp about 100 yards farther down the ridge. He was all excited about something that had made big tracks through his camp along a somewhat circular route. He was scared, or somewhat upset, for the tracks went right along beside his camper truck and he thought the thing had been looking in his windows. From the lack of new snow in the prints they must have been made between the time the snow stopped, at 5 a.m., and when he got up about 7 a.m. Then he went back to his camp, also to cut wood.

Oscar and Jess sawed down-logs and talked about the news. They thought then it was some joke, but about 11 a.m. they took a break and went over to take a look. The sun was out and it was just above freezing, not thawing very rapidly. They saw the tracks, made by something walking on two legs, and measured some with a yo-yo tape. This is Boeing slang for a six-foot roll-up tape measure. (Both Mr. Hickerson and Mr. Carter worked for Boeing aircraft company). The tracks measured 17 inches long and 9 inches wide. Oscar said he could stand with both his boots on one track touching each other and the two boots together did not reach the outside edge of the track across the ball of the foot.

He was impressed by the depth the tracks sank into the snow and by how hard the bottom was. Like packed ice, and very flat, no arch showing at all. At first glance he thought

Tracks in the snow found by elk hunters near Ellensburg, Washington, *November 1970.*

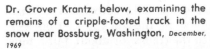

Dr. Grover Krantz, below, examining the remains of a cripple-footed track in the snow near Bossburg, Washington, *December, 1969*

about boards on a man's feet but the stride was far too long and they sank too deep. A second glance showed toe action, snow kicked up as the thing stepped ahead, and some heel drags behind individual prints.

In the men's camp area the tracks were fairly close together, 30 to 40 inches, and Oscar got the feeling the Bigfoot had been looking around, moving slowly, for the stride widened beyond the camp as it strode off normally. They followed and came to a place where Bigfoot had stepped over a down log without disturbing the ridge of fresh snow on top. Oscar and Jess looked carefully for some signs of tracks going around the ends of the log, still thinking hoax, but there were no marks at all in the fresh snow except the one line of prints right over the log. He and Jesse had to crawl up over the log where Bigfoot simply stepped over. Oscar is 5' 9½" and Jesse 6' 1" tall. They figured Bigfoot must have had legs 5 feet long to make that step. His description of crossing the log was "like getting on a horse" and there were several more along the track farther on.

At that time they followed the tracks for about half a mile, zig-zagging through the woods, curving back up behind their own camp. They cut across to the tent, got their rifles and continued following Bigfoot, also looking for elk sign, most of that afternoon, for a total of about three miles, with all the curves and zig-zagging, meandering around etc., to where the tracks finally went down out of the snow into Coleman Canyon. He said that at times Bigfoot idled in one place, back and forth, then would stride out for a way, then idle around some more. The tracks did not circle but switch-backed around contours. Full stride was measured at between 4 to 5 feet, length between tracks varied, some were 4, some 5 and others in between.

The pictures were taken first, in the camp area. Oscar started down to see the tracks without his camera, but went back and got it, fortunately. He took the three shots around 11 a.m. before they tracked Bigfoot into the woods . . .

Oscar tried to match stride where prints were close together, and it took his maximum stride to "almost" reach. When Bigfoot really stepped out it took two of their strides to match one of his . . .

The one thing about those tracks which astonished Oscar most, and impressed him deeply, was the flat bottom of those huge feet, and how hard the snow was packed.

In his letter to me, Mr. Hickerson said that if he had been wearing something on his feet the size of the prints it would have been like wearing snowshoes, and there would have been very little depth penetration in comparison to the tracks. Normally with tracks in the snow there is no way to be certain of the condition of the snow at the time the tracks were made, so that it is impossible to judge how much weight was involved in making them, but there would have been no such problem on this occasion. In any event there would have been no way a man could have cleared the fallen logs without breaking stride or disturbing the snow on top of them.

Another impressive set of tracks in snow was seen and photographed in 1934 by Dave Zebo, who was aviation director for Humboldt County, California, at the time when he was interviewed by a correspondent for the *Humboldt Times* at Eureka in 1960. He said that while living in Weaverville he obtained permission from the Forest Service to spend a night at the lookout on the top of Mount Bally, south of the city, and set out on an 11-mile ski trip to the top of the mountain. The story continues:

Two miles above the timberline, Zebo ran into strange tracks in the snow. There was no animal or human to be seen within range. He stated, "I have never seen anything like these indentations of tracks before or since." The tracks were deep and heavy, but the spacing was what especially drew his interest. The tracks were from 4 to 6 feet apart. Too far for the stride of a normal man, but they were single tracks of a two-footed person or creature.

Pointing to the human element, Zebo said, was the fact that an animal will meander. A human, usually takes a straight path (and sometimes the hardest way) to his objective; while an animal is known to meander to find the easiest direction.

The footprints in the snow, of which Zebo was so curiously engrossed that he took photos of them, went from the bottom of the mountain to the top, from west to east; there was no deviation at all.

"I followed the old trail, and as far as I could see I saw the tracks, making the single line," Zebo said. "There were no other tracks around and I stayed the night in the lookout and came back down the next morning. A heavy snow fell during the night and covered the tracks."

The photos gave Zebo proof that the experience really had happened, and upon returning to Weaverville, he had the pictures developed. He showed these to a number of persons in the vicinity.

Speculation ran high, but no one came up with a solution,

or among those contacted, had anyone ever seen such an incidence. The forest personnel were among those contacted, with no better luck at a solution. Everyone was interested and intrigued, and discussed the event for days without solving the mystery. "In those days," Zebo said, "we had not heard of Bigfoot." He has since wondered if Bigfoot was the answer to the puzzle.

To summarize the experience: The big tracks were definitely there. They were single, as a human's would have been, but too wide apart in stride (they never hesitated but went energetically up the mountain, as if made by a creature with gigantic strength) for an average man's . . .

When I spoke to him on the telephone in 1970, Dave Zebo said that the tracks seemed to him to show considerable weight, sinking four to six inches in hard snow that would have supported the weight of any small animal. No toe prints were discernible, he said. Here again is a set of bipedal tracks of a creature doing something that no man would have been capable of doing — in this case taking strides that averaged five feet in length while climbing mile after mile up a very steep slope in snow. There is nothing in the story or my notes to indicate how he knew the tracks were ascending, but apparently that was obvious. In any event, as I recall, they continued in the same way on the opposite side of the mountain, so they had to be going uphill one way or the other.

Generally there is a great advantage to tracks in snow, since there is no place where the creature making them can get out of the snow it ought to be possible to follow them to the end and find out what is making them. In the forests of the Pacific slope, however, that is not as simple as it sounds. Presumably a man would need a snowmobile or an aircraft to catch up to an animal capable of walking up mountains with five-foot strides, but in the kind of terrain where most of the reports come from tracks would be hidden from the air under the evergreens, and much of the area is too steep or too overgrown for a snowmobile. For a man on foot there is an additional problem besides his inability to travel fast enough. It was stated very well by a farmer who had found huge "human" tracks crossing one of his fields and heading up onto a mountain. He told me:

I followed the tracks back to the hill, and the farther I went the less I wanted to catch up to what was making them, so I turned around and went home.

There is another story, told in a later chapter, in which a man was able to observe two of the animals as a result of following their tracks in the snow, but such reports are very rare. Tracks in snow could also be expected to show something of the activities of the creature that made them, and I do have two reports of such tracks in connection with theft of parts of carcasses left in the open by hunters, but I have not talked to the people involved. Another set indicated that the creatures were eating the soft lower parts of clumps of grass.

Jim Attwell, of Skamania, Washington, described a set of tracks he saw near Port Townsend in the late 1920's that came out of heavy brush and returned to it again after walking for a distance along an exposed water main.

These tracks came out of a heavy stand of timber and downfall, near where the trail was, never once using the trail. I inspected the tracks on the water line and they appeared to be made by a barefoot man. I did not measure the tracks but would guess from memory that they were made by a large man and about a number 10 or 11. I had never heard of the Abominable Snowman at that time so just guessed that it was some nut of a mountain man that one might find around Quilcene. This bare foot track walked down the pipe about 100 yards and on leaving the pipe he or she jumped about 6 feet across the ditch and landed on a 12-inch log covered with snow, something no logger could have done with caulked shoes, and then headed up the mountain through the roughest kind of going, downfalls, brush and rocks covered the snow . . .

The tracks were not too wide apart as I walked down the pipe between them except when it left the line over onto the log. This was a far greater step or jump than I could make, it left the pipeline across a ditch of about six feet and landed on the small log two feet higher than the pipe and did not mess up the snow but landed neatly. This impressed me more at the time than the barefoot tracks in the snow did. He or she did have powerful muscles.

In his letter to me, written in 1969, Mr. Attwell said he regretted that he had not tried to follow the tracks back to where they came from, in spite of the heavy brush, as the distance from the point where they intersected the pipeline to the town was not great. He speculated that the creature must have been in the area before the snow started. That would indicate that after the snowfall it had headed for the mountains, not for the ocean beach. Jack Woodruff, who has found tracks himself on the banks of the Coquille River below his home near Myrtle Point, Oregon, told me that in former years he had met two different people, in 1924 and during the 1930's, who told him of seeing trails of huge barefoot humanlike prints heading east into the Craggie Mountains after an autumn snowfall.

The most important thing about tracks in the snow, in my opinion, is that they are not seen very often. When there is no snow on the ground only a small proportion of the total area is soft enough to show the marks of a foot with a large area and no claws or sharp edges, no matter how heavy the animal. When it snows, an animal cannot move around without making tracks. This change in the tracking conditions should result in a great many more tracks being found, but it doesn't. Tracks in mud, sand and dirt outnumber tracks in snow by four to one.

On the other hand, if one assumes that there is no animal at all, then there should be no tracks in snow— at least not in places where there is snow everywhere. A person faking tracks in snow under those circumstances would have no place to start and stop. Someone would be sure to hit the trail near one end and follow it to that end, finding where the faker put on or took off his false feet, or got in or out of a vehicle, or something like that. Of course some sets of tracks in snow have been followed to a logical ending, such as in a body of water or in a place where there was no snow, and many have not been followed at all. But even though they make up only a fifth of the track reports on file, there have been more than 100 reports of tracks in snow. If they were all made by hoaxers someone should have learned the secret by now.

To me the most logical answer is that there are few sasquatch tracks in snow for the same reason that there are few bear tracks— the animals hole up for the winter. Many different kinds of creatures have adapted to some form of suspended animation during cold weather. I don't know of any such ability in a primate, but primates are mainly tropical animals. The only other primate that made a success of life in a cold climate required the unprecedented adaptions of learning to use the fur of other animals and to make fire.

Evidence for hibernation is entirely negative. No one has ever reported finding a hibernating sasquatch. But if there are such animals at all they have to be somewhere in the winter. The more usual explanation is that they migrate, but there are at least three sound arguments against that. First, the sasquatches of western Alaska, for instance, would be faced with an extremely long walk. Second, there are occasional sightings and tracks in mid winter in very cold areas. Third, in spite of sustained efforts to find it, there is no evidence of any migration. The few reports in severe winter conditions might seem to be evidence against hibernation, but I think it more likely that the occasional individual might come out and move around than that migrants would rush back once in a while, making tracks for hundreds of miles. Certainly there is not a resident sasquatch population that is out and about all winter, or tracks

358

would be very common. It is worth noting that in Russia the creatures are reported to sleep through cold weather but to get up and move around occasionally.

To those who live in warmer climates with occasional brief snowfalls the fact that only a fifth of track reports involve snow may not seem too unreasonable, but in some of the areas where reports originate there is snow on the ground for half the year, or more. Another factor that should also be taken into account is that tracks in any material are far scarcer than might be expected. We are dealing with what is considered to be a very elusive animal and a nocturnal animal at that, yet the thing itself has been reported almost twice as often as its tracks, and reports of sightings outnumber reports of tracks in the snow more than 10 to 1. If I were looking for a cougar, for instance, I would expect to be able to find tracks in snow in a day or so if I were in a place where cougars were known to be. I have no idea how long it would take, on the average, to see one without tracking it with dogs, but I expect it would be a considerable number of years. Bears are not particularly elusive and are often seen. They don't make many tracks in snow either. Yet I am sure that bear tracks are seen a great deal more often than the bears are.

The scarcity of sasquatch tracks is so great that in my opinion it can be explained only by assuming that sasquatches deliberately avoid making tracks, most of the time. There is not much in the way of direct evidence, but there is a little. Clayton Mack, a game guide at Bella Coola, British Columbia, who specializes in grizzly bears, tells of finding some tracks in snow near the Quatna River that he could not identify because the prints were all crossways on top of fallen logs. Curious, he followed them until they reached a place where there was no log to walk on and the whole foot showed in the snow. He identifies them as having been "ape" tracks. Two other Bella Coola men tell of seeing a sasquatch run down a slope where there were patches of snow, but finding that it had avoided stepping in the patches. The sasquatch that Patterson and Gimlin saw made lots of tracks as it walked away from them, but it had avoided stepping in anything that would make tracks in getting to the place where they first saw it. Note that on this matter also the Russians have arrived at the same conclusion, that the manlike animals in that part of the world deliberately avoid leaving tracks.

It has been argued that concealing tracks would indicate great intelligence, but I don't think that necessarily follows. It is generally acknowleged, although I couldn't personally prove that it is true, that animals that hunt and are hunted by scent understand the importance of trying to confuse or interrupt a scent trail, and have various ways of doing it. It would surely take no more intelligence for an animal that hunts by sight, as a sasquatch probably does, to understand the importance of not leaving a visible trail.

As for actually avoiding leaving tracks, it would be no great problem on most of the terrain where sasquatch reports come from. My experience in looking for tracks after a sighting has been reported has usually been one of immediate discouragement because it was obvious that unless the thing deliberately picked out a soft spot to step in there was no likelihood of finding anything. In seven weeks of following the shore of the inlets and islands of the northern B.C. coast in the summer of 1975 I doubt if we found a total of 20 miles of sand or mud that tracks would show in. There were days when we didn't find a single place to stop and look for tracks. At that we found plenty of grizzly and wolf tracks, but any animal that didn't want to leave tracks in that country would have no problem at all. It is the same in most forests. The material that accumulates under the trees is either too hard or too resilant to show tracks.

When tracks are found it often seems as if the creature making them had just decided "to heck with it" and thrown caution temporarily to the winds. In 1975 I received a letter from Robert Franks, president of Raft River Hunting Guides Co. Ltd. at Vavenby, B.C., stating that one night a sasquatch went by one of their hunting camps while they were asleep and

> "Marched with a forty-inch pace for ten miles never stopping once, we saw his tracks in the dust on the road, he did not show any toe nails and he had a big foot like a human being. No bear walks on his hind legs for ten miles like a soldier. The next night it snowed and he was moving out south."

In the Bluff Creek area there were often tracks in places where it would have been just as easy not to leave any. The way the same individual used to leave tracks on one particular sandbar once or twice a year, for no discernible reason, raised the question of whether they were deliberately placed there to be found. It would make sense that a hoaxer able to produce tracks of such quality wouldn't want to go to all that effort without being sure of an audience. But those tracks presented about the most difficulties for anyone trying to fake them or any tracks I have seen or heard of. During the initial excitement over "Bigfoot", and later when the Pacific Northwest Expedition and others were known to be looking for tracks someone might well have been motivated to try to supply some, but the same tracks were also found by accident later on, when there were no construction or logging crews in the area any more, no sasquatch hunters active and no reason to expect the tracks would ever be

360

found. Here is a story from the Eureka *Humboldt Times*, August 10, 1962, describing one such discovery:

An adventurous Naval Academy midshipman, tracking Humboldt county's far northeastern wilderness area for material to complete a research paper, yesterday came back with more than he had bargained for—two plaster casts of the legendary Bigfoot's tracks.

Annapolis senior, D.M. Clark, his younger brother Danny and midshipman T.E. Williams Jr. re-kindled interest in the mysterious monster of the Six Rivers wilderness by discovery of a new set of gigantic tracks.

The two military students, both 21, and Clark's brother, 15, found the series of 16-inch tracks four miles northeast of Onion Mountain, near the Siskiyou county line. The site is about 36 miles north of the Klamath river above Weitchpec.

Clark said the tracks were along Bluff Creek, about 200 yards over a bank from a logging road on which he and the others were driving.

He said the three noticed "a disturbance in the sand" from the road and climbed down to see what it was. "At first we thought is was probably bear tracks," said Clark, "but we decided the tracks could not have been made by a bear, or anything except Bigfoot."

Measuring with a hunting knife and the stock of a 30.06 rifle, the students guessed the length of the prints at 16 inches and the width about 8 inches at the ball of the foot.

The youths followed the tracks for about a half mile in two directions along the creek. Clark said, "no human being alive could get up it on a straight approach."

Clark estimated the angle of the bank at 35 degrees. Attempting the climb himself, Clark said, "no human being alive could get up it on a straight approach."

Danny Clark, a high school sophomore, went back to the car for plaster of Paris to make a print of the track.

An impression of the toes in the hillside was taken in addition to that of an entire footprint on flat ground.

Bigfoot's stride on level ground was measured at 51 and three quarter inches. Where the tracks were going up the hill, the distance between steps was 36 inches.

Clark, who said he entered Bigfoot country with a "skeptical" attitude, related last night he was 94 and 44/100th's per cent convinced of Bigfoot's existence.

"And the remaining per cent is spent wondering how anybody could fake the tracks," he said. "Reading about this,

361

it's hard to imagine it because of all the other things it could possibly be."

After seeing for himself, the naval student changed his mind. "Nobody could pay me to walk around in that country with fake 16-inch feet, especially carrying 600 pounds of rocks to make deep impressions," he stated.

Referring to the depth of the tracks, Clark said they were sunk into the ground about an inch and a half deep. Clark, who weighs 185 pounds, jumped from a five-foot high rock and sunk his heels only a quarter of an inch into the ground.

The tracks were found at about 5:50 p.m., only two hours after the party of three arrived in the area on their first trip.

A steady rain began to fall at about 7 o'clock and continued through the night. Unable to move their wet plaster casts, the resourceful adventurers hollowed out the top of a log and covered the casts with wood and brush. They then built fires around the log so the plaster would dry. Not wanting to leave the area, they went about two miles back down the road and slept in an empty surveyor's camp.

When the plaster impressions were dry enough, the party drove back to Willow Creek . . .

Usually people find tracks of only one, or occasionally two individuals. I have only a half dozen reports on record in which there were three or more. The best is contained in a letter from Jack Woodruff, written in September, 1967, as follows:

On the morning of August 15 I discovered the footprints of four persons along the river bank at this location. I measured these footprints, made casts of four different prints and have these casts in my possession.

The casts are of three different people or sub-humans. Two of them are perfect in every detail and the other two are of the fore part of the feet. One of the perfect casts is of a small child, I should judge about three or four years of age. The other is of a full grown adult and is a foot in length, four and a half inches wide at the ball of the foot and has a three and a half inch heel. This footprint or cast is perfect in detail and distinctly human.

The two casts of the fore part of two foot prints are not identical. One seems to be of an adult, matching the footprint described above. The other cast is of a huge footprint. This print was a full six inches across the ball of the foot. The big toe is huge and well defined and the other toes show up as being big and powerful.

The print was made along creviced bedrock, and although

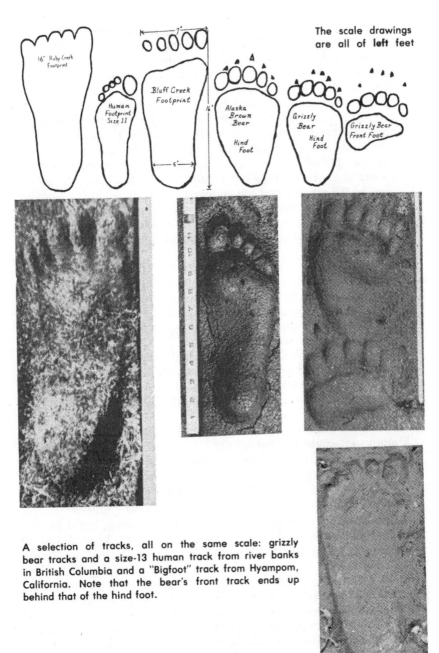

The scale drawings are all of left feet

16" Ruby Creek Footprint

Human Footprint Size 11

Bluff Creek Footprint

Alaska Brown Bear Hind Foot

Grizzly Bear Hind Foot

Grizzly Bear Front Foot

A selection of tracks, all on the same scale: grizzly bear tracks and a size-13 human track from river banks in British Columbia and a "Bigfoot" track from Hyampom, California. Note that the bear's front track ends up behind that of the hind foot.

At right is one of a set of 16" tracks found in 1976 near Terrace, British Columbia and later cast by Bob Titmus.

the entire footprint showed when discovered, only the fore part of the foot was well enough imprinted in a sandy crevice to show in the cast. I measured this print carefully and it was a full fourteen inches long, very wide and appeared to have been made by a very heavy man.

There were six prints left by this barefoot family. Two of the prints were of children but one of these was spoiled by a careless viewer. It was the footprint of a child about eight or nine years old. The other print was of a huge adult and this one was spoiled by a raccoon stepping on the sandy ridge and caving it in. I am confident I have casts of the male of this family, the mother and the youngest child.

These prints were made Monday night, August 14, between seven thirty p.m. and eight a.m. the following morning. I would guess they were about six hours old when I found them.

There was no possibility of measuring the stride of these visitors as I did not find any two prints in a succession that would be made by a person walking. The bed of the East Fork of the Coquille River is bedrock, sandstone. These people are very canny about stepping on sand bars or any place when they would leave prints. All the prints I found were made in shallow crevices filled with sand and fine gravel. I scouted the bed for half a mile in either direction and came up with nothing. This in spite of considerable trapping experience in a misspent youth. These people do not walk where they can be trailed.

The casts of the adult feet show huge callouses. The smaller adult cast could be that of any civilized woman except that she was barefooted in one of the roughest creek beds in Oregon. Her cast does not show anything extraordinary except size but the other cast is unbelievable. The old boy must be tremendously strong, judging from his footprint.

Really good sets of tracks don't turn up all that often, but there was one excellent set found by loggers in Clark County, Washington, in October of 1974, and another in the Skeena Valley in British Columbia in July, 1976. The Washington tracks were found by Cyril Gillette of Amboy, Washington, and his logging crew on going back to work on Monday, October 7. There were 165 prints in stirred-up dirt and dust where they had been working. The tracks were 18 inches long and seven inches across the ball, with a 50-inch stride. They showed excellent detail. At the time, Robert Morgan was just winding up his "American Yeti Expedition" in the nearby Mount St. Helens area. He was contacted, and in turn called in Grover Krantz.

"They were about three times deeper in the dust that my own tracks," Dr. Krantz told the Lewis River *News*. The depth was caused by weight, not impact. Impact tracks would have scattered the dust. There was no sign of that around these tracks."

Those tracks were very similar to some of the tracks from Bluff Creek. Several of them were cast.

The Skeena tracks were found by children near a slough in the Terrace area. They were 15½ inches long and 6½ inches wide, and there were only about a dozen of them. They had a walking stride just short of seven feet. Bob Titmus, who now lives at New Hazelton, made casts of five of them on July 17, and they were in such good shape then that he described two of them as the best tracks he had ever seen. Since he has seen the best that Bluff Creek has to offer, that is quite an endorsement. The good tracks were in a patch of hard moist clay mixed with silt, while others were in ground where there was some grass and brush. Rain had washed some silt into the tracks, flattening the toe impressions somewhat as well as the impressions of what appear to be deep pads behind the toes.

For me the Skeena tracks and the casts made from them supply a long-felt need. In 20 years of investigation I had heard of a number of good sets of tracks in British Columbia, but I had only seen one set of good pictures, of tracks in snow at Tallheo Cannery near Bella Coola in the winter of 1975, and I had never before seen either pictures or casts of good tracks in sand or dirt from anywhere in British Columbia. Just as with the Clark County tracks they appear very similar to Bluff Creek tracks, except that the stride is unusually long.

-20-

Where are the Bones?

There are at least five dead sasquatches to be found in old stories in British Columbia, but that's about the only place where they are so common. When it comes to producing a body nowadays, there are none anywhere.

That is always the ultimate argument against the existence of such a creature, and an excellent argument it is. If there is a real animal, then the remains of one should have turned up somewhere, sometime, and they should have attracted attention. Other forms of evidence for the existence of the sasquatch are ignored partly because they aren't the sort of thing that any branch of science ordinarily deals with, but bones would be something else again. Physical anthropology and palaeontology have always been waist deep in bones.

I would never argue that there could not be some sasquatch bones around that have gone unnoticed. On the contrary, it could probably happen quite easily. A person who finds some interesting bones has no clear-cut course of action laid out for him. He can keep them around the house as a conversation piece, or take them to a newspaper for a story, or ship them off to a museum or university for study. Any of those actions could possibly bring the bones to the attention of someone competent to identify them, but none of them would necessarily have that effect. It might seem that any university or museum should be able to identify just about any bone that might be submitted, but that is very far from being the case. Smaller institutions may well have no one on staff who deals with that sort of material at all, and even where there is some such person there is no guarantee that the bones will come to his attention. That line of thinking is not just speculation. I could site examples. None would be adequate, however, since they all involve cases in which someone has complained of the fact that bones sent to the institution have disappeared, or else the stuff, after being lost for a period, has come to someone's attention. The perfect examples of bones lost without trace have got to be those that really were lost without trace and still are.

Speculation along that line serves little purpose, however. While it is reasonable to point out that there could be some bones that have been found and that have since been lost track of and lost again, it is not reasonable for that to have happened every single time. I am more and more inclined to think that the reason no institution is known to have any sasquatch bones is that none of them do have any.

The same line of reasoning holds good with regard to reasons why bones are not found at all. It is reasonable to suggest that when sasquatches die naturally they may first crawl into holes where they are as good as buried. It is not entirely out of the question to suggest that the survivors may bury them, since it is claimed for both apes and elephants that they sometimes heap things on dead members of their own species. It is also true that bones are eaten by a large variety of creatures; and that in the climatic conditions of the west coast bones disintegrate quite quickly, so that it is rare to find the bones of any animal that has died a natural death. All those things are true and they would account for the finding of bones being rare, but as the years go by they are less and less adequate to account for the fact that no one finds any sasquatch bones at all, anywhere, ever.

It is the same old problem. If you consider this one matter by itself it indicates clearly that there is no such animal. The trouble is that such a conclusion does nothing to provide an explanation for all the evidence indicating that there is such an animal.

After putting that problem aside as insoluble, there are still the stories of sasquatches that were found dead or were killed or captured. Recognizing that none of those stories is very satisfactory, since in every case physical remains should go with them, and there are none, still the stories exist. Let's take a look at what there is.

I have mentioned Bruce McKelvie, who told me that he had a friend who had killed a sasquatch, but would say no more about it. That and the story of Jacko, already told in full, are the only reports from non-Indian sources in British Columbia. Turning to Indian stories, there are two that are well known over a considerable length of the B.C. coast. One involves the late Billy Hall, of Kemano. Like many well-known stories, it varies somewhat according to the teller.

William Freeman, of Klemtu, places the incident near the head of Gardner Canal in the spring of 1905. He says that while hunting from a canoe with another man, Billy Hall saw two animals in a clear spot on the mountainside near the ocean and shot and killed one of them, thinking they were bears. When he reached the spot, however, he found the other creature bending over the fallen one, apparently trying to revive it, just as a human might do. Terrified, he fled back to the shore with the surviving sasquatch in pursuit. There he found his partner asleep in the beached canoe. Terror lent him such strength that he whisked the canoe down into the water, partner and all, without a pause. Mr. Freeman had the story from his father, who was a resident of Kemano at the time.

The other story is older. Mr. Freeman placed it at Klekane Inlet, near Butedale, long ago. At the time there was a village at Klekane. There is none there now, and for that matter Butedale has also been abandoned, but it is still on the maps. The villagers are supposed to have found a humanlike animal with a baby in a cave. They brought the baby home and the mother followed. She used to dig clams along with the Indians and eat them, but she and the child did not live long. People made masks resembling their faces which are still passed on from generation to generation and used in dances at Kitimaat village.

One of the other reports may have been mixed up with "Jacko" in some manner. My informant quoted Chief August Jack Khatsalano as saying that he had seen a sasquatch on display at Vancouver in 1884 at a cost of 25 cents. Later it was taken away up Burrard Inlet, where the people cut its hair off to see what it was like underneath. It then died. Vancouver did not exist under that name in 1884, but there were enough residents on the inlet to provide an audience. However the people who would have taken the creature away up the inlet would have to be Indians, and Jacko was supposed to be in the possession of a white man.

A similar story from Alaska, undated, tells of berry pickers capturing a hair-covered human at Bristol Bay and keeping it in a cage. They didn't shave it, but its hair fell out anyway, and it died.

Rene Dahinden was told that Indians in eastern British Columbia, near Invermere, were supposed to have killed a one-armed sasquatch. In the 1950's I had a letter from a man named Peter James, who claimed that in 1949 he saw two old males and a female sasquatch lay out the body of a dead young female on a rock high on a mountain. I understand he died shortly afterward. In any event I never tried to talk to him, mainly because I didn't put any stock in the story.

I have referred already to a report of a capture in Tennessee in

1878, and to the dead creature from Monkey Cave Hollow in Kentucky. The South has produced several other stories of captures and killings. Earliest was found in the files of the Helena, Montana *Daily Independent*, September 19, 1875. It ran as follows:

A STRANGE CREATURE

A wild boy was lately captured near Austin, Texas. He was first discovered wallowing in a pond of shallow water, and when approached he broke like a quarter-horse, running about a mile before he could be overtaken by men on ponies. Riding up near, the boy was lassoed, when a fierce contest ensued, the strange being striking, kicking and lunging about in a most fearful manner, and apparently being frightened almost to death.

Finally he was overpowered, tied and taken to the house of the man who first discovered him. His body was covered with hair about four inches long and from size and appearance, he is supposed to be about twelve years old. He is unable to talk but possesses reasoning power, and now follows his captor about like a dog.

Another story from Texas, very vague, is that a wild man was caught in west Texas in 1903 and later died and was buried in a cemetery. Yet another story does not start in Texas, but ends there. In the book *Ozark Country*, Otto Ernest Rayburn tells of a wild man, seven feet tall and covered with hair, living in Saline County, Arkansas. Although he harmed nobody, this wild man was greatly feared and a party was organized to hunt for him. The account continues:

A daring young man led the group with a pack of deerhounds. The wild man was tracked to a cave and lassoed with a rope. When the lariat noose fell over his shoulders he emitted a strange sound like that of a trapped animal. He was taken to Benton and lodged in jail—a small building· made of logs. He immediately tore from his body the clothing provided by his captors and escaped from the flimsy jail, only to be recaptured, this time in the canebrakes.

Just what became of the wild man no one seems to know. Old-timers say he disappeared and was never seen in the country again. The following story ties in nicely as a sequel to this.

Soon after the giant escaped the young man who had led the first hunt rushed into the cabin home of his parents, grabbed his gun and called to his mother, "Ma, don't look for me till you see me coming; it may be a day, it may be a year." He had found giant footprints and wanted to get started while the trail was hot.

These tracks were fourteen inches long and four feet apart. The place was Saline County, not far from the county seat of Benton. According to the story, the young man followed the tracks successfully across southern Arkansas and into Texas. Along the way he came upon nine other men who had discovered the big tracks and were following them. In this company, he travelled across the Lone Star State, subsisting almost entirely upon raw meat killed along the way.

It was almost a year before the Arkansan returned home with the disappointing news that not one of the trailing party had caught a glimpse of the giant who made the tracks, although they did find several persons who claimed to have seen him, always travelling in the darkness of night.

Moving up to the 1940's, there are two reports of sasquatches being killed, one in Missouri and one in Georgia. The thing in Missouri, "something like a Gorilla" was supposed to have lived in a swamp in the southeastern part of the state, and to have ripped up full grown cows and horses, abandoning the carcasses without eating them. Finally someone shot and killed it. John Keel tells the story in *Strange Creatures from Time and Space*, but doesn't give the source.

The Georgia story was told to Rich Grumley, of the California Bigfoot Organization, by a man who claimed to have seen the creature after it was dead, when he was a boy in 1943. He said the creature was killing sheep and calves by tearing off a leg and leaving the animal to die. Men tracked it onto a small mountain, cornered it and felled it with some 60 slugs from their shotguns, one of which went through an eye into the brain. Brought to town in a pickup truck, it was so big that its torso was wider than the truck box, and even though the tailgate was down its feet dragged on the ground while its head was propped against the back of the cab. It was covered with reddish-brown hair, with the palms and soles of its feet bare and sparse hair on the head and chest. The smell was terrible.

The man said that the creature was buried under a pile of rocks, and Rich, who knows the town where it is supposed to be, would like very much to go there and look under the rocks. For myself, having been around and aware of what was going on in the 1940's, I cannot credit such events as those taking place without the news being spread all over the front pages. There is, after all, a tremendous difference between a sighting report and being in possession of a dead sasquatch. For the same reason, I can't credit a story told in a letter to Ivan Sanderson in 1971 that a hair-covered humanlike creature found scrounging in garbage cans in the Minneapolis-St. Paul area was at that time being kept in the Rochester State Hospital, where it would wear no clothes and food had to be thrown to it because it was too dangerous to go near. The same goes for a recent report of a dead sasquatch being seen on a truck in Plymouth, California, from which it was removed with a forklift.

The reports of captures and killings seem to bunch up in certain areas, and California has three more, one that no longer interests me and two that do. Until Patrick Householder went to work on it in 1971, the story of Deadman's Hole north of Julian, California was one of the minor classics of ancient sasquatch lore. It is quite a story, telling in detail of the finding the year before of the bodies of William Blair, and, some months later, a girl named Belita. There was said to be no evidence of a cause of death except slight bruises on their throats.

In the story proper two hunters ventured into the "dark and mysterious canyon" and there encountered "an immense and unwieldly animal" that fled upright over the rocks and logs. After killing it they found its lair complete with human skeletons.

The story was written by James Jaspar, proprietor of the Julian *Sentinel*, but appeared in the San Diego *Union*, on April 1, 1888. The date was apparently no coincidence, the only puzzle being why anyone would choose to incorporate two actual and recent murders into an April Fool's Day story.

Mr. Householder found the original written verdicts of two inquests. One, into the death of David Blair on June 30, 1887, showed that he died of a blow on the left jaw by some heavy weapon and two knife wounds in the left side of the neck. The other, concerning the death of Franciscia Ranteria, showed that she died between December 20 and 21, 1887, of "two bullet wounds shot through her back." Those are not quite "William Blair" and "Belita" and there are no places given for the deaths. However the dates correspond approximately with those given in the story and there were no inquests in the county at that time involving anyone who was choked to death.

Being able to relegate the creature from Deadman's Hole to the fiction shelf is quite helpful from the point of view of making sense out of the sasquatch. No other substantial story indicates such totally carnivorous behavior, or gives the creature a home in a cave, or describes so human a face.

The next story is more recent by about 30 years, but is still too old to follow up. It is told in a letter I received from a lady in Pacifica, California after she had heard me on the radio. She wrote:

When I was four years old my mother died. My father placed me in a Catholic Orphanage in San Francisco, and put my brother, who was seven, with the priests of San Rafael . . .

When I was 12 years old, one day a nun called me and three other girls to help her. Up to the 3rd floor she took us which was the bathrooms. I believe there were 15 or 18 tubs, as this was called the 3rd Band, girls 12 and under.

We were told to draw a tub full of water. In came another nun with this creature. Sister guessed her age to be 11 or 12 years. First they drenched her hair with larkspur, as she was full of lice. Her hair was jet black, coarse and down below her hips. It took four of us to hold her in the tub while the nun scrubbed her head. She was hairy from head to toes. She just sneered at us. Once in a while she let out a gutteral sound.

After fine-combing her hair and removing lice and nits the nun took and put a large diaper on her. She tore all clothes off we tried to put on.

She had jet black eyes and was very strong.

This happened in 1916, something I'll never forget.

She stayed four days with the nuns and was taken away in a van to go East to some hospital for study, where we were not told.

We four girls were sworn to secrecy. Of the 500 children no one but us saw her.

I am now 70 years old, but can see that creature as if it were yesterday.

She stood like a human and had fingers (hairy) and toes like anyone else.

Some men who were hunting somewhere in California found her roaming the hills.

I did not give name of orphanage as I have to be sure their name will not be publicized. The nuns would not appreciate publicity. They have now changed the orphanage into a boarding school for girls.

All the nuns who were there in 1916 must be all dead by now. However they must have records of this creature being brought to the orphanage.

Sister Helena was the Superior at that time. A Sister Hortence took charge of this creature. She was in full charge of horses, cows and chickens and was a very strong person herself. She could handle anyone.

P.S. You are the first person I've told my story to.

The first thing you expect when you get a letter like that is that the name and address will be fictitious. This lady turned out to be genuine, but when I asked for more information she wrote that she had enquired and found that records so old would have been destroyed and that no one now at the institution had any suggestion for learning anything further. The waist-length hair is an anomoly, and it is hard to imagine that the strongest of nuns and four 12-year-old girls could safely bathe an animal itself the size of a 12-year-old. Then there is the basic question of how so interesting a creature could just disappear without further notice being paid to it. Still, I don't think this reads like fiction.

Finally, there are some bones that someone should be able to find. In June, 1971, the Salem, Oregon, *Capital Journal* carried a story about a local girl and her friend having seen a dead Bigfoot near Happy Camp, California, in 1967. I didn't learn of the story until several years later, and didn't pay much attention to it. For one thing, it contained the statement, "We had walked about eight miles into the woods," and the idea of two girls being able to find any particular spot that far off in the bush the same day, let alone years afterwards, seemed most unlikely. Still, I never felt quite right about leaving such a story hanging, and I tried a few times to find the girl or to get someone else to do it. Late in 1976 I finally took the matter up in earnest and got lucky. Within 24 hours I had met one of the girls, now married and raising a family near Tacoma, Washington, and had talked to the other on the telephone in Norco, California. It turned out that they had not been wandering in the bush at all, but had been walking on an old forest road. In other words, whether they could find the exact spot or not, they could walk right over it.

Their stories differed considerably. One of them was quite sure she had seen a Bigfoot, lying on its back with its legs spread out. She recalled trying to turn it over by prying at it with a stick, and finding it solid and too heavy. The other recalled only a great deal of hair, like a horse's mane except that there was too much of it and it was spread over too wide an area, with the body all rotted away. She decided she must have seen some bones too, because it was from the bones that she got the impression it was a big animal. She knew of no known animal it could have been.

The first girl could not remember any face—that was rotted away and so was the stomach area, and there were weathered white ribs showing. Ignoring the general impression and comparing the actual details described, there did not seem to be any actual contradiction in the two descriptions, and as far as the surrounding circumstances were concerned the second girl not only did not contradict the first, she cleared up a couple of apparent inconsistencies.

The two of them had been neighbors for a time at Salem, then one had moved to Kemper Gulch, 10 miles north of Happy Camp, in the

Bigfoot country of northern California, and the other had later come to visit her there. One day they took a walk several miles up the road leading to the old Wagner Ranch. The girl I talked to first was certain she had been there in August, yet insisted that the animal was lying in several inches of old snow, an impossible combination at the indicated elevation. The second girl, the one who lived there, said her friend has visited twice and this had happened in the Easter holidays. She thought the remains were a bit below the road. The other girl remembered them as being right on it.

The visitor had the impression that they had been followed that day by some mysterious creature, which the dog kept running after, although it was wagging its tail and did not act at all concerned. The girl who lived in the area and owned the dog said that it kept chasing deer. The two young women had lost touch with each other years ago, although one of them had been able to provide the clues that enabled me, through people in Happy Camp, to reach the other.

I am entirely satisfied that there was indeed what remained of some big-boned creature on or by that road, and I am pretty sure that at least the second girl, who knew the area, should be able to take someone pretty close to the spot. Even 10 years afterward I think there is an excellent chance, in fact all but a certainty, that some bone or tooth from a creature that big will have survived. It might take many man hours to find one, but it shouldn't take even a fraction of the time spent looking for humanoid bones in places like Olduvai Gorge. As to what the bones would be from, there is one very definite indication. The first girl said that the thing that made the most impression on her was seeing a hand with a thumbnail—a nail larger than a human one, and thick, but squared off; definitely a nail, not a claw or hoof. She said it was the color of a nicotine-stained finger. There aren't too many animals that could match the description given by either girl, and almost none that could match both. One that could, and the only one that could have a thumbnail, would be a sasquatch. It is not in Albert Ostman's written story, but he described the fingernails of the "old man" as being copper colored. And one of the Indian names for these creatures in B.C., again not referred to anywhere in print, is supposed to translate as "copper nails"

A sasquatch that died in the middle of a road would most probably have been shot, but if that is what happened whoever did it has not made the information public. There are many reports of sasquatches being shot at—I have made a rough tally of more than 80—but none anywhere near Happy Camp in the mid 1960's. As a matter of fact there are almost no reports of shootings in California at all. On the average, shooting is reported in about five percent of sightings—

374

one in 20. In Oregon somebody is supposed to have pulled the trigger in about one incident out of 12, and several of these stories are fairly substantial. California is at the other extreme, with shooting referred to in only about one report in 50, and only one story that seems likely to amount to much.

In the majority of such reports someone saw an unknown animal and took a shot at it, often with a rifle, but sometimes with a shotgun or a pistol, and the thing took off. It is probably reasonable to assume a high percentage of misses, and it seems reasonable also to assume that most shotgun loads and most bullets from handguns would have little immediate effect on a full-sized sasquatch. Of 49 reports in which the type of gun is specified, five were revolvers, 11 were small rifles and 16 shotguns. As far as I know none of the shotguns were loaded with slugs. (Not including the yarn from Georgia in which the creature was supposedly killed.) Four of the 17 rifles were 30-30's, which may be a bit small for so large a quarry. In nine cases the size of the rifle is not mentioned, and only four definitely involved a 30.06 or some other high-powered weapon. I have no record of any of the large-calibre and magnum rifles used for African big game ever having been fired at a sasquatch.

There is a striking description of the effect of using a small gun on a sasquatch contained in a letter I received from Mike Jay, who was active for several years in investigations in Oregon. The incident had taken place in February, 1972, and the letter was written in September. Location was near Triangle Lake, west of Eugene. Names are left out for obvious reasons. Mike's letter says:

Anyway, it seems that Mr. - and friend were in the habit of spotlighting deer in the vicinity of the lake. It is very rugged country. On this particular evening, they went in to pick up the previous night's kill. They were armed with .22's. All of a sudden, they spotted several deer running down a hillside, "like bats outa hell." They thought it strange, because they'd never seen deer act like that, especially at night. They decided to investigate. They flashed their light around the hillside, until they caught a glimpse of what was making the commotion. They claim it was a Bigfoot. No description, just "a Bigfoot". They supposedly fired 30 to 40 rounds at this creature, which turned and fled over the hill, flailing its arms as if to repel a swarm of flies.

An even more graphic account, involving a 12-gauge shotgun loaded with 00 buckshot, was given in an interview taped by Roger Patterson with a teenager at The Dalles, Oregon. This is supposed to have happened quite near the city on a hillside close to the Columbia River in June, 1967. A whole group of teenagers had been spending

375

the early hours of the morning, for several days, hunting for sasquatches which they say moved to the river early in the night and back up to the hills before dawn. They had a great variety of guns, but on this occasion Roger's informant and the boy with the shotgun were separated from the others and had only the one weapon.

This night they came through the cut and turned right so Dave and I took the 12-gauge shotgun and went down to see where they went.

We were walking back up, we walked by this big tree, and Dave wanted to check inside the tree, because the branches on the uphill side of the tree came clear to the ground and he wanted to look in there. I reached down and pulled up a branch and Dave was standing on my right, and we saw it. It was about eight or ten feet away from us, I don't know how far it was, but it wasn't very far. It was in a crouching position. It was downhill from us, and I'd say it was at least nine feet tall when it was standing up. When it was crouched it was probably seven feet tall, or something like that, you know.

So Dave let him have it twice with the 12-gauge from the hip, just shot him twice. That knocked it down, it rolled over twice, and got up and went right through a four-strand fence and snapped three poles off, big wooden poles, snapped them off flush with the ground. So Dave and I figured, well, we shot it, it was just gonna roll, and when it stopped rolling it'd be there dead, you know, so as soon as he started rolling we started down the bank after him.

Then when he got up and ran, he ran away from us, so we decided well, he's not as hurt as we thought he was. We only had one more shot, and we didn't want to go after him with just one shot, so we started back up where everybody else was.

We decided he was wounded and he was more than likely gonna be mad, so the best thing we could do was get off the hill, go back into town and let him cool off and come back up, you know, in the daytime and scout around, see if we can't find him. We figured on tracking him to wherever he was going.

We figured that, if nothing else, he'd just be running until he couldn't run any more, you know like sometimes you might shoot a deer or something and he'll go for a hell of a long ways and then you'll find he's dead. So we decided if we came up there in the daytime, about noon or whatever, you know, and track him by blood and tracks, we figured there'd be lots of tracks you know.

So we came back up the next morning and we went down and took pictures of where we saw him and the fence he went through and everything, and we tracked him for maybe 80 to 100 yards and then we lost track, couldn't find any more tracks and there wasn't enough blood to follow.

Stories involving shooting are the kind that tend to be given a fair degree of publicity, which means that a lot of them come to light through newspaper stories, and often the stories aren't noticed until long after the event. I have more such reports from east of the Rockies than from the west, but few of the eastern ones were ever followed up. A lot of the western reports are old. In doing the computer survey I don't think that we had even one questionnaire filled out in a current interview with a person who told of shooting a sasquatch, although there were several filled out from old information.

Besides Fred Beck, I can recall talking to only four people who described having shot a sasquatch themselves. Two were Indians living on the B.C. coast. In one case two men in a fishing boat had seen a small "ape" on the beach and wounded it. I don't recall which one did the shooting, but Bob Titmus and I talked to both of them. The animal ran into the forest and they did not try to follow it. The other man, George Talleo, from Bella Coola, had shot at a sasquatch on a hillside above South Bentinck Arm back in 1928. He said he was on his trap line when he saw the thing suddenly stand up from behind the trunk of a fallen tree. He snapped off a shot at it with a small-calibre rifle, and it somersaulted over backwards, out of sight. George also decided he had urgent business elsewhere, and he never went back until 1962, when he tried to show Bob and me the place. There proved to have been logging done in the area in the meantime, so there was no chance of finding it.

The third person was a youth at Richland, Washington, who was one of a group that tried to shoot what they called, "the white demon," a huge, light-colored upright creature that was seen several times in an abandoned gravel pit by teenagers cruising in their cars in the small hours of the morning, in the summer of 1966. They claimed to have hit the creature several times, but all they got for their efforts were some piercing screams.

Finally, there was a man in Eugene, Oregon, who told perhaps the most impressive story of all. I have listened to a taped interview with him and have talked to him on the phone, but that was back in the days before I made much effort to keep records. Recently I obtained his address again, but I was also told that he doesn't care to talk about it anymore, which is certainly understandable.

It happened in the fall of 1957, at Wanoga Butte, southwest of Bend, Oregon, when he and another man were deer hunting. As I recall the story he saw a deer entering a clearing in an unusual way, apparently concerned entirely with something behind it and showing no caution at all about what might be in front. He shot the deer and then sat down on a rock for a minute to be sure it was dead before approaching it. Suddenly a giant, humanlike creature emerged into the clearing from the same place that the deer had entered, gathered the dead animal under one arm and departed with "tremendous strides." Although the thing was about nine feet tall and covered with hair all over, the hunter said his only reaction was "that son-of-a-bitch is stealing my deer," and he emptied his 30.06 into its back. I asked him if he could have missed and he said that he didn't see how he could have because the thing was so big and so close that there was nothing but its back to be seen through his rifle sights. Yet he did not see any indication that it was hit, and it just kept going, making a "strange whistling scream." He did not try to follow it. One thing that he noticed about it was that the hair on its arms was extremely long.

The only comparable incident that I know of also took place in Oregon, only two years later. Late in October, 1959, a story appeared in the Roseburg, Oregon, *News Review* about two boys shooting at a monster near Ten Mile, west of the city.

"It's about 14 feet tall, is covered with hair, runs upright, and screams like a smashed cat," the story said.

That was just about the time we were getting the "Pacific Northwest Expedition" started, and Bob Titmus went to see the boys. Rene and I went there briefly too, on our way south a few days later, but neither of them were around. They told Bob that the smaller boy, who was 12, had seen the creature first, down below him in an open hollow near the site of an abandoned sawmill.

He told his 17-year-old friend, and the two of them went there to look for it, both of them armed. They saw it alright, but it also saw them and started in their direction, so they took off. In an astonishingly short time it was up on the ridge at their level and began to chase them. Both of them shot at it, and the older boy, who could run faster, had time to turn and place several shots with his 30.06. He told Bob that a couple of times the thing slumped until its knuckles hit the ground, but it kept coming. It did not really try to overtake them, however, but moved its arms in an outstretched position as if herding them along. They noticed very long hair hanging from its forearms. Then it wasn't behind them any more.

Bob went to the place and found a set of tracks that supported the boys' story. They were very strange tracks, however, eight inches wide at the toes but only 11½ inches long and 4¾ inches across the heel. They had five toes, and no claws. Coming up the hill they sank as much as 14 inches in some wet ground, although Bob could sink a heel or only two or three inches by jumping downhill. In other places they sometimes sank an inch where he did not sink at all. He also found a bed about 12 feet in diameter in the meadow where the creature was first seen. It still had a very strong smell to it.

The shots could have missed, of course, but Bob learned that the older boy kept his family supplied with venison and had the reputation of never spoiling any meat because he always shot deer in the head. There was no indication of what had happened to the animal. Tracks like that have never been reported since, so it may have died, but they had never been reported before either. Another problem about the boys' story was the size of the creature. Bob couldn't persuade them to modify their estimates below 12 feet, and he found that they were good at estimating the height of other things. Had we known then what we do now, a lot more time would probably have been spent checking into that report, and looking for the animal—which by rights should have been lying dead somewhere around. In those days Bigfoot was making tracks at Bluff Creek every few weeks, and we fully expected to find him ourselves before too long.

Stories like that one are no help at all. The people who want the sasquatch to be from another dimension, or something equally exotic, find them very encouraging, but anyone trying to make sense out of the creature as a normal animal is left with the problem that normal animals could not be shot at on scores of occasions without any of them falling down. All I can suggest is that it is not much use shooting at so big an animal with a small gun, and that even with a heavy gun there is a problem as to where to place the shots. If the last two accounts are accurate, then the chest is obviously not too good a place. I don't hunt, so I am not very competent to speculate on what happens or doesn't happen when an animal is shot. There are plenty of experienced people hunting every year who have heavy guns and confidence in what they can achieve with them, and I would be happy to leave the problem in their hands.

My opinion, based on many years of mulling over the evidence, has always been that there is indeed a normal animal out there. There are those who will argue that it may be something supernatural and extra-terrestrial that cannot be hurt by a bullet, but will disappear, or pay no attention, or do something equally disconcerting. That line of reasoning can safely be ignored, not only because it is almost certainly nonsense, but also because if it were true it wouldn't make any difference. If something can't be hurt by a bullet there is no harm to be done by shooting it. Another thing that apparently need not be considered is the possibility of being attacked by an animal that is infuriated by wounds and cannot be downed. There is absolutely no indication of that type of behavior. It has never been reported.

Wading through dozens of reports of hairy giants being shot at without results lends an air of unreality to the whole business, but that takes place indoors, where it doesn't matter. Outside, looking at a living, moving creature of bone and muscle and hair is where the decisions about shooting have to be made. Let a person think what he likes at any other time, if this book has given him information that will enable him to act calmly and rationally when he actually encounters a sasquatch it will have fulfilled its main purpose.

In what manner might sasquatches be hunted successfully? For other animals, man can rely on techniques that have proved effective in the past, and on knowledge accumulated over hundreds or thousands of years. To deal with some types of animals that he would have difficulty hunting unaided he has the use of dogs trained to find them and hold them in one place until he gets there. Man has better eyesight than most animals, more stamina than many, and can move over types of terrain that some animals cannot negotiate. With the sasquatch he has not a single one of those advantages. In all probability all its senses, including eyesight, are better than his. It is much faster and stronger, can go virtually anywhere he can go, and is on its home ground when he is trying to catch up. No dogs have been trained to track it, and no hunter has left a record of a successful method of getting a shot at one. There are only two advantages that man can apply to the problem, numbers and technology, and neither of them is available in significant quantity to sasquatch hunters.

For a while I thought that the way out of the difficulty was through education. Probably every hunting season two or three men at least have the opportunity for a clear shot at a sasquatch. If every hunter were aware that shooting a sasquatch would be valuable to him and important to science; knew that chest shots had not been successful in the past, and that if he did get one down it was essential to bring out a piece of it, that small store of information should be sufficient to result in a sasquatch being collected within a year or so. What I did not count on was the lack of any authoritative voice to say those things.

I expect that there would be few hunters nowadays who would be totally at a loss on encountering a sasquatch. They would at least know that they were seeing something other people had seen before. That alone would remove the main reason why people who had the opportunity have not shot them in the past. Beyond that, unfortunately, the hunter has probably been exposed to a lot of conflicting information. He may have heard that it is illegal to shoot a sasquatch, or that there is a $10,000 fine for it, or that sasquatch are semi-human, or that they are an endangered species. The first is true in some locations. The others are not true at all.

There are some states and provinces that specify which animals can be hunted, rather than which cannot. In places where only specified animals can be shot, then the shooting of an unknown animal is presumably a technical violation of the game laws, although I don't think it is by any means certain that a court would so decide. How does a possible game law violation weigh against one of the greatest scientific achievements of all time? It is also illegal to exceed the speed limit in rushing a strangling child to hospital, and firing satellites into orbit must sometimes causes junk to fall back to Earth in violation of anti-litter laws. Speaking seriously, it is almost unthinkable that a charge would ever be laid, during the excitement and acclaim that would follow the obtaining of a specimen of a creature physically far closer to man than any other. Even if it did happen the fine would be of no significance. The much-publicized $10,000 penalty exists only in a by-law passed by Skamania County commissioners, in Washington, and even the commissioners would not argue seriously that it could be enforced. It was adopted strictly as a bluff to try to discourage people from hunting for a manlike creature and possibly shooting each other, at a time when there was a lot of excitement about a sighting report. Game laws are not a county responsibility, and although a couple of proposals have been discussed, no state or provincial legislature has yet given the creatures sufficient credence to pass legislation concerning them.

The more serious question, of course, is whether the sasquatch could be human or semi-human. Indian traditions generally pictured the sasquatch as human, even though it was often believed that they ate humans. In British Columbia, which seems to have been the one area where Indian stories found their way into general circulation among the non-Indian population, the story of wild human giants was quite generally familiar, although not taken seriously. Discovery of huge humanlike footprints in areas where no possible explanation for them was known quite naturally suggested giant humans, or in most cases a single giant human. There seems also to be a strong tendency for people whose fancy is caught by the idea of unknown giants living in the woods to want them to be human. The idea that there is a forest race too wise to have wars or get trapped in the rat race, living in harmony with its environment, seems to have tremendous appeal. All those factors combine to produce an initial impression that something human is out there, and there is ample publicity given to stories and opinions that reinforce that view of the matter. Finally, each year that goes by without a sasquatch being caught, or even shot, lends support to the idea that they are, if not human, at least smarter than humans in some ways.

On such a foundation it should be easy to build a case for the creatures' human status, but research produces nothing to build with. Humans walk upright and have a foot of a unique shape. Sasquatches, if they are real, also walk upright and have a foot of a similar shape. There is nothing else but wishful thinking.

Humans are the product of millions of years of dependence on their wits. Naked, they require clothing, fire and shelter for warmth. Sasquatches, like other mammals, have all the insulation their environment requires. Physically weak and without natural weapons, men in a primitive state depend on objects to defend themselves and to kill their food—sharp things, heavy things, things they can throw. Too big to need to defend themselves, strong enough to tear their prey to pieces, sasquatches make no significant use of objects. Too slow to run away, too small to stand and fight, men had to be able to organize and communicate to become a successful species Sasquatches have never faced such challenges.

If at some distant time they were ever similar, one species learned to depend more and more on its brain, eventually losing much of the physical ability it must have had and dulling most of its senses, but developing an intelligence that set it totally apart from other animals, able to dominate all other creatures and even alter its environment. The other species specialized in physical adaptions, strength certainly, and probably speed; vision acute enough to function in the dark; a warm fur coat, and perhaps even the ability to sleep through the winter months.

Each in its own way is an outstanding success, established all over the world and able to survive in most climates and conditions, but they have grown very far apart. The one that succeeded physically took the typical animal route and remained a normal animal. The mental giant became something quite different. That is the only evolutionary script that makes sense, and if the two species are viewed as products of divine creation the answer remains the same. In a separation of man and other animals, everything known about them proclaims that the sasquatch are not and never have been human.

It is true that there are people who represent themselves as experts regarding the sasquatch who claim publicly that the evidence shows them to be semi-human, but there is no such evidence. Some of those people are sincere, but they want the sasquatch to be human and can somehow ignore what does not fit what they want to believe. Perhaps if they were forced to recognize that they were looking for an animal the whole thing would cease to interest many of them. Others are not concerned with the truth, perhaps not even concerned as to whether such a creature exists or not. They have found that it is more popular to proclaim that they would never kill a sasquatch, and easier to raise money on that basis, so they say what suits their purpose. They may say things like "our research proves that" sasquatches are semi-human, but you will never find them citing specific instances that can be independently confirmed, because there aren't any.

Then there is the question of whether sasquatches are members of an endangered species. For that to be used as an argument against killing a single one in order to prove their existence their numbers would have to be very small indeed, only a few dozen individuals at most. Oddly, I don't know anyone who uses that argument who does in fact have so low an estimate of their numbers. Those who take a casual interest may be misled by the widespread use of the name "Bigfoot", which carries the inference that there is only one. In fact, in the Bluff Creek valley alone, where the first "Bigfoot" tracks were found, I have personally examined the tracks of four unquestionably different individuals, plus a fifth that is certainly different, but so small that it could conceivably be a human track.

It is strictly guesswork trying to estimate sasquatch numbers, so perhaps the best that can be done is to consider the only omnivore in North America of comparable size, the grizzly bear. A recent study quoted in the *National Geographic* indicates that in Montana grizzlies range over 50 to 1,000 square miles, which would suggest a tiny population, but there are supposed to be a couple of hundred of them just in Yellowstone National Park. The estimate for the Yukon Territory is more than 10,000. The Pacific Coast states and British Columbia alone have a total area of 700,000 square miles, more than half of which is suitable sasquatch habitat, and there are reports from that entire area. Then there are 57 more states, provinces and territories where such creatures have been reported, plus all the other countries and continents from which there are reports. With so much more territory involved it seems most unlikely that the total sasquatch population could be substantially less than the number of grizzly bears—and grizzlies are still being hunted for sport.

But let us assume for the sake of argument that the sasquatch could indeed be an endangered species. What would endanger them? Are they overhunted? Since no hunter has ever brought one in, it hardly seems likely. Is their habitat being destroyed? A great deal of it is still completely untouched, but there is certainly a possibility that they are under pressure of that sort in certain places. Florida, which now has more reports in a year than any other state, would be a prime example. Most of the sightings there are said to occur on the fringes of woods and swamps on the west side of the peninsula, and subdivisions are rapidly consuming those wild areas.

What steps can be taken, then, to ease man's pressure on a unique variety of wildlife? Obviously none, as long as the authorities do not believe that any such creature is actually there. Where sasquatches are plentiful, the death of a few can be of no significance to the species. If in any area they are threatened by man's activities, it follows that something man is doing must be causing them to die or preventing them from reproducing, so he is already doing things that will eliminate them. If the first step towards changing the damaging activity requires killing one sasquatch to prove that there are such things, then obviously that is what should be done.

What, then, of the hunter who finds himself with an upright, hairy biped within range of his gun? What he chooses to do is entirely up to him, but he should be aware that he is not looking at a mythical monster, or a wild man, or an endangered animal. He has the answer to the question of whether such a creature exists, because he is looking at one, and he may be aware that it will be greatly to his financial advantage, will assure him a measure of immortality, and will contribute greatly to the advancement of knowledge if he can bring it in. He should, of course, make sure that it isn't an idiot in a monkey suit, and if it is too far away for that he should hold his fire. That matter settled, the remaining consideration is whether or not he can kill it. Knowing that several people claim to have shot such creatures repeatedly in the upper body without bringing them down, he might be wise to shoot it in the brain or the spine. There is no need to be concerned about destroying a part of the animal that would be valuable for study. The specimen that proves that the animal exists will certainly not be the only one studied. It is the proof that is important.

The final thing to remember, for any hunter who succeeds in killing one, is that other people claim to have done the same, but have failed to produce any proof. By going for help the hunter takes the risk that the creature may not be dead and may get away, that he may not be able to find the body, or that there may be others around that will take it away. None of those things may seem likely, but there is no point in taking a chance. Before leaving the carcass, the hunter should cut off some part he can conveniently carry, even if only a finger or toe, and take it with him. If he has a camera he should take pictures of the body, but when it comes to a choice between camera and gun it is the gun that counts.

The Brookings, Oregon, *Harbor Pilot*, on October 4, 1973, carried the following story:

A Smith River resident, Rick Blagden, thinks he spotted the legendary Big Foot while deer hunting in the Doctor Rock area up the Gasquet Road near Big Flat.

Blagden was hunting Monday, Oct. 1, with three Smith River companions, Steve Leishman, Frank Richards and Jim Downing when he ran across the huge ape-like beast.

The foursome, Blagden said, had each selected an area where they would hunt. The Smith River man chose the Blue Creek area and went off on the kill.

About 7:30 a.m., he reported, he was resting in the area when the beast apparently picked up his scent.

"Something let out a loud sneeze behind me," said Blagden, "and I looked around and saw this huge, furry, dark-colored, barrel-like creature standing erect."

Blagden said he "took off" but not before he had taken a better look at the beast.

"He was well over six feet tall," the Smith River man added.

"He stood there for awhile and then slowly walked down a steep hill and into the brush."

The hunter said he could hear the animal crashing through the brush for several minutes making a shrill noise until he disappeared.

Blagden began tracking the animal which left prints much bigger than a bear's.

The next day he and two companions, Leishman and Leonard Branton, went back into the area to find Big Foot but the brush was too thick to follow the trail.

There have been a number of reports concerning the animal, called Sasquatch by the Indians, during the past few years. Most of the sightings have been in the Gasquet area.

The last paragraph may be a little confusing, since the name Gasquet certainly does not figure prominently, if at all, in reports of Bigfoot sightings. The observation is correct, nonetheless. Mr. Blagden was hunting in the next valley over from Bluff Creek, but he had entered it from Oregon instead of from the south.

More significant, to me, was an additional paragraph written in ink by the man who sent me the clipping, Jack Woodruff of Myrtle Point, Oregon, who has himself had two experiences with sasquatches. "Why the hell didn't he shoot the thing?"

-21-

Screams in the Night

My basic message with regard to sounds is that they are not much use. Should someone emerge with a good clear strip of sound film in which a sasquatch opens its mouth wide and screeches, or steps close to the microphone and chatters a bit, then we will be getting somewhere, but we are a very long way short of that at present. Perhaps the next best thing, although it would be no more reliable than a sighting report, would be a tape recording made by someone who was watching the sasquatch as it made the sound on the tape. We don't even have that. There are a fair number of people who claim to have been watching a sasquatch while it made a noise, and presumably a lot of them would be able to recognize the same noise if they heard it again. Some might be able to give accurate imitations, but I would think that the degree of ability to do that would vary greatly between individuals. Finally, there are people who have heard noises which, for an assortment of reasons, they believe to be associated with the sasquatch. It is only in this last category— obviously far removed from direct evidence—that a few sounds have been recorded, and all the recordings but one are of poor quality.

The one exception involves one of the most interesting series of events connected with the whole sasquatch question, but it is one about which I have no first-hand knowledge. I do have a small phonograph record of some of the sounds, a remarkable series of growls, grunts and gibberings apparently intended more to intimidate the listeners than to communicate with them. The story is told fully by Alan Berry in the paperback book *Bigfoot* that he co-authored with Barbara Ann Slate in 1976. Alan Berry also markets the records. The sounds were first heard in August, 1971, by Warren Johnson of Modesto, California and his brother Louis. Warren Johnson wrote to Ivan Sanderson about the phenomenon early in 1972, sending him a 10-page account of some of their experiences entitled *Our First Meeting with Bigfoot*. The brothers and a few other men shared the use of a crude log shelter high in the Sierra Nevada a short distance northwest of Yosemite National Park, near Strawberry, California, in an area where there have been several other sasquatch reports in recent years. Al Berry dates the first incident in early July, 1971, but Warren Johnson's account puts it on August 7. Also, Johnson spells his brother's name as "Louis", while Al Berry spells it "Lewis", and he says that things in the camp were "pretty much okay" while Berry makes a point of the shelter having been entered and things moved around. There are other small discrepancies in the two accounts as well, which is probably of no significance but does cause a problem for someone trying to write a short and accurate summary.

At any rate, when the brothers arrived in camp they cooked a meal on a stove they kept a short distance from the shelter. They then retired to the shelter, but had been inside only a short time when they heard a commotion among the cooking things and a torrent of moans, grunts, snarls, snorts, tooth-popping and chest-beating, that thoroughly frightened them. After all became quiet they went out to look around, and in a muddy patch where some tea had been spilled they found some enormous five-toed footprints, nearly 19 inches long and nine inches across the ball of the foot, but only five inches across the heel. The book tells of the creature coming back that night, while Johnson's account says it returned the next night. In any event they heard it again on that trip, and again when five of them stayed in the camp either one or two weeks later. On the later occasion, Louis, peering out through a hole in the shelter wall, saw a shape enter a patch of moonlight and leave again as soon as he tried to call someone's attention to it. There is a variation in the stories regarding the estimates of size, but Johnson's account says 10 feet tall, four feet across the shoulders and as much as 750 pounds. On that occasion tracks were found that were 22 inches in length and had a five-foot stride. Far more remarkable, they were just four inches short of five feet apart from side to side. That was entirely different from all the other sets of tracks they found, in which the prints were pretty much in line.

Other sizes of tracks mentioned by Warren Johnson were 20 inches, 14 inches, 11 inches, nine inches and seven inches. That would indicate seven individuals at least in that one area in 1971. Some of the tracks were found at considerable distances from the camp. In the main the larger creatures, in particular, seemed to try to avoid places where their tracks would show. Tracks had been photographed on the first occasion, and subsequently there were also casts made and the sounds were tape-recorded. Besides making the more violent sounds the creatures seemed to communicate by whistling. And although they were so noisy themselves they seemed to be aware immediately when one of the men made a move to leave the shelter to try to get a look at them.

Al Berry was taken to the site in 1972, and obtained his own tape recordings. Attempts to get photos with trip cameras never succeeded. Besides animal sounds, the men sometimes heard a sharp rapping, like wood striking wood. After the close of the 1972 season nothing happened until late in 1974, when more tracks were found and more calls heard, including whooping sounds. Two rather odd things mentioned by Warren Johnson were that the creatures would take cooked food but never disturbed deer carcasses, and that the men's horses showed no concern about the creatures at all, although one morning the big tracks were found mingled with horse tracks where the horses were tied 50 feet from the shelter.

I understand that the nocturnal visitors have not come to the camp since 1974—and I don't know what to make of the whole story. If there is fakery involved it would appear that Al Berry would be a victim, not a participant, and he tells of making considerable efforts to have the sounds on the tapes studied, without ever learning anything not favorable to their authenticity. A photo of a footprint cast in his book looks most unsatisfactory, showing a heel only about twice as big as some of the toes and a second toe far bigger than either outside toe. Also, the behavior indicated for the creatures differs from the concensus from other sources.

As to the sounds themselves, "tooth popping" is something I have heard attributed only to big bears; there are less than a dozen other accounts that mention growls and grunts at all, and I know of no other mention of chest beating. On the other hand, the sound by far the most commonly described by other people, a high-pitched scream, is completely missing. There is a suggested explanation given for the whole sequence of events, namely that the first eruption of sound was caused when a creature quietly stealing food scraps upset a kettle of boiling water on itself. Having found that during its angry outburst the men did not emerge from their shelter, the creature and its fellows adopted the practice of putting on intimidating vocal displays when there were humans there. Such an explanation would account for the fact that the sounds were so different from those attributed to the sasquatch by other people, and there is no reason to doubt that such animals could make such noises if they wished. But a recording of sounds that are not typical is no use to anyone trying to identify the creatures by the sounds they make.

My files indicate that sasquatches are not particularly silent creatures. Sounds are described in almost 10 percent of sighting reports. That is an average, however. There are very different figures from different areas. When I last made calculations, in 1976, the figure for Canada was only four percent, while for the various divisions of the U.S. the lowest figure was seven percent for California. The central states were the noisiest, at 12 percent. British Columbia came in at four percent and the rest of Canada at a mere two percent. Since it can be assumed that the border is a matter of complete indifference to the sasquatch, could the figure be telling us something about the people? Sasquatches usually scream. Are humans in the U.S. more suitable for screaming at, or do they just hear better?

As a matter of fact there are a lot of odd things about the reports of sounds. Why is it that the only sasquatches heard making moaning sounds did so in the 1800's. It isn't a very serious question, since it is based on only two reports, but perhaps it reflects a slight change in the use of the word. Why do only eastern sasquatches gurgle or cough, while only B.C. sasquatches mutter, only the Oregon variety blow or bark and only the California kind sneeze? Why do only those east of the Rockies cry like babies?

The only sounds that are consistently reported everywhere are screams, 37 all told, and whistles, only 10 reports but well distributed. Grouping somewhat similar sounds together, there are 50 screams, yells and howls; 23 whistles, squeals, blowing sounds and wailing cries; 17 grunts, growls and coughs, eight chatters, barks, mutters, laughs and yelps. Some of the animals that screamed did so after being shot, but it is still the most common sound, even without provocation, in every area. One of the stories that impressed me the most was contained in a letter I received in 1969 from Mrs. Callie Lund, of Rochester, Washington, who grew up on a ranch near Oakville, in western Washington. Here is what she said:

The spring of 1933 was a wet one and also as I remember we had a lot of snow. All the ranches in our area were on the river. We all farmed the bottom land but built our homes on the hillside. This area had been logged by the Balch Logging Company in the period of 1915-1923. By 1933 a lot of the hill area had second growth timber on it. I was attending High School in Oakville at this time and this particular weekend I had been to a dance in Oakville with a group of other young girls. When I returned home I always walked around the house to the back door. All my life I had heard coyotes and cougars. Just as I started to open the back door I heard a very loud noise up on the hill. I had never heard a noise like it before and I was a little scared. I went in and woke up my mother and told her to come and listen as this was a noise I had never heard before. Also I could hear another one answer way over on another hill.

Mother got up and her reaction was so strange that I could hardly believe it was my mother. She kept saying, "It's that terrible Ape again." When I kept trying to ask her to explain she kept saying, "I'll tell you tomorrow, and let's get inside as we are not safe out here." This is the story my mother told me the next day.

After my Grandfather died my Grandmother divided up the old ranch and gave the children a Stump Ranch. About 40 acres and most of my mother's land was the area where our house stood. The timber came down almost to the back of the house. She said in the fall of 1912 my father had gone to Aberdeen to sell potatoes he had raised on the 10 acres they had cleared. He had to stay all night so Mother was alone in the small house.

The house had a porch across the front and a large bay window. She said she had not slept very soundly because she was not used to being alone. She said about 1:00 a.m. she heard a terrible stomping noise on the front porch. She got up to see what it was. It was moonlight outside, and at first she thought it was a bear on the porch, but this animal was standing on its back legs and was so large it was bending over to look in the window. She said it appeared over 6 feet tall and it didn't look like a bear at all in the moonlight. She said in a few minutes it walked over and jumped off the front porch and started around the house. She went into the kitchen so she could get a good look and she said it looked just like an ape. She said the strange thing that had scared her most was the noise it made as it walked around the house.

She said as soon as it was daylight she went over to my Grandmother's place and told them what she had seen. Her brothers all made fun of her and told her she had had a nightmare. When my Father returned home he wouldn't believe her either. She said that they made her feel so ridiculous she said she would never mention it again. This was the first time I heard her tell the story.

Many of the best accounts regarding sounds seem to come from Washington. Here is one from Elmer Wollenburg, of Portland,

Oregon, describing an experience on the afternoon of October 2, 1971, while he was in a park looking at the scenery across the Yale Reservoir, a few mile south of Mount St. Helens.

At around 3 o'clock, I heard a unusual sound of tremendous power coming across the lake at me. It started out from a low-pitched raspy quality and gradually raised to a high and clearer pitch. The bellow may have lasted as much as eight or ten seconds. At first I assumed that it was coming from a human being with a bull horn, trying to get an echo from the mountains behind me. It was only gradually that I concluded it could not possibly have been a human voice.

I studied the opposite shore intently for quite a period of time. The air was clear. The sun was more or less at my back. At first I could see nothing moving. The water level in the lake was perhaps 25 feet low, exposing numerous stumps on the beach opposite. I scrutinized these for signs of movement, figuring that one or more might be humans standing or sitting on the beach. Within about five minutes, I saw a figure move up from the beach, across the logging road that borders the east side of the Yale Reservoir, and disappear on the other side of the road onto lower ground. A couple of minutes later I saw a second (?) figure following the same route up to the logging road, across it, and then down out of sight beyond.

These figures were of a faded light brown, with a grayish tinge, uniform in color all over. They seemed broad in comparison to their height. They moved with an unusual wobbling, almost bouncing, action. It was a movement not at all as humans would walk. I could not distinguish heads. The blobs were all of one color and shape.

At the point where I sighted these creatures on the far shore, there is a canyon leading down to the lakeshore from the highlands above. I saw and heard no logging trucks or other motor vehicles on the east shore logging road all afternoon.

In May, 1972, I went back to Yale Park with binoculars. Looking across the reservoir at the same spot, I saw a yellow Forest Service pickup truck parked at the identical place where the sasquatch (?) had crossed the road and disappeared up the canyon. The truck was almost invisible to me without the binoculars. Is the sasquatch larger than a pickup truck? Lighting conditions were different on the two days. I do not think I should draw any hard conclusions.

Another report containing an interesting reference to sound came from Wayne Thureringer, who then lived at Kent, Washington, in 1970. He said that he and another youth saw a sasquatch on the trail between Sunrise Point and Mystic Lake, 7,000 feet up in Mount Ranier National Park, on July 9 of that year. Here is part of what he told me in a taped interview made six days after the incident:

It was about three o'clock in the afternoon and the climb had been somewhat steep, and as we came to the end of one switchback and about to head around to another one we heard a noise. It was like somebody blowing into the neck of an open empty bottle, and we couldn't figure out what it was. It was loud, and we could tell that it was between us and the switchback we had just come up on, so we weren't really fatigued and we decided we would try and look into it more if we could, so we went part way back down the trail towards the noise and we found a log that had fallen crossways across the trail and wedged itself in between two trees and about three feet off the ground that extended down into the bush. So we climbed out on this log without our packs on and watched at the end of it. We watched for approximately five minutes and we didn't hear anything. We were about to go back when we heard this noise again.

We turned around and we saw this animal emerge from the dense underbrush and walk across, across country from us, probably 30 yards away. We were heading in a southwesterly direction, I believe, and it was heading in a northeasterly direction, completely opposite of us, but parallel at the same time.

Q—What did it look like?

A—It was large . . . heavy built, hair over its entire body except the insides of the hands and the middle of the fingers, and the feet about the middle of the ankle down.

Q—How did it walk?

A—It walked in an upright position, but somewhat slumped over at the waist and bent at the knees.

Q—Any idea how high it was?

A—I would say about seven and a half to eight feet.

Q— Could you see its face?

A—At one point when it turned towards us momentarily to push a bush out of its face, we saw it then . . . We saw a face that was somewhat rounded, or a very flat face and actually all I saw of it was the nose and this is the only thing that really caught my attention for that span of time. It was very flat and pushed in.

Q—Would you say it resembled a person, or an ape, or a bear, or what?

A—I would say more an ape.

Q—How long were you watching it?

A—We watched it for approximately 10 to 15 minutes.

Q—What did it do in that time?

A—It didn't do anything, it just kept on walking.

For all the screaming that has gone on there are remarkably few tape recordings. One was sent to a radio station by someone in Pennsylvania, but since it is completely undocumented it is not much use. Another was taped over the radio of a police car from the Lummi Peninsula near Bellingham, Washington. The sound was picked up from another car that was apparently very close to whatever was screaming, but nothing was seen. There had been several reports of sasaquatches being seen in the general area before the scream was recorded and others afterward, and once several people said they saw a sasquatch approach and then turn away after the taped scream had been played at night at a garbage dump. About all that can be said about that scream and the Pennsylvania screams is that they are loud and powerful but they are not at all the same.

The place where screaming really come into its own is Puyallup, Washington, just east of Tacoma. If there is one fact in this whole strange business that has been established beyond a shadow of a doubt it is that something screams at Puyallup. Unfortunately it isn't doing it as much as it used to. The screams are most often heard in an area of mixed woods and subdivisions southeast of the town, but they have been heard at a lot of other places in the general vicinity as well. They first came to public attention in July, 1972, when Marlin Ayres, a resident of a new subdivision called Forest Green, wrote to the Tacoma *News Tribune* about hearing loud screams one or two nights a month in the woods behind his home. He found that his neighbors had not heard the noise and some of them did not believe him, so finally he made a tape recording of it. In the next year he made several more recordings, but did not make any particular effort to look after the tapes, some of which were later used for other things. The best, and the only one made in daylight, was copied by other people, but the original was not kept. It was made, as were most of his tapes, by hanging the microphone of his recorder out a window of his home.

Whatever does the screaming also causes all the dogs in the area to bark frantically, so the sound of barking dogs competes with the screams on the recordings. The screams are by far the louder noise, but because the recordings are made from houses there are always dogs a lot closer to the microphone. On this one tape, perhaps because it was daylight, the dogs are silent, but it was made at that time in the early morning when all the birds are making as much noise as they can. They can make, fortunately, a lot less noise than the dogs. Many other people have also taped screams, but the Ayres morning tape is the best that I have heard. The noise is not an "eeeeeeee" scream, but more of a long "whooOooOooOoo" or "woopwoopwoop" at a high pitch and with immense volume. It is higher and more melodious by far than either of the other taped screams. Heard from a distance it has been compared to the sound of a siren far off, but it is certainly not that. There is a considerable variation in the sounds, which on the "bird tape" continue immediately after each other for several minutes, each scream lasting about two seconds and the intervals a little longer.

There is no proof whatever that the creature is a sasquatch. That has been inferred from the fact that there have been at least eight reports of sasquatches seen in the same area since the screaming was first noted, plus a good set of 14-inch footprints crossing a new dirt road in January, 1975, a few hundred yards from Forest Green. Two of the sighting reports were by a state trooper on patrol in the area at night. In August, 1972, he watched one walk across the road close in front of his patrol car while he was pulled over on the shoulder with his lights on. The other one ran across the road ahead while he was driving. The trooper, Mark Pittenger, has since become one of the most active and persistent investigators of the phenomenon, in his off-duty hours.

Another man who did a great deal to try to get to the bottom of the matter was Charles Challendar, who lives at the end of a road about a mile from Forest Green. He also taped screams on many occasions, and once when he heard screaming very close to his house he ran down a trail in the woods trying to get closer to it and almost ran into a creature that he estimated at nine or ten feet tall as it stepped across the trail. On my first trip to Puyallup I tramped around in the woods at night with him and two *News-Tribune* reporters, both of whom told me that they had heard screams at close range in the area we were in the very first time they went out there. However I was not so lucky, and have heard screaming on only one occasion, in the summer of 1975, although I have spent a lot of nights at Puyallup listening. When I have been there the frantic barking of the dogs has been almost a nightly occurence. I am told that the screams generally are heard where the barking is going on, but usually, as with the case with what I heard, there are only two or three bursts of sound. Just about all of the sasquatch investigators in the Pacific Northwest have spent some time at Puyallup, and many of them have heard screams for themselves, but because the times when the screaming continues long enough to get a tape recorder going are rather rare it is the people who live in the area who have succeeded in taping them.

Other than the sasquatch, the chief suspect regarding the screams is the coyote. The screams are certainly not normal coyote noises and it is hard to imagine a coyote generating such volume. However there are a number of intermediate noises that some people identify as coyote sounds, while others consider them to be made by the screamer. Some say that on occasion it deliberately imitates coyotes. Mark Pittenger has a tape that appears to confirm that interpretation, as it records what seem to be genuine coyote sounds and other sounds much like them, coming closer and closer to each other, but farther and farther from the recorder, until suddenly there is a distant outburst of frantic sounds and then silence. Searching next day in the area the sounds came from, Mark found the remains of two coyotes that had been smashed against tree trunks.

I don't know what screams at Puyallup, but I know that something does, and I find it absolutely fascinating that it goes on year after year without anyone ever being able to prove what is doing it. Whatever it is moves through the area for a considerable distance, and to do so it has to cross roads every half mile or so. It should be possible to monitor its movements by the barking of the dogs as well as by the screams and intercept it crossing a road.

In the spring of 1976 I decided to try to organize enough volunteers to carry out such a project, before the spread of subdivisions in the area became so great that the screaming stopped, but I found that screams were no longer being heard often enough for such an attempt to be practical. Perhaps it will turn out that the screamer has merely shifted his area of operations, but it seems unlikely that a new area would offer the investigators as good a combination of a well-developed road grid and a scattering of homes in a basically forested area. There are much larger wooded areas nearby, connecting up with the unbroken forests of the Cascade Range and Mount Ranier, but if there were screaming going on there no one would be likely to hear it and if they did there would be nothing they could do about it.

Any reasonable person would probably assume that it would be impossible for anything to scream its lungs out night after night terrifying people in their beds, without law enforcement agencies, wildlife personnel and the like doing something about it. The example at Puyallup proves that assumption to be mistaken. The screams have been going on for years, and no one has yet been able to prove what it is that screams.

-22-

What's that on the Road?

The average person nowadays seldom sees a wild animal, and most of those he does see are either glimpsed beside the road as he drives by in the daytime, or caught in the headlights crossing the road at night. The situation with the sasquatch is the same. A quarter of all sightings are reported by people who were just driving along not looking for anything.

Most such stories are not of any particular interest, and there are quite a few in which the identification has to be considered questionable, since the person involved did not get a good look at the object involved. I say object, rather than animal, because the car is doing all the moving necessary to bring about an encounter and quite often the thing seen is stationary and not fully in view.

On the other hand, some people in cars have recorded particularly outstanding reports, because of what happened, or who the witnesses were, or some related circumstance. Quite a few such stories have come up in chapters dealing with reports from various areas, but there are also interesting ones that took place in the Pacific Coast states and British Columbia, where there are so many reports that I am not attempting to deal with them on a regional basis.

One of my favorite stories involves a British Columbia bus driver. There isn't much that's special about it as a sighting report, but the way I came to hear it tells a lot about the haphazard business of getting information about sasquatches. It started when Rene Dahinden and I went to the Anahim Lake Stampede. At that time I was making my living in a very small newspaper and printing business, and I did not have very many hours left over for sasquatch hunting, but Rene had the bright idea that we should drive to Anahim Lake, which could be done on a weekend, and talk to some of the people assembled there for the annual rodeo. You get to Anahim Lake by driving to Williams Lake, in the heart of British Columbia's cattle country, and then heading west for 200 miles across the Chilcotin plateau that climbs slowly up to about 5,000 feet at the back of the Coast Range. You drive slowly too, because it isn't much of a road, but eventually, right at the back of everything, you get to Anahim Lake. The stampede there is what all rodeos were long ago, before the whole thing became a spectator sport. The contestants are the local ranch hands and ranch owners, white and Indian. Some of them are outfitters and game guides too, and they raise cattle on some of the most remote swamp meadows in the world. If there are sasquatches in British Columbia, we thought, these are the people most likely to know about it. Besides, the stampede is a lot of fun. It's the only rodeo I know of where the spectators still sit on the fence rails.

It's about a 12-hour drive from Harrison Hot Springs to Anahim Lake, but once we got there we soon found we were in luck. Panhandle Phillips, whose ranch was well known to be the most isolated of all—the only thing with a motor that could reach it was an airplane—had his tent pitched at one of the camp grounds. When we got a chance for a quiet word with him we asked, with some trepidation, what he thought about sasquatches. No one can be more scornful of such nonsense than a man who has spent a lifetime in the bush, but not Pan.

"They're out there, alright," he said.

Why did he say that? Well, his brother-in-law saw one down by Harrison Hot Springs.

He had to be pulling our legs, but in this business you learn not to give up easily. We got the man's name. Pan said he didn't know his address because he had recently moved, but he was a bus driver in the Fraser Valley. That meant that he worked for the same company as the drivers that came to my home area. I don't think I really intended to bother following it up, but on Monday morning when I was in the local drug store a bus pulled up outside and I decided to ask the driver if he knew Panhandle Phillips' brother-in-law. The driver was a man I had argued with a few times about how I tied bundles of papers I used to ship on the bus, but I didn't know his name, so I asked the druggist. He only knew the man's first name, which was the same first name that Pan had given us. I skipped the next step and just asked the driver if he was the man I wanted. Yes, he was. Had he ever seen a sasquatch? Yes, he had.

It had happened in 1962 or 1963. He had finished a midnight run to Chilliwack and there were three soldiers on the bus going to the army camp at Vedder Crossing. No one met them at the depot, and the road past the camp was not much out of his way, so he gave them a lift in his own car. That left him driving home in the middle of the night on the road from Vedder Crossing to Yarrow. It was pouring rain:

I was going quite slow because of the severe curves in the road and as I came around the final curve I saw what I thought was an extremely large man standing along the edge of the road. And then as I got close to him I realized it wasn't a man or anything I had ever seen that looked like one, and when he looked at me my headlights hit his eyes and they glared. Then he slowly kind of half walked and half loped across the road, and I kind of turned my car and followed him with my headlights. Then he stopped along the edge of the road and looked at me for a second and then jumped up onto the bank.

Q — How tall would this thing be?

A — I would say he was well over seven feet.

Q — Any clothes?

A — No, none that I could see whatever, just quite long hair all over his body.

Q — Would that be similar to an animal's hair?

A — Yes, it looked very similar to that, like a large ape, hairwise, except much taller . . .

Q — How close did you get to him?

A — Twenty-five feet, twenty to twenty-five feet . . .

Q — How fast would you say you were driving?

A — I was going quite slow, no more than 30 miles an hour.

Q — Did you continue at that speed?

A — No, I slowed right down. Actually I stopped completely.

Q — You stopped until he disappeared?

A — Yes, until he disappeared.

Q — You stopped and watched him?

A — Yes, until he disappeared.

Q — You say his eyes shone?

A — Yes, they had an animal glow to them.

Q — Did you notice any particular color?

A — No, I would say it was a kind of a reddish kind of glow. It would be hard to describe exactly. At that moment I was kind of excited and I didn't get too good a look at his eyes . . .

Q — Did he look heavy?

A — Yes, I would say he was well built, very athletic-type build actually, and very well proportioned from what I could see of him. I imagine he weighed well over 400.

Q — Was he watching you all the time?

A — Not at all times, he was looking up the hill . . . actually he only looked back directly at me twice.

Q — He wasn't really paying that much attention to you then.

A — No, no he wasn't . . .

Q — Were you ever at any time able to see any of his features, any part of his face?

A — Yes, I saw part of his face when I was right close to him . . . a very apelike, flat-nosed, some of it quite hairy and very thin-lined mouth as far as I could see. The one outstanding feature is, I didn't notice any ears . . . the only other thing was the lack of neck. He had little or no neck at all.

He did not hear the thing make any noise. When it jumped up the bank it just took off without apparent effort and disappeared above the area lighted by his headlights. After a while he cautiously marked the spot by poking a foot out of the door of his car and scraping some rocks together with his toe. Next day he found that the thing had to have risen six feet on its first jump, to land on a large rock at the bottom of a slide.

The interview was taped some years after the event, but before there were any published photos of the Patterson sasquatch. The man said he had told only two people about the incident. The first one was alright, but the second one thought he was crazy, so he gave up telling the story again until I talked to him. He didn't want his name used.

There have been two outstanding road sightings by deputy sheriffs in Washington state. That is, one of them was a deputy sheriff when he had the experience but ceased to be one soon after, while the other was a teenager at the time of the sighting, but is a deputy sheriff now and hasn't changed his story. The first incident, involving the youth who is now a deputy, took place in mid September, 1966. Here is part of my tape-recorded interview:

I was on the Fisk Road about nine or 10 miles west of Yakima about 11 o'clock on a Tuesday night and I had taken a friend of mine home that lives out there.

I was on my way home and I came round the corner and it

was raining that night, and lightning flashing and everything, and this thing was standing in the middle of the road. I slammed on the brakes and my car skidded right up to it and stopped about three feet from it. It just stood there and looked at me right through the windshield of the car.

It killed the engine on my car when I slammed on the brakes. He walked around behind my car and then turned around and came right back to the window and just bent over and looked in the window at me. I got my car started and took off.

Q — What did it look like?

A — I imagine it was a bit over seven feet tall, pretty good size, had long hair all over. It was just in the form of a big man.

Q — Would it be a heavily-built man . . .

A — Oh, yes, awful heavy. He was awfully broad.

Q — What color was he?

A — Kind of a light greyish white color.

Q — How could you judge the height of it?

A — By my car. He towered over my car by a long way . .

Q — When he was looking at you I guess you got a look at his face. Could you describe that?

A — He had a real flat nose and his lips were real thin. He had hair on his face too, there wasn't any skin other than his lips and just around his eyes that I could see. I couldn't see any ears. He had long hair on him, all over his chest and everything.

Q — What color were his eyes?

A — Well when the car lights hit him his eyes kind of had a red . . . just a . . . almost flourescent, like a rabbit's eyes or something when you shine a light in them at night.

Q — What did they look like when he was looking in the car?

A — Just eyes . . . I was so busy trying to get my car going I wasn't concentrating on staring him in the eye.

Q — Did he seem to be at all menacing?

A — No, it didn't try to touch my car at all. It never made any noise whatsoever. Just looked through the window.

Q — Were his lips open?

A — Well kind of, I could see his teeth. His teeth weren't very big in front except these (pointing out eye teeth) and they kind of protruded more than the others. They weren't

400

fangs, they were slanted more . . . forward. And his face was slanted an awful lot. He had a real low forehead.

Q — Did he have a big jaw?

A — Oh yeah, a great big square jaw.

Q — Did the nose protrude like a human nose?

A — Well not very much. It was real flat and wide.

Q — As if a human nose was squashed?

A — Mm hmm.

Q — Did you see him walk?

A — Well just back to the back of my car. He only took about two steps, just plunk, plunk and that was it. He took a stride about half the length of my car.

The car finally started and he left the sasquatch standing in the middle of the road.

I had never noticed until I transcribed them for this book how close to being identical the descriptions given by those last two witnesses are. The creature seen by the bus driver was dark brown, however.

The other sighting by a deputy sheriff took place in July, 1969, near Hoquiam in Gray's Harbor County. Unlike the other two, this incident was publicized at the time, and I talked to the witness within a few days after it happened. The deputy had not intended the story to become public, but he was overheard discussing it with another deputy in a cafe. After the report was published the sheriff became concerned about a horde of monster hunters invading his county and issued a statement saying that the deputy had decided he had actually seen a bear. He also ordered the deputy not to talk to reporters. The sheriff was very co-operative with me, however, calling the man in while he was on duty to show me the place where the incident occured. His name was Verlin Herrington. He had only a summer job with the sheriff's department, but he had been a summer deputy for a number of years. He was already pretty sure he wouldn't be getting the job the following summer. Here is part of the interview I taped with him:

Q — Now the incident we've been discussing, could you give us the date when that took place?

A — Yes, it was July 26, on a Sunday.

Q — And the year?

A — 1969.

Q — And what time?

A — It was 2:35 a.m.

Q — What were you doing at that time?

A — I had been on an incident at Humptulips and I was

401

en route by way of Deekay Road, by Grass Creek Road, to the beach and into my residence.

Q — You were working at that time.

A — Right.

Q — What was it that happened while you were making that trip?

A — As I was going down Deekay Road I rounded a corner, and my first impression was of a large bear standing in the middle of the road. I either had to stop for the bear or hit him so I decided to stop, put on the brakes, came to a screeching halt, and coasted up the slight grade as far as I could without startling the animal as I was looking at it as I was going towards it. This animal in my opinion was not a bear, because you could see by the way it was standing that it had no snout. It had a face on it. Its eyes reflected, and when I came to a complete stop I could see in the headlights of my car that it had feet on it instead of paws, and it had breasts. When I centred my spotlight on the patrol car on it, it walked to the edge of the road. It didn't fall down on all fours like a bear would. It walked to the edge of the roadway and stopped, turned, still looking at me. I re-adjusted the spotlight on my car so I could look at it better. Its feet had hair down to the soles but you could see the outline of the foot. It did have toes. Its hand was in a position where it was spread out, and it did have fingers. After I'd re-adjusted the spotlight, I rolled the window down, pulled my revolver and crawled out of the door. I aimed my revolver at it and as I cocked the hammer on it the animal went into the brush. I got back in my car and drove off . . .

Q — How far away were you when you were looking at it?

A — Seventy-five to eighty feet . . .

Q — It was standing erect on its hind legs?

A — Yes it was. As it went off into the brush it was still on its hind legs.

Q — Where it went into the brush, that's down quite a steep little bank isn't it?

A — Yes, it is.

Q — And could you see that it remained erect as it went down the bank?

A — Until it went out of the spotlight, yes . . .

Q — Just to sort of go over it from top to bottom, can you tell us anything about the shape of the head? For instance your first impression was a bear. Did it have a snout?

A — No, it didn't have a snout. I couldn't say that it has a

402

nose like a person would have. I believe that there was no hair on its face. It had a dark leathery look.

Q — Did you get any impression of the length of hair on the head or on the rest of it?

A — I would say about three to four inches on the head.

Q — It didn't have long hanging hair at all?

A — Longer on the head than on the body, yes.

Q — What sort of a neck did it have?

A — No neck.

Q — You mentioned, I think, that it had breasts?

A — Yes.

Q — Where on the body were they located?

A — Like a human person's. They were also covered with hair except for the nipples, and they were skin.

Q — Did you mention the color of the hair?

A — No, I didn't. It was brownish black, dark colored.

Q — You say you saw a hand with the fingers extended?

A — Yes.

Q — Could you tell how long the arms were?

A — I would have to guess at how long they are. The arm that I was looking at was in a bent position like you catch someone startled.

Q — You couldn't at any time see how far down the body the arm would come?

A — No.

Q — Comparing it to a human, would you say that it had long legs or short legs?

A — Long legs, I would say, long muscular legs.

Q — Did it take long strides?

A — No, it seemed like it took small steps as it walked. It was watching me as it walked to the edge of the road. They would be large steps for a human, but for this animal they were short steps. About three steps from the center of the road to the edge.

Q — Did you notice any difference in the way it walked and the way a human would walk?

A— Same type of position and same type of walk as a human being . . .

Q — Did it taper at the waist? Did it have a narrow waist?

A — It was a big body, a big stout body, that's the best I remember it was. It could have been thinner at the waist. I don't remember.

Q — It was pretty heavily built?

A — Yes.

I don't really think there is much point in trying to explain reports like those three in terms of hallucinations, or mistaken identity, and I don't think reflections about man's supposed need to have monsters are either useful or intelligent. You can call those men liars, or you can believe there are people who put on ape suits and stand in front of cars in the middle of the night, or you can face the prospect that there is an animal around that has not yet been identified.

Verlin Herrington's report is the first I know of from that part of Washington, which is close to the Pacific beaches. It is well south of the Olympic mountains, but much of the area is heavily wooded. There were several more reports there in the next few years. I have not heard of any recently, but I have not been there to inquire. Quite often when one story has received a lot of publicity there will be other reports from the same vicinity for a while and then they will stop. I suppose it could be that one or more of the creatures comes into an area for a time and then moves on, but there are other possible explanations involving human rather than sasquatch behavior, so I don't think any such conclusion can safely be drawn. It would be easy to argue that other people make up the later stories because they want to share the limelight, and that could be the explanation in some cases, but it would hardly do for all. It isn't a kind of limelight that many people find enjoyable, and a lot of the people who report subsequent incidents are careful to keep their reports from the public. I think it is more a case of heightened interest that stimulates good communication for a while but that gradually dies away.

The story that starts it all does not need to be a spectacular one. In March, 1969, a man who was on the road early in the morning going fishing reported seeing a hair-covered, upright creature run across the highway on the north side of the Columbia River near Stevenson, Washington. He had no chance to study it and it didn't do anything except bound up the cutbank and disappear in the trees. The story reached the media, however, and all sorts of people combed the adjacent hillsides looking for tracks in what remained of the winter snow. They found a few, one of which was good enough for the sheriff to cast. It was this incident that led to the adoption of the famous Skamania County by-law setting a $10,000 fine for shooting one of the creatures. There were two more sightings on the same stretch of road that year, both by women driving at night. Within the next year or so there were two more reports right there, and some others close by.

The year 1969 was the busiest in Washington, with 39 reports, the most I have on record for any one state in a single year. It was also the busiest year for reports from British Columbia, with 14, and for sightings, though not total reports, from California. There has never been another year like it on the Pacific Coast, where there were 87 reports altogether. The closest to that total in any other year was 54 in 1970, followed by 50 in 1968. Part of the explanation lies in the fact that those are the three years following the public release of the Patterson movie, but the big impact of that was in 1968, not 1969. Assuming that the movie was the major factor involved, it could be because increased interest in the subject resulted in more reports reaching someone who could record them, or it could be because the movie stimulated a lot of people to imagine things or to make up stories. If the latter were the case, however, there would be no apparent reason for the imaginings or fabrications to take account of the seasons, and the first two months of 1969 hardly produced any reports at all.

In general there have been more reports in the years since the movie than in the years before, but that is only to be expected, since the movie got a lot of people started at digging up information and must have made it much more likely that witnesses would realize that they had seen something real, instead of wondering if they were losing their marbles. The overall total of 119 reports that I have for 1969 was not exceeded until 1975 and 1976, by which time a substantial proportion were coming from states east of the Rockies, while reports from the west coast average little more than half the 1969 total. One interesting point about the reports from the eastern United States, is that they were less numerous in 1968, 1969 and 1970 than in the two years before or any year since. If the movie publicity stirred up a lot of imaginary reports in the rest of the country in those years, it didn't do it on the eastern seaboard.

Besides the flurry of activity in Skamania County in 1969 there was the Bighorn Dam excitement and the reports from the Kootenay Plains Indians in Alberta and the track and sighting reports involving the cripple-foot at Bossburg, Washington, and also a flare-up of activity around Oroville, California.

Getting back to road sightings, they usually do not involve observations of interesting behavior, but there are a few cases in which the creature seen was doing something out of the ordinary. In 1974, as a result of a radio interview, I received a letter from a man at Rockaway, Oregon, who told of seeing a pair of sasquatches on old Highway 99, south of Shasta, California, while driving home from San Francisco one afternoon in the spring of 1947. The road at that time was apparently no super-highway, because he said that on rounding a curve and seeing a tall object beside the blacktop he thought it was a post indicating a washout, and stopped to study the road ahead before proceeding. Neither he nor his wife could see anything wrong with the road, but before he got going again the "post" stepped out onto the road, starting to cross and then walking at an angle towards the car. However there was still another object of the same size and shape standing where the first one had been.

My informant estimated the creature's height at eight or nine feet, saying that he is six feet tall and could easily have walked under its arm. It came right up to the car and looked in at them, then went over and looked at the bank on the other side of the road, returned to the other creature and took it by the arm, coming back to the same spot that it had inspected. It then helped the second animal, which they could see was a female, up onto the bank. She waited until the male also stepped up and then they each put an arm around the other and walked about 100 yards over into the trees. He said that both creatures were covered with long hair, reddish brown at the ends and lighter near the body. They stood very straight. The eyes of the male, which had peered into the car, were black, with no whites showing. The female never turned her head their way.

That is the only report that I am aware of involving solicitous or affectionate behavior between two such creatures. Another story of particular interest was told to me by a highway maintenance foreman in British Columbia, who reported to his superintendent that he had seen an unusual animal on the road north of Squamish on an afternoon in January, 1970. The superintendent contacted the provincial museum, and word was passed along to me. Here is part of the transcript of my interview with him a few days later:

I came around a slight curve in the road and there was this animal on all fours facing away from me, a large hairy animal. I thought at first it was a bear. As I drew closer it turned sideways and it got up on its hind legs and ran across the road in an upright position. I knew then it was definitely not a bear, it was a human like animal but a monkey type appearance.

Q — How big did it appear to be?

A — It was seven feet tall, at least seven feet and 250 pounds . . . very prominent stomach . . .

Q — What colour was it?

A — It was reddish brown pretty well all over the body from the shoulders down, the head was a darker brown hair and not quite as long as the rest of the body . . . On the head was not much more than a medium type crew cut. It was not flat at the top or anything, it was all the same length, like the back of its head and the top of its head it was like a rounded type haircut, and on the body was very shaggy, about four inches long, like a bear.

406

Q — You said it was a monkey type thing, just what were you referring to?

A — Well, just the general appearance of the thing, a large, large monkey, and definitely not a bear. It didn't have a bear's face, it had more of a flat face like a person or a monkey. Its face was much lighter colored than its body hair, and definitely lighter than the head hair and it appeared to be almost hairless I would say but a lighter color.

Q — How far away would you be?

A — When I first noticed it I would say probably two to three hundred feet and when I slowed down to finally a complete stop to have a look at this thing that I thought was a bear . . . I wanted to have a good look at it . . . I would say approximately 100 feet when I stopped to watch the animal.

Q — How long do you think it stayed before it started across the road?

A — Not long at all, it either fell off this bank or it fell on the shoulder of the road and it picked itself up immediately and just scurried right across the road and up the bank here on the other side, in the upright position like a man would run, and it did run, it didn't waste any time at all.

Q — Did it move very fast?

A — Quite quickly. Also it looked at me when it was standing up and pretty well running all the way across the road, but the first impression I got was it was either terrified or it was very angry, the expression on its face, and then I was terrified, actually, because it was something you don't expect to see.

Q — Could you see the arms or forelimbs distinctly?

A — Yes, it definitely had arms . . . quite long arms. In proportion to its body I wouldn't say they were longer than a man's to his body. It definitely had hands. I'm not sure how many fingers but it did have hands and in the right hand it was carrying what appeared to be a fish, a 10-inch fish.

Q — How was it carrying it?

A — It had the hand wrapped around the fish and the head was on one side and a length of the back and the tail on the other side, the front and back of its hand. It had its fist . . . closed around the fish.

Q — Could you see it well enough to be sure that's what it was?

A — I would definitely say that that is what the animal was carrying.

Q — Did it use its hands at all as it was moving?

A — Its hands swung along when it was running across the road, and when it was climbing the bank it did use its left hand a couple of times, it bent slightly forward like a man would do and assisted itself or balanced itself up the incline with its left hand, still holding the fish in its right hand.

Q — How did the length of the legs seem in proportion to the rest of the body?

A — They were about comparable to a man's legs. Its legs were not overly long or short for the size of the animal. It did have a very prominent stomach, that is what I did notice.

Q — Were the limbs slim or heavily built?

A — Very sturdy type build, but hairy as well, so there would be no way to tell whether it had terrific muscles, but sturdy, a very sturdy animal.

Q — What was the neck like?

A — It didn't have a great deal of neck at all, the head just kind of sat on the shoulders . . .

Q — Could you tell if it was male or female?

A — There wasn't any sign, or at least there was no evidence of sex that I noticed . . .

Following the taping of the interview, we tried re-enacting the incident, with the foreman in his pickup where he had stopped before, while I ran across the road at the point where the animal had been. He then had second thoughts about his estimate of size. I am over six feet and he said the animal had looked half again as big as I did. That would also indicate a much more substantial-sized fish. Nearest water is a river running far down below the road in a steep canyon, so if the creature was indeed carrying a fish it seems probable that it was not for its own consumption.

Another observation that probably indicates something about feeding habits was made by Don Stratton, near his home 12 miles east of Estacada, Oregon, in April, 1973. He was driving home about 6 p.m. when he saw something thrown out of the woods onto the road in front of him, so he stopped and got out of his truck to see what was going on. Just about 25 feet away on the other side of a barbed wire fence he saw a hairy creature that he took for a bear digging at the bottom of a rotten stump. Then it stood up and he could see that it was only about five feet tall but very heavily built, with very broad shoulders and no neck. A chunk of wood from the stump was what had landed on the road. He could see skin in the creatures throat and chest area, the color of a sunburned man, and he could see muscles there and on the chest. The rest of the body, arms and legs, was covered with thick hair. It let one arm hang and held the other slightly raised. The top of its head was hidden by a branch, but he could see the lower jaws and teeth. There were no fangs, and the face was flat. He had not looked closely at any part of the animal except the upper body when the thing side-stepped unhurriedly out of sight among the close-packed trees. It looked as if it might be too wide at the shoulders to move straight ahead between the trees. The hair was brown with a silvery tinge. A search for tracks in the area was unsuccessful.

There are a number of people who claim not only to have seen a sasquatch on the road but to have run into one. I don't recall ever talking to anyone with such a story to tell, but I do have a tape that I was allowed to copy while I was in Los Angeles, on which a truck driver tells of such a collision in mid-January, 1973.

I was hauling a load of logs out of Grant's Pass, Oregon, to Eureka. At that time we had to go down the Avenue of Giants. It was about 7:30 in the evening. I was on the highway by myself so I had all my lights on . . . I was doing about 40-45 miles an hour when I went around a curve and as I got approximately to the center of the curve this being, whatever it was, stepped out from the right hand side and I hit it with the truck. The top of the hood is six foot four inches high and the top of his head was about six inches above the top of the hood. I don't know why it didn't see the lights or notice me. I didn't notice whether it was male or female. The upper torso was turned away from me at possibly 45 degrees rotation.

I hit it with the truck and it flew off to the left hand shoulder of the road. I didn't hear any noise it made. Of course I couldn't hear anything over the exhaust . . .

After I hit it I drove down the road about five or six miles and then stopped to check on the damage, which was quite extensive. I didn't see any blood or hair on the front end. The front radiator shell had been smashed in, away from the fibreglass hood. The glass was cracked in several different places, and we had to replace the whole thing.

Q — You say you have had trouble with bears, you've hit bears before?

A — No, I've hunted bears before. So I know it wasn't a bear, I've hunted bear too often to make a mistake . . . if the body was in proportion with the torso the legs would have been too long for a bear. It did not walk like a bear, it walked almost erect. There was just a little bit of forward angle from the waist. The shoulders didn't look hunched at all. The head, from what I could see it didn't look like there

409

was a neck at all, the head sat right on the shoulders . . .

Q — Did you notice the shape of the head?

A — From the back side it seemed to be round, not like I would picture a gorilla or an ape or something with a crown on the back of the head. It was more humanlike . . . The arms and hands were long enough to . . . they stopped about an inch above the knee . . .

Q — What color was it?

A — It was, I would call it auburn . . . almost a russet color.

The driver reported the incident to his supervisor, who asked him what he had been drinking.

-23-

Walking in the Woods

I have said that sasquatches are more often reported seen on roads than anywhere else, and that they are frequently seen around houses, but obviously that is because those are the places where there are most likely to be people to see them. The vast majority of their time is spent in the wilds, where only a person on foot, or perhaps on horseback, would have any chance of an encounter.

The man whose occupation takes him into the bush the year round, or who does a lot of hunting, hiking or mountain climbing, is often absolutely certain that there could not be any such animal around, because he has never seen any sign of one and he is sure that none of his friends has either. That attitude is quite understandable, but the fact is that such people do indeed have a much better than average chance of encountering a sasquatch and there are quite a few of them who publicly claim to have done so. Many others, quite possibly including some of the sceptic's friends, have seen something but will not talk about it except to someone they know will take them seriously.

Because such encounters are usually at close range, and in daylight, and because the witnesses are usually people with a fair knowledge of wildlife, such reports often have far more than the average content of specific information. Quite a few of them have been quoted already, but some of the best have been saved for the last.

One of the most interesting accounts I know of involves two men who went on a hunting trip south of Mount Ashland in October, 1943. Mount Ashland is in Oregon, but I think the exact location was a little way south of the border in California. Neither of them said a word about what had happened at the time, even to each other. It was just too unbelievable. It wasn't until almost 20 years later, after seeing published reports about the same sort of animal that he had seen, that the late O. R. (Red) Edwards, who by then lived in Fresno, California, wrote to his former hunting partner, Bill Cole, in Nebraska to see what he remembered about it. After an exchange of letters, Mr. Edwards wrote out for his friend the full story of what had happened as he could recall it. Basil Hritsco, an associate of Ivan Sanderson's, later made a tape recording of him reading the written account. Here are some excerpts from it:

It was just about sun-up now, just slightly hazy but the sun was burning it away fast and visibility was practically unlimited. The valley opened out . . .with a flat floor about a mile long and at least a hundred yards wide . . .This flat valley floor was covered with slick-leaf brush about shoulder to head high for the most part . . .The right hand slope was clear of any big trees, but there were numerous patches of brush . . .We sized up the terrain and I said, "Shall we try the brush patches up there?" and you said, "As good as any," so we angled up the hill to our right for 200 yards or so and came to the lower end of this particular brush patch. It was a good six feet tall, I couldn't quite see over it, oval-shaped, maybe 25 feet at the widest and 36 feet long up and down the slope.

You were to my left, so with just a nod by each of us you went left and I went right around the brush. We were both moving slowly and quietly. I was sweeping the area ahead with my eyes. On one sweep I caught a glimpse of what seemed like an apelike head just above the brush at the upper end of this patch. By the time I got my eyes back to focus on the spot it was gone. Then I heard the "pad pad pad" of running feet, heard the "whump" and a grunt as your bodies came together.

Dashing back to the end of the brush I saw a large manlike creature covered with brown hair, about seven feet tall. It was carrying in its arms what looked like a man, I could only see legs and shoes, straight down the hill on the run. I was about 30 feet away and the opening in the brush was only 10 to 15 feet wide. At the speed he was going it did not leave me much time to make observation.

I, of course, did not believe what I had seen, so I closed my eyes and shook my head to sort of clear things up, and looked down the hill again in time to see the back of the shoulders

412

and the head of a manlike thing covered with brown hair disappearing into the brush some 70 or 80 yards below.

Bill, I was stunned. Basically I was okay. I checked myself over but I certainly did not believe what I had just seen. I went to the other end of the thicket where I thought I had seen something at first, found fresh scraps of leaves on the ground as if something had been pulling them off and eating them. The dry ground under the short grass was dusty as if it had been trampled, but I could make out no tracks that I could recognize. Perhaps if I had known what I was looking for I could have.

I walked on around the brush to where you should have been. I saw those dusty scuff marks and nothing more, stood there real still for quite a while, then turned and went up near the top of the ridge, found a little outcropping of rock and sat down on it. I was in plain sight from below and had a good view of the area where you had just disappeared. I lit a cigarette and watched and listened.

There is no need to say that I knew that something was wrong. Plenty was wrong. My hunting partner had just disappeared and there was no logical explanation. What I had just seen I did not believe. I waited and watched, lit another cigarette, then something else began to bother me. It was too quiet .

I sat there and smoked for over half an hour. Now, Bill, it seems to me that this thing either packed you for quite a ways or you were out for quite a while. I guess I should have started looking for you. I don't know why, but I didn't. Maybe I was afraid of the creature and maybe I was afraid I'd find you dead. Anyway I climbed up to the ridge. I followed it up to the head of the valley . . .

He then spent the best part of the day working his way around the ridge above the valley and returning on the opposite side towards the place where they had entered it. There was little brush on that side and just a few scattered trees. He moved along very slowly and quietly, well up the slope, and then angled down towards the mouth of the valley:

Approaching at this angle, I reached the edge of the brush-covered valley floor with about 200 yards more to go to the end. While the brush in the center of the valley seemed quite dense, the stream of water ran through there of course, there were clumps along the edge with room to walk between them. I proceeded to do this as it was much easier than following the steep slope. Suddenly I realized that I was

413

following a beaten path. The grass here was taller, about eight inches and dense, and had been pressed down to form a soft, noiseless path winding though the clumps of brush. The grass was not cut with hooves, so it was not a deer, sheep, cow or horse path. Something with a big soft foot, I thought, probably a bear.

I had nearly reached the end of the valley and the last clump of brush was about 30 feet away when from it came a very human "Shhht!" At the sound I froze for about a minute but nothing happened. I proceeded another few steps to within about 20 feet of the brush, and peering hard could make out a dark object or objects in the center of the small clump. The outline was so much like two men in dark clothing sitting close together that I spoke out with something like:

"Okay, this has gone far enough, I have a loaded rifle trained on you and I don't want to hurt anyone."

Not a sound, not a movement. Could be a black, burnt stump, but I had never seen anything quite like it. I moved slowly within ten feet and bent over to see into the brush patch. When I did this the right half of the stump moved with great speed toward the pine tree on the slope to my right and about 15 feet away. I thought I caught the outline of a great, long-legged man through the brush, but by the time I had jumped back to see around the brush it had disappeared behind the tree.

I stood still for quite a while hoping that old bear would come out from behind the tree so I could get a shot. All was quiet. Just a puff of dust from behind the tree drifting slowly in the quiet air. I then moved to the right around the brush four or five steps, brought me almost directly between the clump of brush and the tree that something had disappeared behind. I bent down to peer into the brush from this angle and the other part of the stump went out the far side of the brush clump and running bent over, back into the brush patch approximately the way I had come. I could not see the head or feet, but I could see the back and shoulders, which were flat and broad like a man's, and the rocking motion was exactly that of a man running bent over at the waist.

I would like to interject here that this second half of the stump had been definitely shorter than the one that went out first, and as I saw it go out the back side of the brush it seemed to be clasping something tightly to its chest with both arms. A young one, perhaps. I'll never know. I did get a good look at the back. Going into the brush bent over, the bending had opened up cracks in the fur so that I could see through the tips of the outside of rather light brown. The fur or hair underneath was quite dark like chocolate.

Since I was hunting I had instinctively brought up my rifle to my shoulders as I watched, but since I was not sure just what it was I did not draw a bead. Then came the damndest whistling scream that I had ever heard, from right behind me. My hackles went up and I whirled to face the tree. Just in time to see a flash of something brown disappear behind the tree about six feet from the ground. More dust drifted from behind the tree.

I was less than 12 feet from this tree, and it wasn't any squirrel cutting off pine cones. I stood still covering the tree with my rifle. A full minute or more passed. Nothing happened. Then I moved a few feet to the south, past the tree, putting myself between those two things and the only exit I knew of, but still within 15 feet of the tree. Still covering the tree I stood real still for several minutes, probably four or five, straining my senses.

Then I noticed a knot on the side of the tree about six feet from the ground and to my right as I looked there seemed to be an eye in the middle of this knot. I had been staring so hard and long that my eyes were beginning to water, so I shut my eyes real tight for a minute to clear out the tears, and when I looked again the knot was gone.

By this time I had had enough. I started walking slowly down the path or trail towards the car, looking back at every step. As I looked back at a distance of 60 or 70 feet I saw the head, part of the back and one outstretched arm disappear behind the brush, as if it had made a flying leap from behind the tree to the cover of the brush. I also heard plainly the "flop flop" of two feet landing on uneven ground at the end of the leap.

That was all, Bill. A very few minutes later I was back in your car, where you were waiting. You asked if I had seen anything unusual and I said, "Noooo, did you?" You said, "Oh no."

Then you asked me if I'd heard a scream, and on my negative answer you said that you thought you had, but were not sure.

Bill Cole's recollection of the incident was a bit different. He didn't think that the animal had carried him after it ran into him and he thought he had been conscious the whole time.

415

"Funny, neither of us had the guts to say what happened to us," he said in a letter to Edwards.

Another story that was both informative and convincing was told by the late John Bringsli, a long-time woodsman and hunter who lived at Nelson, in the Kootenay district of British Columbia. I first learned of the incident from a newspaper story, but later met Mr. Bringsli and talked to him on several occasions. At the time I knew of no other reports from that southeast corner of the province, which made his story hard to accept, because my father had grown up nearby at Kaslo, my grandmother still lived there, and they had never heard a rumour or even a legend of such a thing. Since that time, however, there have been more than a dozen additional reports from that vicinity. Here is how Mr. Bringsli told me the story, on tape:

Q — Do you recall just when that happened?
A — It was on the seventh day of August, 1962.
Q — What were you doing that day?
A — I was picking huckleberries.
Q — Where did you go?
A — I went up Six Mile Road to Lemon Creek . . .
Q — Is that into Kokanee Park?

A — Yes, the vicinity of Kokanee Park . . . It was just about 7:30 in the morning when I went up there and started to pick berries . . . I was roughly about 100 yards off the road, following an old logging road . . . I was down on my knees picking berries and the berries were plentiful . . . This bush that I was picking on finally gave out so I got off my knees and rose up and turned around to look for another bush and I saw this animal standing there, at least I thought it was an animal.

At first I thought it was, like, a bear, but then I looked closer at it and realized it wasn't an animal, it was more like a human being. It was coated in steel-grey hair and it was very tall, I would say between seven and nine feet. An enormous size.

Well naturally a man like me hadn't seen a thing like that before, so I dropped the bucket of berries and I took off. I come out this logging road, you know how old logging trails are, trees felled across it here and there, I never even saw them, I went over them, and got into my car and took off. I don't remember much till two or three miles down the road that I really come to.

Q — About how far away from you was it?

416

A — About 40 feet . . . It was watching me, and coming towards me. Coming towards me slowly.

Q — Did it walk like a person?

A — Walked upright, just upright. It had an enormous set of arms on it. Believe me, I was so close I could see the nails on its fingers.

Q — Did the way that it walked differ in any way from a human being walking?

A — No, well I wouldn't say any different, only it was so big it was more or less of a lumbering walk, if you know what I mean . . .

Q — Heavier built than a man?

A — Oh yes, its arms were just like a thigh on a man . . all of that. It looked terrible to me, like a terrific human being . . . it was hairy all over, all over the whole body, and I think myself it was a male. You could tell a male from a female.

Q — Could you tell how long the hair on the body was?

A — I would say about four inches . . . it was fairly smooth, not too shaggy, but it had this sort of a steel grey bluish tint to it . . .

Its head was cocked to the side like it was trying to figure out what I was doing, who I was or what I was doing.

Q — It looked curious more than anything?

A — That's right, I wouldn't say it looked menacing at all, that way, but it sure was curious to see what I was doing.

There were several good reports from places in Oregon in the summers of 1972 and 1973. Barbara Ann Slate and Alan Berry in *Bigfoot* quote from a letter from Thomas E. Smith, Portland, as follows:

In late June of 1972 a friend and I went fishing in the Cascades at the base of Mount Jefferson. I don't know the name of the lake or if it has one. I do know that we had been able to reach it only after descending a very steep, boulder-strewn cliff for perhaps a thousand feet.

As we sat in our two-man raft on the lake, enjoying the great fishing, my friend mentioned something about having visitors, a couple of 'guys' standing near our packs on shore. As we rowed closer, my friend declared, "Those are apes!" My back was to them as I rowed.

I turned to see them. They weren't huge. The largest was perhaps six foot. The other was maybe four feet tall and a darker color, almost a dark brown. We thought them perhaps a female and her offspring. They didn't run. They

417

watched us as we watched them. Only when we started to drift closer did they seem nervous and begin to move about. They were graceful and not at all fierce looking. Finally they just walked away toward the boulders and we lost sight of them about halfway up the cliff.

That summer Ron Olson, at Eugene, Oregon, had some money to work with, and was doing research and publishing a newsletter under the name North American Wildlife Research Association. He learned of a report that on June 27 three timber fallers clearing the right of way for a new road up the McKenzie River valley east of Eugene had encountered a sasquatch as they arrived at work one morning. He quoted them in his newsletter as follows:

It was 6:30 a.m. as we began to cross the final knoll to where we left off the day before. As we topped the knoll, we looked down to where we had finished working, and standing about 100 yards from us was this creature. It was standing sideways to us in the middle of the downed timber we had felled the day before. The creature was upright and covered with dark-brownish hair. It seemed to be just looking around in total amazement of the construction taking place. It caught wind or heard us shortly after we saw it. Moving very quickly, the creature walked off into the timber . . .

Ron put a man in the area to live with the crew and work as a watchman, in hopes that something more would happen. Sure enough, on July 20 the chief surveyor on the project saw a similar creature about 8 a.m. as he was walking along a trail cleared ahead of the route of the new road. Coming to the edge of a clearing he saw his dog standing motionless looking at something. The quote in the newsletter tells the story:

I walked out of the thicket and looked in the same direction. Squatted down in the deep grass was a creature staring at my dog. The minute he saw me, it exploded into the alder thickets . . .The thing was dark brown. It had a pointed head, no neck, and was extremely muscular. I heard it for about 10 steps as it ran through the thicket. It moved so fast that if it would have been a deer in season, I would have never got a shot.

Ron was on the scene within two hours. He noticed the dog going into the thicket just where the surveyor said the creature had run, so he followed it, and found some huge scuff marks as if made by a two-legged creature running hard and landing heavily. It was decided not to try to follow the creature with dogs for fear of scaring it out of the area. However it was not seen again.

In July, 1973, I received a letter from a man living in Oregon City, containing a most interesting account of something that had happened in the preceding month. Here is what he had to say:

The encounter took place in the following locality—along the Collowash River, a small tributary of the Clackamas River, about 40 miles above the town of Estacada. This region is extremely rugged, heavily timbered, and very seldom visited by man.

The sighting took place about six weeks ago. I and my nephew decided to go on a weekend fishing trip to this locale. I have fished the Collowash for 12 years, and it is one of the most beautiful, totally unspoiled places I have ever been.

We had fished all day Saturday, and were very happy to return to our camp early that evening. We proceeded to build a fire circle on a large gravel bank close to the stream. We built a fire, prepared our meal, and settled back for a much needed rest and conversation about the day's activities.

My nephew proceeded to fall asleep rather early, but I remained awake until fairly late, perhaps 11:00 p.m., feeding the fire, and feeling quite content and at peace with the world.

This peaceful attitude was rather quickly shattered. First of all, I heard a sound exactly as if someone was walking along the gravel bank close to the river. I could distinctly hear the crunch and grinding of each step.

I strained to see who or what could be making such a noise, and at that time a very large form came within view of the fire light. It was walking erect, with a very determined gait, almost fluid in motion. The creature swung its arms as it walked, almost as a man would, but slightly more pronounced. As it passed directly in front of the fire, about 20 yards away from me, it paused slightly, and gave almost an *indifferent* glance towards me and towards the fire.

Needless to say I was virtually thunderstruck. I rose slowly, turned my flashlight on the creature, drew my .22 pistol and fired three shots. Upon my execution of this action, the creature emitted a high-pitched scream, and moved with really astounding speed over a huge log, and into the heavy timber.

I was so upset that I could not sleep the rest of the night. Early the next morning my nephew and I searched the bank for signs or tracks but found only a few depressions in the coarse gravel which could have been foot prints. We also examined the point at which the creature moved into the timber. There we also found no trace of its movement. The log which the creature literally stepped over, was waist high, and I had some difficulty climbing over it.

In a later letter he noted that the creature had a "truly nauseating, acrid odor," and that it was breathing very heavily, almost wheezing, like someone with asthma or emphysema, although it was not exerting itself at all.

Another sighting took place in June, 1973, this time near Sechelt, British Columbia. The Sechelt-Peninsula *Times* started things off with a story quoted here in full:

Reports of a Sasquatch have been bandied about. This time the huge hairy monster was seen in Jackson Bros. logging camp by an employee.

Joel Hurd was visiting me at the time and expressed an interest in looking into the matter, which he did with characteristic thoroughness. He reported that a cat skinner-powderman working on a logging road extension job had made the sighting near Mile Nine on Chapman Creek at about 7:30 in the morning. He had just driven a drilling rig to the end of the road and climbed on top of a huge log to look over the grade ahead. Across the clearing he saw a shaggy-haired creature about 150 or 200 feet away, standing on two legs. Besides being covered with hair, it seemed to have a goatee. It appeared to be larger than a man, and it was jumping up and down, with its arms bowed out at its sides. He couldn't see all of its legs, but thought it was standing on a log.

The witness looked for a place to run to, but glanced back and saw the creature "do a somersault off the log to the lower side of the road cut." Then the man leaped onto the drill rig and started to drive it away in reverse, but soon got stuck in a mud hole. He then ran down the road and cut across to where he had left a tractor parked on a lower switchback. There he had a clear view all around him, so he waited for about an hour until someone came, and then told his boss that he would not work alone on the road anymore. When the men returned to the site they found some footprints in mud, only one of which was much good. All the witnesses agreed that it looked more humanlike than bearlike. The man who saw the creature did not say that it was a sasquatch, but he insisted that it was not a bear.

The location of that sighting is only a few miles away, although across an inlet, from where a prospector reported seeing two such animals in the fall of 1971. He told me that he surprised them beside a small lake, and they ran. They were the color of a collie dog, and bigger than a man. Tracks in the mud did not show a clear shape, but appeared to indicate that the creatures had been pulling up lily roots.

Considering how rare it is for a person to see a sasquatch once, it always seems suspicious when someone reports that it has happened to them twice, although there is certainly nothing impossible about it. Unfortunately, from that point of view, four of the most informative of all sighting reports come from the same man. Nor does it help much that all those sightings took place within a period of a little more than a year. But the man had been spending every available hour since the first encounter trying to find the things again, and he lived where he could be in good country within minutes after leaving his home. All his reports seemed entirely reasonable, and there was no indication that he was not a reliable person. He did not gain anything by telling the stories, in fact he did not tell them unless asked, and was most insistent that his name be kept secret for fear his neighbours would find out what he was doing. While the rest of us were wondering how he could possibly have seen so much, I expect he was wondering how we could have been fooling around for so long without seeing anything. But his most recent sighting was more than eight years ago now, and he has been hunting most of that time. It is my personal opinion that he is telling the truth. If he isn't, I don't know who is. Here is how he told the story of his first encounter, in the mountains west of the Clackamus Valley, southeast of Estacada:

I was supposed to be watching a catskinner as he was fire trailing, but it was awful cold, and I walked a mile or so down that trail, because he had no need of anyone at that time, and I thought I'd warm up and see the country. Up where he was, it was cold east wind blowing; a little farther down, it was a west wind coming in. It was late fall, the last weekend in deer season I think, in 1967.

It was just a mountain trail, they have several of them up there, footpaths, and for horses. The elevation was about between four and five thousand feet. I came out lower down, into the fog, before I saw anything, and the fog was freezing on the trees because it was so cold, but if the wind would blow, the fog would break and fall off. That made it kind of noisy, it sounded like walking.

I came around a bend . . .well, first I noticed some rocks that were turned over. All the other rocks were wet, because of the fog, but these rocks were dry. Then I looked up, about forty or fifty feet up on a ridge of rock, and I saw these animals there . . . looked like human or just about. Large male, the female wasn't so large, and a small baby . . .well not really small, it was moving with them. It was standing up, mostly. The two older ones were squatting down and sort of bending, as they picked up rocks and smelled them. They were kind of careful. They moved on for a few minutes, and then finally the male found possibly what he was looking for and dug real fast down into the rocks, which were large boulders . . . not the round type of rock, but the flat, sharp kind.

I could not explain why those rocks were there; there hadn't been a slide or anything. They were on top of the ridge, so they couldn't have come down from anywhere. They are loose, quite a few holes underneath them, and they are as if they had been broken up. Definitely not the round, river-type rock. But they would pick them up, and after they smelled them, they would lay them down, on top of each other. They didn't just lay them back where they had picked them up, they stacked them up, in piles. And when the male found what he was looking for he really made the rocks fly. The big rocks weighed fifty, sixty, or even possibly a hundred pounds; he just jerked them out with his hand. He didn't seem to take any precautions for his safety. Later on I looked, and there was some rock there that could have fallen on him, but he wasn't concerned.

He brought out what appeared to be a grass nest, possibly some stored hay that small rodents had stored there. He dug through that, and brought out the rodents. It seems they ate them. The rodents appeared to be in hibernation, or asleep, or something. There were about six or eight rodents. The small animal, I noticed, only got one, but the others got two or three apiece.

But about that time they became aware of my presence and well, just became alert, I was alongside of this trail that follows the ridge. I didn't remember getting there, but I was squatting down behind a small tree when I became aware where I was. As soon as they realized I was there they suddenly began to move, real quiet, behind some low-hanging limbs on a tree there. I didn't see them again after that.

I tried to follow their tracks in the direction I thought they would have to go, but I couldn't find any, although there was heavy frost there. But the next day I found two tracks, a heel print, and the front part of the foot, the toes, but they were in a different direction, the direction from which I had come, and I never did get to connect them up with exactly which direction they had gone, or know anything about them.

The footprints, I would say, were about twelve to fifteen

Jim Green, then 5"10" tall, standing in the hole the sasquatch dug.

inches, but there wasn't enough of the track to tell exactly. They were possibly five inches wide, I don't know, at the widest part. I don't think they could have been six. I didn't know if it was one of the animals I had seen that made the footprints . . .

When I left the catskinner, he was on Low Creek, but I had walked to Jim's Meadows, possibly a mile or more. I saw the footprints between where the catskinner was and where I had seen the other animals . . .

The only time I saw their faces was when they became alert. They gave me the impression of having a face a little like a cat, without the ears. It seemed like the nose was much flatter, it didn't stick out like a human's. The upper lip was very short, and seemed to be thin. I couldn't remember that it had a chin, like a human has. So somehow or other I felt that it was a face more like a cat than a human.

The male was darker than the female, a dirty brown, where the female was a buckskin or fawn-colored animal. The male had much longer hair on his shoulders, head and neck, and hung in strings, like you see on an Angora goat. He was much heavier in the shoulders than the female. From just above the hips, the male got larger; he had a very wide "small" of the back. From there on up he just got bigger and bigger. They had very rounded or stooped shoulders. The head was set lower on the shoulders than a human's. They don't seem to have the neck stand up as we do.

Most of the time they were not standing, but were squatting down and leaning forward to pick up the rocks . . This one that was digging just seemed to go right on down. I don't remember seeing him get up, but as he was down here, he was just digging and kept going on down, and, well at that time I couldn't see exactly where he was, because I was down, and they were up a little bit on the side of this rock, which kind of levels off some, and he went down, and so I couldn't see exactly what he was doing in there, but I did see when he came out.

At that time I was a little bit nervous. I'm not sure, now about half of it . . . seemed like a bad dream for a while. I just couldn't believe what was happening. It just couldn't be, but it is.

Q — Did you notice the hands at all?

A — I noticed that it had hands. I did not notice if it had thumbs. I couldn't tell from the way it worked . . . it didn't seem to use the thumb . . .

I didn't see any knees projecting when it squatted. They were in an awkward position because of the rocks, and they couldn't just squat down like we would on a floor. They would be on different levels, and off too far to be comfortable. That's as close as I could explain it.

When they went from place to place they would shift in position, according to the terrain. The male, well, actually both of them, seemed to be moving in a certain direction, possibly from tracing the small rodents. I thought possibly it was the scent left by the rodents coming up through the rocks, because it was not a runway that they could have been picking, because they were just picking the rocks up anyplace, and as they picked it up, they'd turn it over and smell it, and then lay it on the stack. They left it very definitely in a pile. They would leave anywhere from three to fifteen or twenty in one pile, as they would reach back, and then, oh, six or eight feet farther, they would leave another pile.

In other conversations I had with him he mentioned that the small creature seemed to be afraid of the male, and always kept on the other side of his mother, away from it. He described their method of eating the rodents as being the same as a person eating a banana, except without peeling them. Neither of the big ones helped the small

424

one, it had to find one in the hay for itself. Rene and I arranged to have him show us the place where this was supposed to have happened and we did not allow him any time to get anything ready for us. Sure enough, the piles of rocks and the deep hole were there. The hole went down about five feet and was as steep-sided as a well and just about as straight down. I won't say a man could not have made it, because men can do surprising things, but I don't think it would have been easy.

The other sightings were all in the same general area. In the spring of 1968 he was looking for a place to take some sighting-in shots with a gun when he noticed at a distance of about 100 feet an animal eating leaves off a willow bush. He climbed up on the bank by the road to get a better look. The creature was a female, with breasts like a woman's, but in a lower position on the chest. He estimated it to be only five or six feet high, but it was very heavily built and covered with short, dark brown hair. As he watched it strip the leaves and shove them in its mouth he had the impression that its thumbs were not used. They seemed to be farther back on the hand than a human thumb. He had watched it only a few seconds before it noticed him and ran off into the trees. Although not many miles from the location of the first sighting, it was at a much lower elevation.

For the next half year he hunted without success, but in November he hit the jackpot. Crossing a ridge at about 3,500 feet, he found two sets of tracks, both about 16 inches long, and followed them for several miles in the snow, going down a logging road. They were not always on the road but kept coming back to it. Eventually they reached a level where the snow was petering out, and he lost them in the woods. Next day he went back to where he had left off, and while casting about for more tracks he saw something dark on the snow across a small open area.

Using binoculars at a distance of less than 200 yards, he found himself looking at two sasquatches sleeping out in the open, with their backs to the sky and their knees and elbows drawn in under their bodies. He settled down for a long vigil, and they slept for about an hour, with very little movement, then one got up and then the other. They went to a creek a few feet away and began pulling up and eating water plants. Both were obviously females, with more pendulous breasts than the ones he had seen before, and one appeared to have a swelling in the genital area and kept rubbing itself. It also gave an occasional loud call, which he described in a computer interview as "like a scream in an echo chamber." He also saw one defecate in the creek. It stepped up on a wide, low stump that was in the water, "bent forward about 45 degrees with its knees

425

slightly bent and let fly." It then wiped itself with one hand, and licked the hand briefly.

After feeding for about half an hour, working their way up the creek from him, they lay down again in a new location for about an hour, then both got up and crossed the road. The female with the swelling climbed a few feet up in a dead yew tree and wailed, then both moved off into the timber above the road.

He estimated them to be only about six feet high, but very heavy. They were both dark brown, covered with shaggy, dirty hair. When they lay down they did not seek shelter, although there were trees nearby, and at one time there was snow falling. After they left, the observer went home. He said he was "too spooked" to go over where they had been. Next day there was new snow.

His final encounter took place only about a month later. He was following some elk tracks, walking on top of old, deep snow, when he happened to look back and saw right behind him, only about 10 feet away, a scruffy looking, dark brown sasquatch at least nine feet tall. It had both arms raised in what he took to be a threatening way, and he scrambled to try to get a revolver out from under his rain clothes. However the animal quickly ran and ducked behind the roots of a blown-down tree, and since it was leaving anyway he did not shoot at it. Instead, as the computer form notes, he "left the area running." His impressions of that individual are pretty spotty, but he realized afterwards that it had a mild smell "like an old outhouse," and that it had a long upper lip that fluttered as it blew through its mouth. He also noticed that its hands were very large and long, but with the thumb "not up where a human thumb is."

He decided at that time not to hunt in the woods alone anymore, but I don't think he has kept to that decision. The thing certainly could have harmed him if it had meant to. He even speculated afterwards that it might have been a very old animal with failing senses and might have mistaken him for one of its own kind.

That was his last sighting report, and in recent years he has not seen any evidence that the creatures are around the area, although there have been sightings by other people, including the one on the Collowash River in 1973 quoted earlier in this chapter. The last thing he did find, in February, 1969, was the tracks of two creatures, one with a 14-inch foot, the other 11-inch, that came out of the forest into a field, where they appeared to have been eating the lower stems of clumps of grass. Tracks were made during the night, more than once, but then he found them taking off across country, eventually getting out of the snow. He then back-tracked into the woods and found that the animals had been sleeping just a short distance out of sight. What particularly interested him was that the two tracks were always near each other but never right together.

The winter during which all those observations were made was an unusually cold one in northwestern Oregon, with an exceptional amount of snow.

Bearing in mind the hazards of basing too much on reports from a single source, there is a wealth of information in those observations. They picture creatures that do not have an effective opposable thumb, or if they do, don't use it much. There are three accounts of sasquatches actually eating something — rock rodents, willow leaves and water weeds—and circumstantial evidence that they eat grass stems. The observation of the infant avoiding the male suggests that the trio was not a family group, and the tracks in the field suggest possibly a juvenile still hanging around its mother, but not welcome. The other pair was made up of two adult females, again suggesting that family groups are not the norm. Female chimpanzees have to put up with large sexual swellings when in heat, but other apes do not, and there is no other report of a sasquatch in that condition. That, plus the calling, indicates reproductive arrangements unlike those of any other ape, but perhaps suitable for a species very thinly distributed. On that point, however, note that he describes a total of seven individuals, all different, and tracks that appear to belong to two more, all in an area of about 200 square miles.

Finally, one of them defecating into running water, at the same place where it was eating, might indicate a complete lack of concern about such bodily functions, or it might possibly be a set pattern of behavior. If so, that would certainly cut down the chances of collecting sasquatch droppings. It would also give the people who safeguard domestic watersheds from pollution something to think about.

-24-

Apes Under Water

Would an animal built like an upright gorilla be at home in the water? Obviously not. If you were making up a semi-aquatic monster an ape should be the last thing you would think of. They can't even swim. And if you were dreaming up a way of life for a monster ape, about the last place you would put it would be in the water.

Why, then, does water figure so prominently in so many sasquatch stories? It goes back a long, long way. When Beowulf pursued Grendel's mother to her lair he had to dive deep in a lake to reach the entrance. Bernheimer, in *Wild Men in the Middle Ages*, tells of a whole tradition of aquatic wild men, including one described as tailless and hairy but bald, caught in 1161 in the sea near Orford on the English coast, and dumped back in again when no one could make him talk. He goes on:

> The wild woman, too, may be a maritime creature, and may then be referred to in the glossaries as lamia vel Meerminne (woman of the sea) just as the hairy man (pilosus) may occasionally be rendered in German as alpe (elfe) vel merewonder (sea monster). In Denmark, where the presence of the sea on three sides may have suggested the mythological change, the wild woman very often takes up her abode in the water.
>
> In all these instances the change of habitat implies no change of habits, which are the same as those of corresponding demons above water level. Indeed the writers

of medieval epics, when they consign their creatures to an aquatic life, make sure that their kinship to creatures on land is well understood. The wild woman and mermaid in "Wiamur" has two brothers living in the forest.

I am not aware of any Indian legends that make an aquatic creature of the sasquatch, the dsonoqua or the boqs, but oddly enough I have talked to many modern Indians who ascribe aquatic activities to them at the present time. Chief Tom Brown at Klemtu thought he had seen one that had climbed out of the ocean at night into the stern of a rowboat. As I recall, he hadn't been able to tell for sure what the creature was. The point is that he considered it quite normal for an ape to come out of the sea. He and several other men also told of seeing a whole set of tracks, both large and small, crossing the sand at the head of Kitasu Lake, from which Klemtu gets its water supply, and entering the water. There was no sign that they had waded out again, but most of the lake shore is rocky.

At the mouth of the Nooksack River, near Bellingham, Washington, there have been numerous recent reports of sasquatches seen in the river or in the sea. Chief Joe Washington told me that "the river is their highway", and to reach that area across the many miles of farmland that separate it from the mountains, the river would indeed be the logical route. In the fall of 1967, when there was a big run of sockeye salmon, a sasquatch was seen on more than a dozen occasions, and tracks were found and photographed coming up out of the water. Although the area is less than two hours drive from my home I did not learn of those things until years afterwards. However I later did computer interviews with many of the witnesses, and one of them has since passed a lie detector test, which Wolper Productions had him take while they were making the movie *Mysterious Monsters*.

One woman told me that while returning from the store one morning in September, 1967, she saw a dark hairy figure, in broad daylight, wading up out of the sea. Another, on the Nooksack in a fishing skiff later the same month, had a sasquatch stand up suddenly beside her boat, also in daylight. It rose up out of about five feet of water, which came only to the top of its legs. When it took a step towards her she spun the boat and left at full power, ruining her fishing net in the process. Both she and her husband had seen one a few days before that, in the early afternoon, standing up to its knees in the river. It simply bent down and slipped under the water and they did not see it again. The next day tracks were found at that point coming out of the water onto a sandbar.

In the most dramatic incident, and the one regarding which the main witness passed a lie detector test, a fisherman was drifting down one of the channels of the river delta with his gillnet set across the current, at night, when the back of the net seemed to catch on an obstruction, making it trail back upstream behind the boat. Shining his spotlight back along the net, he saw a sasquatch standing in the water, gathering it in. He shouted, and several other fishermen who were waiting their turn to drift that stretch of the river came down at high speed and got their lights on the creature, which then dropped the net, waded to an island and vanished in the trees.

There were several other sightings on land, some of them beside the water, and one of a creature wading up the river at night. We rather expected a recurrence of that visitation in 1971, as the heavy salmon run occurs every four years, but there was none. In the fall of 1975, however, there was another round of sightings in the same area, several of them by patrolling policemen. Nearly all of the sightings were on land, but a girl sitting on the beach in front of her home early in the morning said that when a mist lifted she saw a sasquatch in shallow water, in the sea, and she watched it for some time as it stood and swam, apparently trying to catch fish.

Two of the reports involve the sasquatch either disappearing under water or appearing from it, and the Indians I spoke to said that the creatures could not only swim a long way under water but could do so at tremendous speed. They said that when one passed it made a big swell in the river that surged along the bank—I have seen for myself that a big salmon can do that in a small river, though not in as impressive a manner as they described for a submarining sasquatch.

Obviously it would be unlikely that anyone would be able to observe such a creature in the act of swimming under water, and it would be impossible in a muddy river like the Nooksack, but there is one account, from Alaska, of exactly that. It came to me in a very roundabout way, but I have followed it back as far as a person who claims to have heard it directly from the person involved. I have also written to him, and my letters presumably reached him, since they were not returned, but I have not heard from him. Here is the story as I originally received it, in a letter from Nick Carter at Bellevue, Washington.

She had a tale to tell second hand, about a possible sighting her cousin (actually a cousin's husband) had had in Alaska, 14 or 15 years ago. She saw the TV show of a month or so ago, went out and bought all three of your books and read them in proper sequence. When she came to the part about sightings in the Queen Charlottes and the bigfoot swimming, she suddenly remembered the story her cousin had told her . .

The story goes like this. Errol's father is or was a fisherman in Alaska and the boy spent his early years on boats with him. The summer Errol was fifteen, the fishermen were having a lot of trouble with something ripping up their nets and stealing fish and so on. Nobody knew what was doing this. Mrs. N. thought it was the summer of 1960 and the area near Ketchikan, not exactly sure. One night, about eleven o'clock, the men were all in a shack on a dock, playing poker, when Errol's father remembered they had not pulled in their skiff and sent the boy to haul it up the side of their boat. Errol took a big flashlight and went along the long dock, apparently an L-shaped affair which ran parallel to the shore. He had to cross two other boats to get onto his, the three were moored side to side, with their boat the closest to the beach. Just how far off it was, she had no idea, but not far, apparently, and the water was not too deep between the dock and the shore. This was salt water and very clear.

The night was dark so Errol was using his big flashlight to work by as he set about hauling the skiff up the side of the big boat. He laid it on the deck, pointing inshore and was bent down, hauling on ropes when he glanced up to see a humanlike figure standing in the water half way between shore and the boat, just standing and staring at him. The boy froze for a few seconds and remembered the thing was not exactly a person but it had arms like a man and a head. It was probably wet and looked greyish color all over the head and body. It had round eyes, not big, beady like. It just stared at him, quite close, standing up to its waist in the water.

When the boy unfroze, he screamed bloody murder and ran blindly, over the tied-up boats, back up the ladder to the dock and toward the shack, still yelling his head off. The men came running, some with lights, and about 30 or so of them saw the thing. They shone several lights on it as it dived under water and swam away. They could see it under water swimming like a frog, arms forward over its head but not doing a crawl stroke. The legs kicked, the best description was like a frog. The men could see legs and arms as it swam out of sight. Nobody in the crowd had ever seen anything like it.

Errol was so shaken by it that he never spent any more time on the boats or fishing, he wouldn't go near the place again. Mrs. N. said he would not talk about it much but he had once given her the whole story, with a sketch of the dock layout.

The last paragraph is not altogether accurate, as the lady has since told me that Errol, who is no longer married to her cousin, is now back in Alaska fishing for a living. There is, of course, no proof that this story is true, or even that there ever was a man who told it, but I don't have any real doubt on the latter point.

A story almost as impressive was checked as far as possible by Bob Titmus, who was in the area when the incident is supposed to have taken place, in July of 1965. The man concerned lived at Butedale, which then had a few year-round residents, while Bob was sasquatch hunting by boat out of Klemtu, the next town to the south. I am not sure, but I think that they already knew each other before this happened. The Butedale man was a shore worker, but he fished for recreation—and he had his own outboard runabout. He had a friend who also had a boat, and the two of them usually went together, but on this afternoon the friend was not able to leave so early, so each was taking his own boat.

Butedale is on a very large island, facing the mainland across a narrow channel of the Inside Passage, but there is a small island in front of the village. The place where they fished was near the mainland shore on the far side of the small island. On this occasion the first man stopped his motor when he noticed something in the shadows on a tiny islet about 75 yards away and something else in the water. They were well worth a second look. On the islet were two gigantic bipedal creatures, very dark in color, very heavily built and covered with hair. The thing in the water, surging forward with tremendous power but with no apparent arm motion, was another of the same creatures. It was swimming towards the islet, which meant it was also swimming towards his boat. In panic, he got the motor going and sped off, but not before he got the impression that there might be two more of the creatures on the mainland beach.

Bob said that he talked to both the men, the second one's contribution to the story being that as he was on his way to their fishing spot his friend, whom he was going to meet, tore by him at top speed headed back to Butedale and showed no sign of being aware that he was there.

Seals and sea lions are common in that part of the world, and bears can also swim, but all those animals are easily recognizable. No one who lived in that country could possibly mistake any of them in broad daylight for what the man described. Sea lions can be that big, but they can't stand up. Grizzly bears are big enough too, but they are a fairly light brown and they cannot be much inclined to swim, since they are common on the mainland but unknown on islands that are sometimes less than a mile away. On the other hand the "apes", as the Indians generally call them, have been reported on virtually all the islands. Bob himself found incontestable evidence of their swimming ability in the fall of 1961 when he discovered tracks on the beach of a tiny island located in a bay on the offshore side of a much larger island. Here's how he reported the incident in a letter to Tom Slick:

When I started out that day the weather and water wasn't too awfully bad; but a couple of hours later a first class storm suddenly bore down on me from the west. The wind was blowing so hard and the seas running so high that I didn't dare turn broadside to it and run for a bay a couple of miles up ahead. I continued burrowing almost straight into it and made for a couple of small islands less than a mile ahead.

I had been on these islands before and knew that they were joined by a large sandbar, inhabited with clams, which formed a nice little bay that was a good anchorage, especially if the tide wasn't too high.

That was a long trip. The dinghy was in tow and was soon swamped. One seat and both oars were carried away; this being the second set of old oars to be lost in a month. Upon arrival at the little bay you should have seen the inside of that boat! It looked like—well, it was just scrambled, everything awash in a large portion of the ocean that had come aboard. The waves were breaking wild over the sandbar and I wasn't even certain whether the tide was on the rise or fall.

The anchor seemed to be holding alright for the moment at least so I went below to clean up the mess. When this was finished I climbed atop the cabin for a look at the water and weather outside past the bar. It seemed to be quieting down just a little. It was while I was standing up there that I scanned the sandy beach on the island to the right and saw this string of large tracks with a long stride. I felt certain that these were what we've been looking for and an inspection with the binoculars confirmed it.

Although I was little more than a hundred feet off the island there seemed no way to get ashore without swimming; —eventually I screwed up my courage, stripped off and did just that. The tracks came out of the water angling toward the timber and the undergrowth, paralleled the growth line for about 125 feet and then entered it. The tracks would measure about 13 or 13½ inches long and were approximately 6 inches wide at the ball of the foot. The stride would have been just about an even four feet. Some of the impressions were quite deep although the creature was only walking. It would be impossible to estimate with any degree of accuracy the age of the tracks since it was raining hard at the time and had been more or less steadily for days on end. Nor was there any clue offered where the tracks abruptly started at what had been a tide line. From what was left of the impressions and for all practical purposes I would say that these tracks were very similar to the fourteen-inch track that we are familiar with in California.

I did not find any other tracks on the beach of this island nor where he made his exit from the trees. Was not too successful in following these tracks in the undergrowth; however, I must confess I did not spend any great amount of time on the island as I was literally turning blue with cold and was in constant fear that the boat would pull its anchor and be carried away. Needless to say there were no measurements, casts or photos taken.

Quite a few accounts of aquatic activities have been dealt with in other chapters, such as the creature hit by the boat's propellor in Louisiana, and the tracks on island beaches in South Carolina. The Lake Worth Monster was reported seen swimming from the mainland to Greer Island, and the tracks of the Sister Lakes Monster led into a pond and didn't come out. Several of the very old "wild child" stories associate the creature with water. The "What Is It" at Cincinatti came up out of the Ohio River. The Saginaw fisherman referred to in the Indiana Folklore Archives saw a manlike monster climb out of the Saginaw River and then go back in. At Bossburg the tracks of cripple-foot were always by Lake Roosevelt, and when they were followed in the snow they both began and ended in the river. A person faking them would have to have worked from a boat. Tracks that I saw myself near Woodland, Washington, in 1963 came up from the Lewis River, and people who had seen them when they were fresh said that at the beginning they could be seen quite a long way under water. A 10-year-old boy phoned me from West Linn, Oregon, recently to say that he and his aunt and uncle had seen a big hairy animal standing upright in the Detroit Reservoir, in Oregon, and had watched it reach down and catch a fish, which it ate after biting off the head. When it noticed them it ran off and left 16-inch footprints in the mud. A logger in northern Oregon saw one standing in a small lake. There are dozens of such stories.

Perhaps the most startling account of aquatic activity was contained in a publication called *The Yeti Newsletter*, that was put out for some time by Gordon Prescott and Frank Hudson at St. Petersburg, Florida. It stated that a shrimp boat crew from Placida told of seeing such a creature swimming in the Gulf of Mexico 20 miles from shore, and said a captain had seen one swimming under water. Such behavior by a land animal is not unknown. It would be

434

no problem for a polar bear. If sasquatches can swim like that their presence on the Queen Charlotte Islands ceases to be a problem. However it is not only a most unlikely attribute for an ape, it would also raise the question of why none had ever been caught or killed from a boat, as even the best-adapted of air-breathing marine animals is at man's mercy when encountered swimming in open water.

Another type of activity fairly common among mammals that use both land and water would go a long way towards clearing up some of the mysteries about sasquatches if it were not so improbable in itself. That is the use of a den dug into the shore of a body of water with the entrance below the surface. Precisely that is described in Beowulf, but the only other possible reference to it that I know of is in a humorous best-seller of the early 1940's, *Anything Can Happen*, by George and Helen Papashvily. The book is about the adventures and misadventures of a Georgian immigrant in the United States, but contains a chaper titled "Where Sleep the Giants Still" that tells how George and his friends in the Caucasus when he was a boy had discovered a natural cavern with an opening beneath the surface of a small lake in which there were several skeletons of what appeared to be enormous men. I have no idea whether the story was intended to be serious—it is certainly not included in the book for any humor in it—and it apparently cannot be checked. It mentions in the book that the little lake was later obliterated by a landslide.

If sasquatches do use dens of that sort, it would go a long way towards explaining why their bones are not found, and how they can disappear under water, and where they are when the snow is on the ground. But it would open up a new set of questions, because such holes would be there to be blundered on by excavators from above or encountered by skin divers from below, and no such thing has happened as far as I know.

A story that could possibly indicate that sort of activity came to light in 1967 when the B. C. Provincial Museum issued a public appeal for information about hairy giants and huge tracks. It doesn't prove that the sasquatches were ever actually in the water, but there was certainly something interesting going on. Two brothers who are professional prospectors reported that they were travelling on foot at about the 4,000-foot level in the rugged mountains northwest of Pitt Lake, British Columbia, in June, 1965, when they encountered some fairly fresh tracks in the snow. The prints were enormous, twice as long as their own boots and as wide as a boot is long. They were perfectly flat and showed four clear toe impressions, with the big toe on the inside of the foot like a man's. The stride was double a man's stride. Snow in the bottom of the prints was tinted pink.

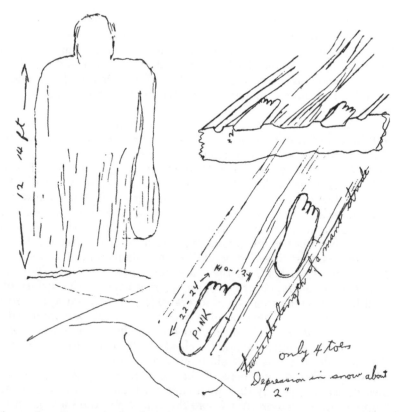

only 4 toes

Depression in snow about 2"

Above, the prospector's sketch. Below, a cast of overlapping grizzly tracks. The result is about the size and shape of an average sasquatch track, but the extra toes in the middle mar the effect, and the claws show clearly.

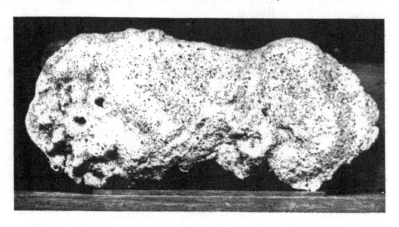

436

Parallel to the tracks were three grooves in the snow. The prints were widely spaced from side to side, and there was a wide but shallow drag mark running along between them. Outside the line of prints on either side, but close to them, were narrower, deeper grooves. The snow was old and hard, but both the prints and the outer drag marks sank in about two inches. One of the men sketched all this in his notebook on the spot. They followed the trail up the valley until they came to a small lake that was still frozen over. The tracks led out on the ice, to a place where a large hole had been made, with the broken pieces of ice lifted out and piled around.

Baffled, the men went on around the side of the lake, until they noticed in the trees on the other side a figure standing watching them. The thing was auburn in color except for its hands, where the color lightened gradually almost to yellow. As sketched on the spot it had a human-shaped head set directly on very square shoulders and its forearms and hands bulged like canoe paddles. It was swaying slightly as if shifting from foot to foot, and its hanging arms swayed too. They couldn't make out its face because of the distance, but the features seemed flat. It was just noon, and they sat down and had a cigarette and a chocolate bar while they watched it and tried to estimate its size. Counting the sets of branches on the evergreens where it stood and comparing them with those on their side of the lake, they decided it was between 10 and 14 feet tall. It just continued standing there, so finally they went on. When they came back later there were more tracks around, but the animal was gone.

The following day after climbing over a ridge onto a plateau they came to some very small lakes where there were a lot of smaller tracks, so old that all that was left of them was compacted snow sticking up above the level of the melting snow around. Some of them were 18 inches long, and they led out to a place on one of the tiny lakes where the snow had been pushed back and a hole more than five feet wide made in the ice. Other smaller tracks, about 10 inches long, did not go out on the ice.

A few days later one of the brothers went back to the spot with a newspaper reporter, in a helicopter. They photographed the big tracks in the valley, which by then were very badly melted out, and saw some fresher-looking ones on a ridge but did not land. Those tracks led to the edge of a cliff, with no snow down below it. I have talked to one of the prospectors and seen his notes and sketches. I have also talked to the reporter, a man I had known years before, but not until about 10 years after this event. My enquiries have left me in no doubt that the men saw what they said they had seen, but as to what it means I don't know.

Plainly the creature they saw had dragged some big object to the

437

lake, and presumably, since no such object was in sight, it had dumped it through the hole, but there is no clue as to what the object may have been. Only the two prospectors know exactly where the lake is, and the last I heard they had never had an opportunity to go back to it. There is an additional possibility, that the things that made the tracks were entering the water through the holes they had made in the ice. However holes in ice are the easiest thing in the world to spot from a plane, and both Rene and I have tried that and failed to find any, checking every lake within 50 miles of the area of these observations, so there is certainly no group of animals doing such a thing with any frequency.

Another story linking a sasquatch to a body of water came to light only recently, although the events referred to took place in the 1880's. The Red Bay, Alabama, *News*, May 6, 1976, quoted a letter written by Fred Collum, a resident of that area, as follows:

I thought maybe someone in this area might be interested in hearing of an incident I heard about many years ago. It happened in the Horshoe Bend area of Bear Creek. The man involved was Mr. Jade Davis, he told me this when I was a small boy.

I later confirmed the story by his son, Mr. Arthur Davis of Red Bay, this happened around the 1880's. At this time, many families lived in this remote area on small farms and supplemented their food supply by hunting, fishing and trapping.

One morning Mr. Davis went fishing and carried his gun along in case some game showed up. He was fishing at the lower or west end of the big bluff near where the Haithcock branch runs into Bear Creek. While fishing he became aware that he was being watched. He turned and looked behind him. About thirty feet away stood a hairy creature that looked like a man, his body was covered with reddish brown hair, his head and shoulders had turned grey. It seemed to be very old.

Mr. Davis said its eyes were reddish like an old person. Not knowing what the creature might do, he reached for his gun, which was lying on the ground beside him, but as soon as he moved, the creature jumped into the creek and disappeared under the rocky banks. Mr. Davis told me he thought it was a man that had gone wild. He said he never went near the place again.

A few winters later some other people were hunting in the area. It had been snowing and they were walking the bluffs where the snow had not collected. When they reached the lower end of the bluffs where the snow had drifted they saw what they thought were barefoot tracks of a man in the snow. It had been under the bluffs and had moved out ahead of them. They followed the tracks and found where it had stopped and looked back at them. If they had been looking ahead they would have seen it.

They followed the tracks to the creek, where they disappeared over the creek bank at the same place where Mr. Davis had seen it disappear. He believed there was an underwater cave that led back under the mountain . . .

At this time no one had ever heard of a 'big foot'. No doubt that was what it was. No one ever knew if there was more than one of these creatures in the area. From what we have heard later, these creatures are very shy and seem to be harmless. Maybe some day, someone may find their remains.

Walking in creeks is often about the easiest way to travel in mountain country with steep sidehills and heavy underbrush. It would also be a good way to move around without leaving any tracks, particularly with snow on the ground. Steve Adams, of Albany, Oregon, says he and his wife found tracks by Clearwater Creek on Mount St. Helens in April, 1969, that came out of the creek onto the snow on one side of the road, crossed the road and went back in the creek on the other side. There was a pool of deep water under the bridge, and there was still ice in the water, so perhaps the track maker considered it too chilly to get all wet.

Tracks are very commonly found on beaches and sand bars, and in swamps, in fact very nearly 30 percent of the tracks for which I have a record are from one of those locations, two thirds of them from beside rivers and creeks. British Columbia and the eastern U. S. also show a high percentage of tracks on ocean beaches, with the result that in B. C. 44 percent of the tracks reported have been near water, and in the eastern U. S. 55 percent.

Sightings are somewhat less likely to involve water, but again British Columbia and the eastern U. S. show a different trend from the U. S. west. Nearly a quarter of all the creatures reported in B. C. have been on beaches, and a full third were either beside or in the water. In the central and eastern U. S. nearly 20 percent of sightings were in or near water or swamps. Reports from swamps are largely concentrated from the Mississippi east, plus Arkansas, and on the eastern seaboard they make up 10 percent of all reports. In that area the totals are small, however, so the percentages can easily change.

One thing that undoubtedly distorts the B. C. figures is the fact that Bob Titmus spent several years patrolling among the islands and inlets of the northern B. C. coast in a boat, finding quite a few sets of tracks on beaches himself, and gathering a lot of information from people who spend much of their time on the water. In the early 1960's reports of sasquatches seen at the water's edge by people who were in boats or even on the beach themselves seemed to be the most common of any—so much so that Bob spent his time hunting there although it would have been much cheaper, easier and safer to hunt somewhere on dry land. In recent years that has not been the case at all, and when I made a trip all along that coast in the summer of 1975 I did not learn of anything I had not known about in 1970.

Taking in all reports of both sightings and tracks, everywhere, 22 percent are beside or in water. The percentage rises to 24 in summer, drops to 17 in winter, and is 21 the rest of the time. I expect the seasonal difference is as likely to be accounted for by the presence of more humans on or near water in summer as by any pattern of sasquatch activity. In this connection, however, I should mention the one indication I have ever found of what does look like a pattern. In my records 66 percent of all reports are in the six months from June to November, and February and March have among the least reports. The situation on the northern B. C. coast is the reverse, with sightings on the beach most common in February, March and April, and 66 percent of reports in the six months from December to June. The numbers are small, only 18 reports altogether, but I think the difference is great enough to indicate beyond doubt that sasquatches do tend to be down on the beach on the northern B. C. coast in the early part of the year. One reason why I find this convincing is that there is hardly anyone around in boats to see them at that time, while in summer there are fishing boats and cruising yachts around in good numbers, yet not so numerous nor so close to shore that they could be expected to scare all the animals away.

The commonest thing for a sasquatch to be seen doing in connection with water is standing in it, which has been reported 13 times, the water most often being a river (six occasions) or a swamp (four). Ten have been reported swimming. Of course the number actually in the water is small, most are just on a stream bank or beach. One curious anomaly is that I have recorded about three times as many sets of tracks going into bodies of water as coming out. Again the numbers are small, and certainly of no significance.

In a lot of instances the presence of water near where a creature or a set of tracks was seen is presumably of no significance either, nor are my files set up to record that type of information systematically. I am just picking it up from the brief summary on each card. Still, I think there is sufficient information to establish that there is a much greater association between sasquatches and water than anyone would be likely to imagine.

440

-25-

The Giant's Portrait

There is a poem about several blind men trying to describe an elephant. The one who got hold of its tail said it was like a rope, the one who encountered a leg said it was like a tree, the one who bumped into the side of it, said it was like a wall, and so on. The business of trying to learn about an animal that is usually seen for only a few seconds is a lot like that, in two ways. First, it is necessary to put together information from many different sources before there is much chance of getting an accurate picture. Second, there is a tendency to build a complete picture on the basis of a limited amount of information, and think you have it all.

From time to time when I go to places that have had a flare-up of sasquatch reports I meet people who think that they have everything figured out, on the basis of what has gone on in that one area. It is an understandable attitude, but experience suggests that the more they learn the sooner they will end up with problems and doubts like the rest of us. There is more to the story than is likely to be learned at Bluff Creek, or Basin Gulch, Antelope Valley, Lummi, Puyallup, Greensburg or Harrison Hot Springs. When the excitement has died down what is left is another set of observations to go into the file. They come out of the file again either as stories—of which the other chapters of this book are full—or as statistics.

Anyone who likes to think about things in terms of general impressions should have plenty to go on, if he has read this far, but there are other people who prefer to work with totals and percentages and such. This chapter is for them—not so much an attempt to describe the sasquatch, as to paint it by numbers.

It should be obvious that the numbers can not be terribly accurate, since they originate in a great variety of stories, some of which are sure to be mistaken, or made up, and since they have reached me by a variety of direct and indirect routes. But the same can be said of statistics compiled by governments, and they use them anyway, so why shouldn't I?

If it is any comfort, I once did a study of a wide variety of points, comparing percentages obtained while using only those stories which I thought should be reliable, and those obtained using every report in the file. The differences were insignificant, in almost every case. I can't say whether that proves that the doubtful reports are as accurate as the good ones, or the other way around. In any event, it relieved me of any feeling of responsibility for trying to assess accuracy before compiling statistics, and the figures I now use include everything in the file.

The detailed study was completed in June, 1976, when I had 1,350 reports, and the figures used are from that date, unless otherwise indicated. Of course, there has been more information coming in since—almost 300 additional file cards by November, 1977—and in some places in the book I have used more recent information, but not in this chapter. We are dealing, by definition, with creatures that walk or run on two legs, and are covered with hair. What else do we know about them?

I have not tried to sort out the largest, or the smallest estimate of size given, but they range from under four feet, to over 14. In 465 reports in which size was estimated, nine percent were smaller than an average man; 17 percent about man-sized; 27 percent about seven feet; 27 percent about eight feet; 16 percent nine to ten feet; three percent 11 to 13 feet, and one percent 14 feet or more. Average estimate was 7.55 feet, and just over half of all the estimates were either seven or eight feet. The only places with an average height estimate above eight feet were California and Oregon, at 8.21 and 8.45 respectively. The only place with an average under seven feet was the western United States, at 6.82. The only place where the commonest estimate was not either seven or eight feet was California, where 30 percent of estimates were nine to ten feet. The Oregon average was high because 16 percent of estimates were 11 feet or more.

The way the states were divided in my file, when these figures were compiled, is as follows:

Western—Alaska, Montana, Idaho, Wyoming, Colorado, Utah, Nevada, Arizona, New Mexico.

Central—North Dakota, South Dakota, Nebraska, Kansas, Oklahoma, Texas, Minnesota, Iowa, Missouri, Arkansas, Louisiana, Wisconsin, Illinois, Michigan, Indiana, Ohio.

Eastern—Maine, New Hampshire, Vermont, Massachusetts, Rhode Island, Connecticut, New York, New Jersey, Pennsylvania, Delaware, Maryland, Virginia, West Virginia, Kentucky, Tennessee, North Carolina, South Carolina, Georgia, Alabama, Mississippi, Florida.

Since that time I have made a separate section for Florida and another for the balance of the Southern states, but I can't readily convert the figures.

Dealing with the size of tracks, there is a complication in that those small enough to be human can't be counted. That wipes out any possibility of a valid relationship between the average height and the average track size. On the other hand, the actual sizes given for footprints should be fairly accurate, since nearly all were either measured or estimated in comparison with something of known length. Of 349 track sizes given, 28 percent were 11 to 14 inches; 39 percent were 15 to 16 inches; 25 percent were 17 to 19 inches; eight percent were 20 inches or more. Largest average was for the eastern states, at 16.87 inches; smallest for the central states at 13.84 inches, but there are so few track reports from east of the Rockies (42 in all) that averages don't mean much.

On the west coast Oregon had the highest average track length, at 16.09 inches, and British Columbia the lowest, at 15.32. That doesn't indicate much variation, but in fact there were a lot of differences. Commonest size in British Columbia, 43 percent of estimates, was 11 to 14 inches; in Washington it was almost a dead heat between that size at 33 percent, and 17 to 18 inches at 34 percent. In California 62 percent of estimates were 15 to 16 inches, and in Oregon all three sizes were about even. The order of size for both height and foot length is the same; Oregon the largest, then California, then Washington, then British Columbia. I don't bother with weight estimates, as I consider them to be universally far too low.

Colors described are almost identical to the hair color range in humans, although it seems probable that there are sasquatches with hair that is actually grey rather than just a mixture of some other color and hairs without pigment. Of 407 animals described, 126 were said to have dark hair, 79 were black, 49 brown, 30 white, 29 grey, 23 light, 23 dark brown, 23 red brown, 13 light brown, seven silver-tipped, four light grey and one dark grey.

Grouping similar shades together, 50 percent were black or dark, 27 percent were some shade of brown, 15 percent were light in color and eight percent were some shade of grey. The percentages on the west coast were about the same in each area, except that there were 27 percent light colors reported in Oregon, and only six percent in

California. Biggest variation from the average was in the eastern states, where 69 percent were dark or black and only four percent light or white. In the central states, on the other hand, 39 percent were dark or black, and 30 percent light or white. Only 15 percent of the animals reported in the eastern states were brown. No grey sasquatches were reported in Canada east of British Columbia, and very few black ones, yet in the eastern United States black was the common color, and there was a higher percentage of greys there than anywhere else. Most of the white ones were in Washington and Oregon. There were none in California.

About all that can be concluded, in my opinion, is that if someone claims all sasquatches are some particular color he hasn't done his homework, and if he has seen a purple or a green one, he is color blind, or worse.

Then there is the question of eye color. It would be easy to get the impression that the eyes always glow, but in fact I have only 56 reports of glowing eyes, of which 12 were just said to glow, with no color mentioned. Of the remaining 44, there were 23 that glowed red, 11 that glowed green, nine yellow or amber, and one white. I don't know of any significance to that either, but it is interesting that blue, which is a common color for animal's eyes to reflect, is not mentioned. When no mention is made of glowing eyes, that doesn't mean much. There may have been no light to reflect, or the animal may have been looking the other way, or the eyes may have glowed, but the informant didn't mention it. I don't have an exact figure for the number of night sightings, since sometimes no time is known, but it would be at least 400.

Smell is another item that seems more prevalent that it is. Even in Florida, home of the "skunk ape", only 14 percent of sighting reports contain any information about smell. In Canada the figure drops to one percent, and the overall average is 5.6 percent. Most of the emphasis on smell probably results from the impact of the descriptive words used. Anything described as "putrid, overpowering, intolerable, rank, old outhouse, dead, foul, terrible, vile, rotten, rancid, powerful, horrible, sickening, awful, skunk putrid, dead fish, rotten eggs, nauseating" must really smell when it smells. There are almost no descriptions of mild smells, which suggests to me that the smell is much more likely to be caused by the creatures rolling in something than to be their own body odor or the result of something they ate. Of course the percentage of smells would be much higher if there were any way of leaving out of the calculation all reports in which the witness was not in a position to notice a smell if there was one.

When it comes to physical details, I don't have much information

Dr. Daris Swindler at the University of Washington, with casts of the lower jaws of Gigantopithecus, left, and a large gorilla.

Dimensions of the creature in the Patterson movie, scaled from the known length of the footprints.

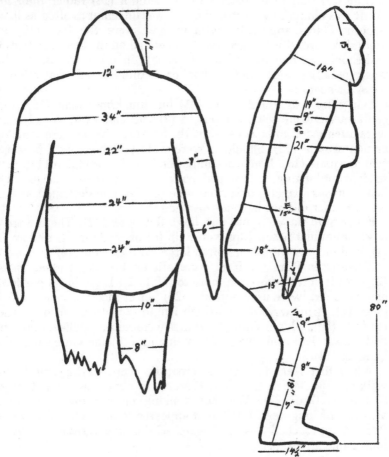

on the file cards. The best record I can go to is a preliminary run from the unfinished computer study, which included only 153 sightings. Since a great many of those involved something seen very briefly, or at a distance, there isn't much to be learned, but I will summarize those items about which there was a substantial number of answers.

Comparing the build of the creature to an average man, most called it very heavy (37), or heavy (22). Only two were thin and none very thin.

Posture was usually erect (55) or slightly stooped (17). No other posture was indicated more than three times. So we are dealing with something that is of heavy build, and that stands either erect or with a slight stoop.

Hair was said to be short (26) or medium (13) rather than long (3). It was almost always straight, but was unkempt as often as it was smooth. Places apparently without hair were the face (19) and around the eyes (5). Longer hair was noticed on the head (7) arms (3) and shoulders, midriff and legs (2 each). In other words, the hair was generally straight and of short or medium length, but might be messy or neat.

Skin color was usually dark (35) but sometimes light (12). Long canine teeth were seldom reported (2) and ears even less (1). Ears were often noted to be covered by the hair (17). Necks were short (35) or non-existent (12). Shoulders were wide (48) but might be sloped (17) or square (13). Foreheads might be low (11) or high (6) but never straight or bulging, just sloped back (11) and there was usually a heavy brow ridge (11) or a small one (3). No one described seeing a creature that did not have a brow ridge.

Faces were flat (30) and had large flat noses (11). There were no large prominent noses. Lips were thin (6) or medium (4) seldom full (1), and mouths were more often large (6) then small (2).

Torsos seen from the front were often wide (29) seldom narrow (2); and from the side even more so (22 to 1). Legs, in proportion to general build, were usually heavy (20) but of medium length (25) or long (16). Arms were of mid-thigh length (23) or down to or below the knee (19), seldom short (4). Hands were never pawlike, and never had claws. The arms were seen swinging (26) far more often than hanging (14).

When first seen most of the creatures were staying still (39) or walking (19). When they left, there were almost as many that ran (22) as walked (23). Whether walking or running they almost always took long steps. Of 72 that appeared to notice the observer, 20 fled, 14 approached closer, 22 watched the observer and 11 avoided him. None attacked.

On the subject of food there is very little information. Including absolutely everything that sasquatches have been described as handling or carrying there were only 64 reports in the file that could indicate what they eat, and there were 33 items on the list. Thirty-two reports mention types of meat, 26 mention vegetables and one a mineral, salt. In five instances the creatures were seen taking garbage or things from under rocks, but the items could not be identified. The only things mentioned more than half a dozen times were deer and berries, seven each, but in all but one case the sasquatches were just seen carrying the deer, they weren't seen eating them. Fish and roots were mentioned five times each, sheep, cattle, corn, rabbits and garbage were mentioned three times each. Not much of a case can be made out of figures like that.

Some of the things that might have seemed likely to lead the menu hardly appear on it at all. Oranges, grubs and apples were mentioned only once; clams, small rodents, leaves and water plants only twice. It seems probable that all those things would figure prominently in the diet of a big omnivore in one region or another, but actual observations have not been reported.

When it comes to reports by identifiable witnesses who tell of seeing something actually eaten, the list is very short. Albert Ostman mentions certain roots. William Roe mentions leaves. Three men describe seeing the creatures eat berries. All but one of those are British Columbia reports. I have no such reports from Washington at all. In Oregon, there are reports of creatures eating grass stems, leaves, water weeds, small rodents and a fish, but all except the fish are reported by the same man. In California there is a specific report of a sasquatch eating raw bacon and eggs, and in Manitoba one was seen eating berries. In Florida, a couple have been seen eating garbage. I might have missed one or two, but I think that is the lot.

Including things taken, or carried off, as well as those actually eaten, I have tried to find some pattern relating use of certain types of food to certain periods of the year. Vegetables do come out slightly ahead in spring and summer, animal matter in fall and winter, but the totals are too small to mean anything. To sum it up, there are indications that sasquatches are interested in a great variety of potential foodstuffs, but they have very seldom been reported actually eating anything.

As to things sasquatches do, there is a lot more information, but I don't know if it is much more helpful. They have been reported seen down on all fours 10 times, which would seem to indicate a thorough commitment to bipedalism. They can step and jump over things, and pick up and throw things that would be too big for a man to deal with in the same way, but that is to be expected from their size. There are five reports of them being clocked running at specific speeds, ranging from 35 miles an hour up to 80 miles an hour. The latter figure seems ridiculous, but some men can sprint at 20 miles an hour, and I don't think it would be unreasonable to expect a sasquatch to be able to double that. A hunter in British Columbia who watched one cross the tangled mess left behind on a logged-off hillside said it took about 10 minutes, where he had taken two hours:

> He seemed quite big, because I was crawling under logs and trees that were cut down by the loggers . . . I had to climb over, and crawl under some, but he was just walking over them.

The vast majority of the creatures reported seen are solitary and the same is true of the tracks, but slightly more so. Only 5.4 percent of total reports involve more than one individual. In the central and eastern United States, the percentage drops to 2.6. It is highest in the western states, at 10 percent, and British Columbia at nine percent. None of the counted groups consisted of more than four individuals, and there were only two groups of four. However there were a half dozen instances in which the estimate was "several", which would likely be more than four, since most people can recognize four of anything without counting. In 48 of the groups reported there were two individuals. Only 25 included more than two. The groups were not usually recognizable as families. Only 27 included individuals that appeared to be female or juvenile.

In most cases sex can not be determined. Out of 967 sighting reports, which would involve at least 1050 individuals, only 25 creatures were identified as females, only 14 as males. It is generally assumed that the ones that are not obviously females are males, but that is perhaps not an entirely safe assumption. Many are seen at too great a distance to tell. There is also a possibility that only lactating females have prominent breasts. Still, it seems most probable that many more males are seen than females. Only 10 groups were described as including a male and a female, and only five of those included one or more juveniles.

There is even more confusion when it comes to identifying juveniles, since there would seldom be any clue except size. Assuming that all the small ones accompanied by large ones are juveniles, 22 have been identified, in 16 groups with one juvenile and three with two. Nine of the 19 "family" groups included only one adult. Without knowing the size range of adults, there is no way to tell for sure whether the ones seen alone were juveniles or not. A total of 41 were reported to be smaller than man-sized, but that doesn't help too

448

much when you consider that Ostman's "young male" was seven feet tall.

One thing that can be said with certainty is that small creatures are very seldom seen. There is also a high probability that females are not seen very often. A reasonable explanation for that imbalance in the reports would be that females and young stay away from places where they are likely to encounter people, and that although all sasquatches seem to be fairly elusive, the males could be a lot more so if they chose. It could be that most of the individuals seen are footloose young males out to explore the world, or possibly old ones driven away by a group and not exercising the usual care.

The figures also indicate clearly that sasquatches do not normally associate in groups larger than would be expected for a family unit, but I don't think that constitutes evidence that they live in families. My chief sources of information on the other apes are *The Apes* by Vernon Reynolds, *Year of the Gorilla*, by George Schaller, *In the Shadow of Man*, by Jane Goodall and *In Search of the Red Ape*, by John MacKinnon. It seems agreed that gorillas form groups containing several adult males as well as females and young, while chimpanzees belong to much larger groups, various members of which associate haphazardly. In neither case is there any link between fathers and their offspring. Only the gibbons live as families. It used to be assumed that orangutans were seen in families too, but recent studies indicate that is not usually the case. Instead, the males are apparently territorial, each one occupying a definite area of forest and proclaiming his ownership with loud calls. The females do not stake out territories, so they can go where they please, and what appears to be a family group is more likely to be a mother and child in company with the male whose territory they are temporarily sharing.

The orangutan model seems to me to be the one that fits best with what has been observed regarding the sasquatch. If the males tend to be territorial, then in order to find unoccupied areas many of them might have to frequent areas also occupied by humans, and in finding territories for themselves they would have to explore unfamiliar places. That would go a long way towards explaining why the creatures sometimes show up in inhabited areas for a time. Something of the sort seems the most likely explanation for a series of incidents reported on Lulu Island, just south of Vancouver, B.C., in July, 1969.

Lulu Island is in the delta of the Fraser River and is not merely near Vancouver, it is a part of the urban complex. However some areas of the island are composed of peat bog, covered with little evergreen trees in some places, and there are still some farms on it as

449

well. The first thing that happened was that two brothers who own a farm at the southeast corner of the island noticed that something was scaring their cattle, making them stampede from a distant field up into the farmyard. That was particularly puzzling because they were range cattle brought in for finishing, and normally stayed at a distance from the buildings. One brother said he had seen something dark chasing behind the cattle while he was up on the barn roof doing some repairs, but he couldn't tell what it was. A few days later the other brother was working among the trees, building a fence, when he heard the cattle stampeding. He rushed out to find out what was going on and saw a dark animal running "on all twos" behind the cattle. It apparently was not after meat, however, because a dead calf thrown on the farm dump in the same area was not touched.

About a week after things started happening on the farm, an artist driving on a crossroad about five miles farther west on the island saw a manlike thing, covered with hair, standing in the open in a small area of woods. He stopped, and drew a picture of it, depicting a dark, rather thin form, which he said was about seven feet tall. That night, a mile farther west, a girl who was baby-sitting at the next-door-neighbors heard some horses making a great fuss in their stable about 10 p.m. On her way home a couple of hours later, she saw the head and shoulders of what looked like an enormous man above a row of raspberry bushes that were over six feet high.

The following morning Rene Dahinden and I were on an open line radio show, and someone phoned in about this thing on Lulu Island, which apparently had been mentioned in the local paper, and had been discussed on another radio program the night before. Then another caller said they had overheard a man in a cafe telling about seeing it that same morning. When we got off the air we went to investigate, and found that all of the reported witnesses stood by their stories. The man who had made the sighting in the morning had driven out onto the dyke road beside the Fraser River and parked for a few minutes before going to his store. As he sat in the car he saw an upright, hairy giant walk from behind a clump of trees across a dirt road that led from the dyke to a farm, and go into some bushes beside a big drainage ditch. We found no tracks there, but did find a place in the bushes where something big had been lying down.

The reports suggest that the thing had probably crossed to the north side of the river onto the island from a large bog on the south shore, where a fair variety of wildlife is still reported to live, although it certainly wouldn't harbor a sasquatch population. It had apparently hung around in a similar bog near the farm for a few days, then gone exploring farther west. All the places where it was seen were close to the ditch that ran inside the dyke, which was almost big enough to be called a canal. By moving west the creature was getting into areas more and more solidly inhabited. The final sighting was a little to the east of where the baby sitter saw it, so it was apparently headed back to where it had come from. It was moving in that direction anyway.

If, as do orangs, sasquatch territorial proprietors announce themselves vocally, that could provide an explanation for some of the screaming, although screams could also be the means by which individuals keep track of each other when they are moving independently. I am not suggesting that sasquatches scream all the time for territorial reasons. There are not nearly enough screams heard in the woods for that to be the case.

Jane Goodall's observations of chimpanzees established that family ties are very long-lasting, so that a pair keeping together when the female was not sexually receptive would usually be a mother and son or even sister and brother. Sisters, or unrelated females, with or without offspring, might also move around together. Similar, non-family combinations of adults have been observed with wild orangs. Pairs, or groups of male chimps also move around together, something territorial males would not be expected to do. The potential for groups of apes that are not families is almost unlimited.

Creatures living in a temperate climate usually have a particular breeding season, assuring that the newborn will have time to build some strength before winter, so I have checked to see whether there is a tendency for more pairs of adults to be seen at any particular period of the year, which might indicate mated pairs, whether identified as male and female or not. Summer would be the logical time for pairs to be together if infants were to be born in the spring but there is no indication of it. Groups make up 5.7 percent of summer reports, 6.6 percent of fall reports, four percent of winter reports and 4.8 percent of spring reports. I don't consider the number of group reports to be large enough for a variation that small to mean anything.

That wraps up the statistical information about the creatures themselves. Other matters that can be dealt with statistically include when the creatures are seen, where they are seen, and who sees them.

The time of day when tracks are made is not generally known, but in those cases when it is known, mainly in the Bluff Creek area, almost 90 percent were made at night. Sighting reports, overall, are divided almost evenly between day and night, but there is a big variation in different areas. Sightings at dawn make up only two percent of the total and at dusk 5.4 percent. If dawn is considered to last half an hour, it would occupy 2.1 percent of the day, so that figure would be exactly average, while there would be more than twice as many dusk sightings as could be expected. The half light lasts longer the farther north you go, and British Columbia has by far the highest percentage of dawn and dusk reports, 19 percent. But people differ in their impressions of when day becomes dusk, and so on, or so I would assume, so the whole thing is extremely approximate. It is clear, however, that sasquatches are not among those animals that tend to be about mainly at dawn and dusk.

The statement that reports are evenly divided between day and night may surprise people who have read something on the subject in the past, because it wasn't always that way. In 1970 I had 144 day reports, only 59 at night, and in 1973 I had 242 day reports, 149 at night. In June, 1976, there were 349 in the daytime, 332 at night. There has been a trend all along for the proportion of sightings at night to increase, and I have no explanation for that, but the sudden acceleration of the trend results from the addition of so many reports from new areas. In southern California and the eastern United States the vast majority of sightings are at night. That is one matter on which figures from the various areas are much different, with no consistent pattern. Not counting dawn and dusk reports, 73 percent of the remainder in British Columbia are in the daytime, only 46 percent in Washington, 68 percent in Oregon, and 49 percent in California. Why there should be such a fluctuation in places that are basically similar I have no idea.

Without having any statistics on the subject, it is obvious that there are far more human observers covering far more territory during the day than at night, and more at dusk than at dawn for that matter. The observers can also see a great deal more in the daytime than at night. I have always considered the existence of a substantial proportion of night sightings to indicate that the sasquatch are largely nocturnal, and the trend continues to reinforce that view.

So much for the time of day. Time of year is a considerably more consistent matter. Dividing the year into seasons, with March, April and May considered to be spring, and so on, 16 percent of reports are in spring, 36 percent in summer, 31 percent in fall and 17 percent in winter. There are two factors that would obviously tend to put those figures out of balance however. One is the presence of thousands of hunters in the woods in the fall. The other is the presence of snow to show tracks in some periods of the year, and not in others. However those things don't influence the result as much as might be expected. Subtracting all reports by hunters and all tracks in snow, the percentages become 15, 40, 28 and 16. If there were any way to make allowance for all the extra observers who are outdoors in summer because they are on vacation and the weather is nice, it would presumably cut the summer percentage considerably and increase the others.

Florida is unique in having the greatest proportion of reports in winter. Percentages there are spring 26, summer 22, fall 20, winter 33. The difference in climate would seem to be ample explanation. In every geographical division the greatest percentage of reports is in either summer or fall, and the least in winter or spring. The extremes are 47 percent of reports in summer in the central states and seven percent in spring in western states. On the Pacific Coast the highest percentage is 39, for summer reports in Oregon, and the lowest 12, for spring reports in Washington.

The most reports in spring anywhere are in British Columbia, 21 percent. The least in summer, not counting Florida, is 29 percent, in the eastern states. The most fall reports are in Washington and the western states, 36 percent, and the least are in Oregon, 22 percent. The most for winter are in Washington, 23 percent, and the least in the central states, 11 percent. While the variations are by no means as great as those concerning time of day, they are just as short of explanations.

The big puzzle, to me, is the shortage of reports in spring. Even when tracks in the snow are not counted there are less reports in spring than in winter, and there is no geographical division where there are substantially more reports in spring than in winter except British Columbia. If winter is considered to be from December 22 to March 21, and spring from March 22 to June 21 then there are probably a few more reports in spring than in winter, since there are more reports in both December and June than there are in March, if that makes anyone feel better, but there is just no explanation as to why there are so few reports in March, April and May.

The only encouraging thing about this anomaly is that since there is no apparent explanation for it in the ways of humans, with which I am familiar, then the explanation is presumably to be found in the ways of sasquatches, which are unknown. In other words, if the sasquatches are imaginary the explanation for this anomaly has to be found in the human mind, and I don't think it is there. If anyone knows a reason why people would imagine monsters more often, and fake their tracks more often, in summer, fall and winter than in spring, let's hear from him.

For quite a few of the older reports no time of year is available, and there are also a lot of reports that give a season but not a month, or say something like "June or July". Using only those reports in which a specific month was named, the totals in June, 1976 were as follows:

Jan.	Feb.	Mar.	Apl.	May	Jun.	Jul.	Aug.	Sep.	Oct.	Nov.	Dec.
51	41	42	51	40	58	113	103	89	83	60	56

The puzzling thing in that table is the low total for May, which is quite consistent geographically. There are substantially less reports in May than in either April or June from British Columbia, Washington, Oregon and California. East of the Rockies the totals for the three months are about even. If there is a month when sasquatches head for the high country or the deep bush, May is apparently it. Perhaps they are shy creatures and May is the mating season. I don't know anything on the human side of the situation to explain it. Possibly it is just accidental and will disappear when more reports are in, but it doesn't seem likely. The first time I did such a count, in 1968, there was one more report in May than in April. By 1970 there were 19 April reports, 15 May reports. By 1972 there was almost a balance, 30 in April, 29 in May. Now April has moved a long way ahead.

I won't list the numbers of reports by states and provinces, since they can be shown much more effectively on a map. Another way of studying the places in which sasquatches have been reported is to consider the type of location. Including both sightings and tracks, by far the greatest number were reported on roads—37 on major highways and 245 on lesser roads including dirt roads. Total reported in the woods was 118; on stream banks 111; in people's yards, including those close to farmhouses, 105; in wild open areas 78; on hillsides 75; on farms 72; on beaches 61; on trails 32; in swamps 31; in running water 18; in still water 15; in berry patches 11, and in the sea seven. The trouble with a list like that is that it can't include more than one thing at a time. If the creature was on a trail in a berry patch on a hillside, you just have to pick one answer arbitrarily.

Another basic question is what people are doing when they see sasquatches. Dealing with sightings only, not tracks, the witnesses were in moving cars 170 times; on foot 164 times, and at home 134 times. Of the drivers, 44 were on main highways and 126 on lesser roads. The lucky pedestrians were mainly hunters, 69; or just out walking, 43; or prospecting, 24. Of the people at home, 105 weren't doing anything special, while 29 were farming. A lot of other categories can't be lumped together with anything else. There were 40 campers, 22 loggers or road builders, 32 boaters, 30 fishermen, 13 groups of children playing, 26 (couples mostly) in parked cars.

In approximately 30 percent of sighting reports there was more than one witness, and that figure is consistent in every area, varying only from 27 percent to 32 percent, except for the central states where there was more than one witness in 45 percent of sightings. I have no idea why that figure should be so regular almost everywhere or why it should jump out of line in that one area.

Anyone trying seriously to make some sense out of this is probably fuming by now, because you obviously don't learn much unless you can relate things like time of day, time of year, type of location, geographic location, activity of the witness, etc., and find consistent patterns. If the weather, the altitude and the phase of the moon would fit in too, that would be even better. That is the sort of thing we tried to do with the computer, and I have also spent many a day at it with pencil and paper. I have made charts that will tell what the witness was doing, the type of locale, the season, the time of day, whether it was forested or open, what decade it was, approximately how high, in what state, whether there was snow on the ground, and whether there were tracks found, all from a single entry. I also have maps marked to show what was seen, what time of day, what time of year, by whom, how long ago, how high, and whether there was snow on the ground. They don't show any useful pattern at all. What they do show is that the patterns some people claim to have found—migration routes for instance—just aren't there.

Just for examples: the most common type of report in British Columbia is a sasquatch seen on the beach from a boat in the daytime. In every other area the most common type of sighting is by a driver on a side road at night. In the central and eastern states and in California, however, sightings by·people at home at night are almost as common. By far the most likely place to find tracks is on a dirt road, with stream banks a solid second, but in B.C. the most tracks have been found on beaches.

Taking the seasons into consideration, the most common form of sighting is by someone driving on a side road, in the fall, in California or Washington. They second most common is by someone sitting at home, in the fall, either in the eastern states or in Washington. Of course there are many times more people in the eastern states than there are in Washington, and the odds against such an experience are many thousands to one, for any person's lifetime, even in Washington. One thing that is of interest is that there are more sightings by drivers in the fall than in summer, yet there are a great many more drivers in summer. If one takes into consideration the number of potential witnesses involved, the people with the best odds are on boats in British Columbia, in the winter or spring, followed by hunters in British Columbia or Washington in the fall, but even for them the odds are too high to estimate.

Actually the people with the best mathematical chance to see a sasquatch are almost certainly those who sit up nights where skunk apes have been reported garbage hunting in western Florida, and after that would come anyone who regularly drives the road to Easterville in northern Manitoba—but I didn't learn those things from statistics.

-26-

Sasquatches, Humans and Apes

Science tells us that humans are animals, but we do not think of ourselves as such. The differences between any human and all the other animals are so many and so obvious that we make the distinction automatically, with no need to think about it.

It seems natural to us to be so different, and yet there is no obvious reason for it. If all of the higher primates of which we have found fossil remains were alive today we might have to give very considerable thought to the distinctions between Homo sapiens and some of his relatives. We might find ourselves examining each distinction by itself, and the results might surprise us.

The most obvious differences are that we walk upright on only two legs, that our bodies are not covered with hair, that we have speech, and that we make and use things. The first two are the most noticeable, but are they really important?

We recognize a human at a distance or from a fleeting glimpse because of the upright posture, so in that way bipedalism can be considered to be a basic distinguishing feature. That applies even if the person is sitting, on all fours, or lying down. The way we normally stand and walk upright has determined human proportions and shape to the point where they are readily distinguishable even when not in an upright position.

Actually, it is a particular type of bipedalism that makes the difference. Many dinosaurs used to be bipedal, and birds still are, but their bodies balance at an angle across the top of the legs instead of stacking everything straight up. Many other mammals are capable of standing upright on their hind legs and some can walk that way, including bears and apes, but the only one that customarily walks upright is the gibbon, and he does it with his long arms touching the ground. None of them look very human while doing it.

Does our posture really make that much difference? If a gibbon's legs were longer and its arms were shorter that would obviously fail to make it a person. There is a strange chimpanzee named Oliver that stands and walks more upright than most humans, and that does not associate with other chimpanzees, but no one would mistake it for a man. A trained bear can not only walk upright, it can perform feats of balance that the average man would not dream of attempting, such as standing on a rubber ball and walking it backwards, around corners, up an incline. Still, no one would consider it a human. On the other hand, people too crippled to stand up are still people. Our upright posture may well have contributed to other differences that are more decisive, but plainly there is no reason why there could not be creatures that stand and walk like humans without having developed the other attributes of humanity.

Nakedness, the other difference that strikes the eye, is not really a basic distinction at all. Whole classes of animals are without either fur or feathers, and even among mammals there are some, like the elephant and the hippo, much less hairy than humans. Lack of hair serves mainly as an easy way to distinguish us from our close relatives, the apes. There are, however, some very hairy humans. They are still humans, while an ape kept shaved would only be a chilly ape.

It is not really our physical attributes that distinguish humans from other animals so much as the things that we do. Most creatures use sounds as a means of communication, but not in anything like the way that people do. A few animals make and use things, but again there is little similarity between what they do and what humans do.

Members of the most primitive of human societies have the ability to communicate complicated information by sound. A person can tell others about things that they have not experienced themselves; plans can be presented and discussed, and arrangements can be made for concerted action by many individuals. Information can even be transmitted by sound, and remembered in the form of sounds, so that people can call upon the experience of others to guide them in situations that are new to them, and in places where they have never been before. No animal is known to be able to do anything of the sort.

Field studies of wild animals have produced a wealth of new information in recent years, and have disproved a lot of what everyone thought was known of animal abilities and activities, but there has been no suggestion that any animal has a language. Chimpanzees have shown an unexpected ability to learn how to use words, even combining them correctly in simple sentences, but they deal with them in sign language or with printed words on computer keys. No one has been able to teach an ape to speak, or a parrot to understand what it can say.

When it comes to making things, animals have some truly remarkable abilities, as anyone who has watched a spider spin a web must realize. Many mammals and birds construct elaborate nests or burrows. A beaver builds an actual house, and that is just one of his accomplishments. He can also build dams and canals that would be beyond the ability of an untrained human to engineer. Perhaps the most remarkable thing about a beaver's ability is that it is entirely instinctive. In *African Genesis*, Robert Ardrey reports that the few beaver which have survived throughout the centuries in Europe got along without building anything, until those in the Rhone Valley in France were protected and began to increase in numbers. They proceeded to build dams and lodges with a skill exactly equal to that of their North American cousins who had been doing it all along.

There are also animals which use things. Hugo van Lawick has photographs in *Innocent Killers* showing Egyptian vultures and mongooses throwing rocks to break ostrich eggs, and in the *National Geographic* showing wild chimpanzees soaking up drinking water with a sponge of pre-crushed leaves, and trimming twigs to poke into holes in termite mounds. Those are remarkable accomplishments for animals, and they have forced abandonment of two attempts to define the difference between man and animal—that man is the only tool user and that man is the only tool maker. The fact remains, however, that man is the only creature that depends on tools or that uses them in more than a very rudimentary way. He is also the only creature that uses or makes fire. His technological progress has proceeded unevenly, and there are still groups found occasionally that have not even discovered the use of metals, but a baby born into any cultural group has the inherited capacity to master any other culture if he or she is raised in it.

The differences between humans and all known animals are so plain that no confusion is possible, and this would remain the case if there were animals with many times the mental ability of any now known. Unfortunately as our mental abilities have grown our physical gifts seem to have atrophied. We have retained good eyesight, but our hearing is second rate, and our sense of smell all but useless. Our muscle tissue is poor stuff compared even to that of our closest relatives. It doesn't matter whether the development of the human brain was a compensation for physical deficiencies, or whether human physical ability declined because brain power made it unnecessary for survival; in either way, and almost certainly in both, our bodies and our brains have been tailored to each other. No species without such a brain could have survived with such a body, and no animal that remained physically the master of its environment would have had to evolve such a brain.

Those considerations and a host of others have to be taken into account in assessing any creature that appears to be something between man and animal, whether it is something known only from fossil evidence, or something reported seen in the present day. It certainly makes no sort of sense to say, "It walked upright, therefore it was human." Yet that is exactly what a lot of people are doing.

In 20 years of active participation in the investigation of reports of giant, hairy, humanlike creatures, I have constantly looked for indications of whether the things described were human or animal or something in between. The initial impression, from Indian traditions, was that they were some sort of wild humans, a tribe that kept aloof from its smaller relatives, but nevertheless lived in villages, spoke human languages, used fire, and even carried off human females for breeding purposes. Indeed the non-Indian community generally had the idea that the Sasquatch were hairy only to the extent of having long hair on their heads, something almost unknown among Canadian males at that time. I have seen a drawing in a high school yearbook depicting a handsome long-haired Sasquatch wearing a breachclout. At the same period a sasquatch costume was made by an Indian living in the same area and it was a complete fur suit.

Some of the earlier reports, particularly Albert Ostman's story of being kidnapped by a sasquatch and observing the activities of a family group, gave support to the idea of the near-human giant, although there was no suggestion of villages, clothing or fire in any report. Ostman's physical descriptions of the individual creatures, however, contained little hint of humanity. In fact there are no eye-witness descriptions that do. A sasquatch walks in much the same way as a man, and therefore has considerable physical resemblance to him. It is also like man and unlike other primates in its omnivorous diet, its ability to swim, and its successful survival in areas that differ widely in climate and vegetation. That practically completes the list of resemblances. The list of differences is much longer.

· A sasquatch has a fur coat. It has the size and strength to discourage most predators, if not all. It may have the speed to run down its prey, and certainly has the strength to kill most other animals. It can see in the dark. It requires no shelter and no clothing. It does not depend on tools or weapons, or need or use fire. In all those things it is the opposite of man. It cannot be proved until a specimen is dissected, but it appears that a sasquatch has a ridge on the top of its skull to which powerful jaw muscles attach, enabling it to eat types of food man's puny jaws can deal with only if they have been softened by cooking.

Most significant of all, the sasquatch is a solitary animal. It appears most likely that the largest groupings, normally, consist of a female and young. At most there may be single families including the adult male. Mankind, for probably millions of years, has depended on numbers for survival. His societies always include groups of families. The interaction within such groups probably had as much or more to do with making him what he is as did his upright posture or his use of weapons. It was surely essential to his development of language. Speech, co-operative effort, and the use of weapons and tools go with the big brain, which is what really distinguishes man from all his animal relatives. All those things the sasquatch never needed.

To put the matter in perspective, consider the implications if the sasquatch were ruled to be human. Not only would they be free of fear of hunters and of zoos, they would be entitled to welfare. In the U.S. they could have food stamps, while in Canada they would have baby bonuses and free medical care. Not only would they be entitled to send their children to school, they would be required to send them. So instead of being in zoos they would have to be put in jails, or perhaps homes for the retarded. It would not help them to take to the bush. The census takers would have to find and count them. Civil rights workers would be after them to make sure that they registered to vote. Activists would point out to them that since they had signed no treaties their aboriginal rights should be worth billions. Fair employment laws would require that every business have its token sasquatch. How would they look in the mail-order fashion ads? There is more to being a human than walking on two legs.

Why, then, is there an apparent tendency for people to want to see in the sasquatch some kind of human—to insist on its humanity in spite of all the evidence? Partly it must involve the initial reaction that anything sitting, standing and walking like a human is a human, since in all our lives everything we have ever seen that looked like that was a human. Partly the problem is caused by people who claim that their studies of the sasquatch indicate that they are human—a clear case of telling the public not what they have learned about the creature, but what they have learned the public wants to hear. Partly, I believe, it is a yearning for our own innocence, for the noble savage we never were.

Many people seem to have a need to believe that man in the wilds, without civilization, is purer, better, happier than they. Since studies of primitive societies have provided no consistent support for such a belief, the sasquatch is asked to fill the role. It has to be out there avoiding contact with its unsavory cousins, living in harmony with its environment, because it has looked us over and found us wanting; not because of its lack of weapons and inability to act in groups having rendered it unable to compete for more desirable territories, leaving for it only the mountainsides and swamps where man does not choose to live. It is a beautiful thought, but only a thought. It exists only in the mind.

Lacking physical evidence, there is a question whether sasquatches exist at all, but if they do we know a lot about them and all of it says one thing. They are all animal. Magnificent animals, completely self-sufficient on their physical endowments alone, but no more than animals. As higher primates, and huge ones, sasquatches undoubtedly have bulky brains, but until their smaller cousins acquired technology they never faced a challenge requiring that the big brain be used. When that day did arrive it was too late.

How, then, should we treat this animal that walks like us? How do we treat other animals? How should we treat other animals, and why?

I do not subscribe to the belief that nature exists solely for man's use, nor do I believe that all of Earth that can support a human population should be used for that purpose. In my opinion our species is a blight that the world would have been better off without. Had we realized a century or two ago that we were becoming too numerous, and had we been able to halt human multiplication, everywhere, man might have been the crowning glory of the planet. With his present numbers and continued growth, man can only be considered as a cancer that is destroying the planet. We are already far too numerous to hunt for our food, the world could not support us that way. We are also too numerous to be supported by the normal growth of such normal plants as we can eat. Our dependence on huge acreages of specially-bred single crops, on herbicides and on favorable weather patterns, has already made us very vulnerable, so that the problem of our numbers may contain its own solution.

I don't have any objections to animals being killed for food, or for research, or even for furs, provided that the species concerned is sufficiently numerous to recoup the loss, but I am opposed to the killing of animals merely for recreation, and most strongly opposed to the elimination of animal populations or of animal habitat, both of which are inevitable as long as the human population continues to grow. I consider this alone to be sufficient reason why the proliferation of humans should be stopped.

In short, I am entirely susceptible to arguments suggesting that

461

man reduce his interference with other animals, but what I think, or what the reader thinks, is of little significance in this matter. The pattern of mankind's treatment of animals is thoroughly set and subject to only gradual change. The question of the sasquatch must be considered in relation to things as they are, not as we might wish them to be.

How, then, should we treat an animal that walks like us? No differently than we treat anything else. To give special treatment to one type of animal because it reminds us of ourselves obviously reflects concern for ourselves, not for animals.

Leaving ourselves out of consideration, let us consider the animal. Does it perform some beneficial function for which it should be protected? There is no present evidence one way or the other, but it seems unlikely. Is the species endangered? Certainly not by hunting. There is no record of one being hunted successfully. By destruction of its habitat? That could quite possibly be the case in a few areas, but in general there does not appear to be any pressure on the mountain forests where most of the reports originate. Logging does disturb those areas from time to time, but it increases their ability to support animal life by letting the sunlight get down to ground level. Logging roads can have a very adverse effect on the population of game animals by providing easy access for hunters, but there is no evidence that sasquatches have been shot in appreciable numbers, or ever killed, and there is ample evidence that logging activity in an area does not drive them out of it.

Is there any other reason why sasquatches should be treated in a special way? I am not aware of any except for their resemblance to men.

Should sasquatches be hunted for sport? Fortunately there are many precedents for denying hunters the right to hunt one type of animal or another, so it should not be difficult to have this species protected from trophy hunters. Should they be hunted for food, or for their hides? They are presumably slow to reproduce and slow to mature, so they could not be a significant source of animal protein, fur or leather. Nor is it likely that many people would care to eat one. Descriptions suggest that their pelts would not be particularly attractive, but it is possible that they might have other valuable qualities. Should they be hunted for scientific purposes? Definitely yes. To begin with, one must be presented to the scientists in the flesh in order to establish that such a creature exists at all. Until that is done there is no possibility of having them studied effectively, or of preserving their habitat in areas where it is being destroyed. It would be difficult enough to hold back the tide of real estate development in western Florida, for instance, on behalf of any animal, no matter how manlike. To attempt it on behalf of an animal considered to be imaginary is obviously impossible.

Following that, should they be captured for public display and for study? The same considerations apply to them as to other animals. If there are sound reasons for having zoos, then there are the same reasons for having sasquatches in them. Should they be killed for dissection? Each question is more emotionally charged than the one before. A few people are beginning to protest the imprisoning of animals for man's benefit. A considerably greater number object to killing them for experimental purposes.

Dr. Geoffrey Bourne heads the Yerkes Primate Centre at Atlanta, Georgia, where large numbers of chimpanzees, gorillas and orangutans are kept for scientific study. The centre is one of a chain maintained by the U.S. government to raise primates for medical research, and in many cases the animals must be killed. Thousands die that way each year, which is why the government uses taxpayers' money to breed them. Dr. Bourne told me, however, that public objection to the killing of any of the great apes for research purposes has risen to the point where projects involving it are not approved, even though there are sufficient animals available.

The situation of the sasquatch differs considerably from that of the other great apes in at least three ways. First, there is no shortage of wild sasquatches. They cover such a tremendous area that there must be many thousands of them, and there is nothing to indicate that their numbers are declining. On the contrary, their appearance in more and more places where they were not previously known suggests that they are steadily becoming more numerous.

Second, the sasquatch as an experimental animal is of unique importance. Primates are expensive and bothersome to raise. They would be little used for medical research if other animals would serve as well, but for some purposes their close relationship to man makes them essential. As the only other primate that walks in the same manner as man, the sasquatch is the only potential experimental animal for the whole range of human ailments that are linked to upright posture. Whether it will be worthwhile to breed them for experimental purposes remains to be seen but before the answer is known they will have to be studied, and study will have to include dissection.

The third difference is that sasquatches are not available for study without killing them. With the other apes a great deal of information is already available as a result of earlier research, and if new cadavers are needed there are enough natural deaths among captive animals.to provide a fair selection. Thorough study of all the various systems—muscles, nerves, glands, blood vessels, digestive organs and so on—requires dissection of quite a few bodies. The only way they can be obtained, in the case of a sasquatch, is by hunting.

Many people will find those statements unacceptable, and I sympathize with their feelings, but there is no point in ignoring reality. Mankind is still killing the animals that are truly endangered, and doing it merely for sport or for profit. Where there are genuine and important scientific reasons to collect animals for study it is going to be done, especially when no case can be made that the species can not afford the loss. Death is a basic element in nature, the inevitable end of all but the simplest forms of life, and the majority of animals die violently. It seems certain that the sasquatch do their share of killing. There are simply no grounds for concern if a few sasquatches are killed for research purposes.

It is often proposed that there is no need to kill a sasquatch to prove their existence, that one should be captured instead or that good photographs would constitute proof. I don't suppose there are many people who have concerned themselves with the matter at all who have not gone through a period when they held those opinions, but they turn out to be in conflict with the facts. For photographic proof to be decisive, it would have to be studied, but an adequate film already exists, and no such study has taken place. Everyone knows that films can be faked, and there is nothing to suggest that additional footage, or better footage, would be treated differently from the film that has already been available but ignored for a decade. Even if studies were made and the verdict was favorable, photographic evidence would always be open to challenge and the more of it there was the more pointed would be the question, "If people can get pictures of it, why not the animal itself?"

As to capture, that would be decisive, certainly, and I don't doubt that when he applies enough resources to the problem man will find ways to trap sasquatches. The difficulty is the usual one—the resources will never be available until the animal is proved to exist. To trap an animal you must first have a trap that will hold it. Then you have to be able to put the trap in a place the animal frequents, and you have to have some way of attracting the animal to the trap. In the case of the sasquatch suitable traps are not available, unless one would be obliging enough to crawl into one of the large culvert traps used for bears. The knowledge of where traps should be placed is not yet available either, despite years of effort towards that end, and there is no saying when it will be. Finally, no one has yet demonstrated an ability to attract a sasquatch, even in the most general way, let alone persuade one to get involved with a contraption that could imprison it. If anyone wants to try trapping sasquatches there is no reason why he shouldn't, but it won't rate as a practical possibility until a lot more is learned about the quarry than is known now.

The other method of live capture, more frequently proposed, is with a tranquilizer gun. That sounds easy, and one of the favorite practices of some of the publicity-seeking sasquatch hunters is to talk about how they would never kill one, or even have one in captivity, but would just tranquilize one for a while and then let it go. Again, it is an idea that appeals to almost everyone, but it has to be abandoned by realists because it conflicts with the facts. If tranquilizer guns were accurate at long range and could be loaded with something that would drop a sasquatch in its tracks but do it no harm, they would obviously be the thing to use. In fact they are good only at very short range, and for practical purposes the animal should be fenced in, or at least in the open where it can't get out of sight.

Experts do use tranquilizer guns on wild animals in the woods, but they know how to hunt them and what quantity of drug to use for each species, and it isn't terribly serious if the animal gets away or dies. The odds of getting a shot at a sasquatch at all are so poor that it makes no sense to try to do it with an inaccurate weapon of doubtful effectiveness. If the dose of tranquilizer in the dart is too heavy it may be fatal, and if it is too light it won't work, but to use the correct dose you need to know both the weight of the animal and the tolerance of its species for the drug. If you do hit it, and the dose does happen to be correct, it may go a long way before it falls down.

Improvements are being made, and there may come a time when tranquilizer guns will do all the things people imagine they can do now. There is, however, no point in waiting. Even if, through some remarkable chain of circumstances, the first sasquatch to be collected should be a live one, there will still be just as many killed for dissection. Any contribution that anyone makes towards advancing knowledge of the subject helps to speed the day when sasquatches will be studied on slabs. A lot of people involved in the hunt would like to believe that isn't so, and a few who know better still pretend publicly that it isn't, but there is no escaping reality.

Another appealing dream is that the sasquatch could be left unmolested, but studied in the wild as has been done with other great apes. That would be very desirable, but probably not very practical. Gorillas travel in groups, leaving plenty of evidence of their passing, and they travel slowly. It is possible for an observer to find them readily and hang around long enough so that they get used to him. Orangutans present a different problem, since they are generally up in trees, but they too are generally slow moving, and recently have been successfully studied in the wild.

Chimpanzees can easily avoid humans who try to follow them, but they are numerous enough in some areas so that it is not too difficult to make observations. Like orangs they can readily be watched when they congregate in trees to eat fruit. Even so, for intensive observation it has proven necessary to attract them to the observation point with bananas.

Sasquatches are probably far more mobile than chimpanzees, or men, as well as being infinitely harder to find. Present experience suggests strongly that it will never be possible to find or follow them with sufficient regularity for any sort of study. The head of the U.S. Inter-Agency Grizzly Bear Study Team has been quoted as saying that the bears' ability to travel 20 or 30 miles in a night over mountains makes them almost impossible to study first hand in the field. The situation with the sasquatch will probably be the same or worse. Perhaps eventually some food will be found that will attract them, but bananas have already been tried without success, as have a great many other things. Vocalization may also help make observations possible, if it is first established just what sounds sasquatches make. In any event such studies would complement, not replace, those that will involve captive specimens and dead ones.

Probably no student of wild primates has ever had a closer relationship with or affection for the subjects of the study than Jane Goodall with her chimpanzees. Near the conclusion of her book *In the Shadow of Man* she makes the following comments:

> I should make it clear that I am not trying to say we should never use the chimpanzee as an experimental animal . . . He is as closely related to us in some respects as he is to the gorilla. Because of this, he is probably the only really effective substitute when for ethical reasons research cannot be carried out on humans. Kuru, a strange trembling illness of New Guinea, was a mystery of the medical world and claimed countless victims. Research with chimpanzees established it as a slow-acting virus disease and made the present dramatic cure possible.

My contention, in brief, is that there is not the slightest possibility that sasquatches can be considered human or near-human, neither are they an endangered species, and no other reason is known giving them any unique claim to total protection. On the other hand, they offer unique opportunities to learn things of value to man through the study of an animal. The appropriate action under the circumstances is to collect a sasquatch and get on with the study. Since no scientific institution is attempting to do so, it is perfectly reasonable for some private individual with a gun to get the ball rolling.

It is a normal and common reaction for people suddenly confronted with the prospect that such a creature actually exists to assume that on no account should one be shot, and sometimes people in positions of prominence will support that idea publicly, but they are acting in ignorance. Nor does it make any difference whether the person doing the shooting is some responsible, even reluctant individual who has reached the correct conclusion after due deliberation, or one of the "trigger-happy" types that sheriffs tend to worry about. The most likely person to have the opportunity is a person out hunting for something else. Whoever it is, the man who first succeeds in killing a sasquatch and bringing some part of it back for identification will be doing the right thing.

Barring the unlikely eventuality that governments or major scientific institutions will involve themselves in the investigation, that seems to be the only likely way for the matter to be brought to a successful conclusion.

In summary, I hope that I have been able to convey adequately the main points of a rather simple message:

There is evidence that another erect primate shares this globe with mankind.

The evidence may not be conclusive, but it is certainly ample to establish that the matter should be further investigated.

In the meantime, the person who finds himself in a position to obtain a specimen should do so, in the knowledge that it is important, and that such creatures are neither rare nor human.

Finally, don't worry about them. They are big, but they are nothing to be afraid of.

BIBLIOGRAPHY of books and articles cited in the text, indexed.

Allen, Donald K. "A Report from Southern California," *Bigfoot Bulletin*, Oakland, July, 1970. Page 307.

Annabel, Russell. "Long Hunter-Alaskan Style," *Sports Afield*, 1963. Page 336.

Ardrey, Robert. *African Genesis*. New York: Atheneum, 1961. Page 458.

Barker, Gray. "Invading West Virginia's Saucer Lairs and Monster Hideouts," *UFO Report*, December, 1976. Page 224.

Benton, Thomas Hart, and Schoolcraft, Henry. Letters to *American Journal of Science*, 1822. Page 324.

Bernheimer, Richard. *Wild Men in the Middle Ages*. New York: Octagon Books, 1970. Pages 18, 19, 428.

Bird, Roland T. "Thunder in His Footsteps," *Natural History*, May, 1939. Page 324.

————. "We Captured a 'Live' Brontosaur," *National Geographic*, May, 1954. Page 325.

Bourne, Geoffrey H. and Cohen, Maury. *The Gentle Giants*. New York: G. P. Putnam's Sons, 1975. Page 332.

Buffon, Georges Louis Leclerk, Compte de. *Natural History*. Paris. Page 34.

Burroughs, Wilbur. Article in *Berea College Bulletin*, Berea, Kentucky, October, 1938. Page 324.

BIBLIOGRAPHY

Caesar, Gene. "The Hellzapoppin' Hunt for the Michigan Monster," *True*, June, 1966. Page 201.

Clark, Jerome. "Are Manimals Space Beings?" *UFO Report*, summer, 1975. Page 178.

——————. "On the Trail of Unidentified Furry Objects," *Fate*, August, 1973. Page 205.

Clark, Jerome and Coleman, Loren. "Anthropoids, Monsters and UFO's, *Flying Saucer Review*, January-February, 1973. Pages 194, 205, 257.

Coleman, Loren. "Mystery Animals in Illinois," *Fate*, March, 1971. Page 203.

——————. *The Occurence of Wild Apes in North America*, unpublished, 1973. Page 211.

Colp, Harry D. *The Strangest Story Ever Told*. New York: Exposition Press, 1953. Page 303.

Coon, Ken. *The Antelope Valley Bigfoot Reports*, unpublished, April and June, 1973. Page 317.

——————. "Monsters in Our Midst: New Clues to the Growing Bigfoot Mystery," *Saga*, July, 1975. Pages 311, 312.

——————. *Sasquatch Footprint Variations*, unpublished, January, 1974. Page 322.

——————. *The Sasquatch in Southern California*, unpublished, 1971. Pages 305, 306.

Crabtree, Smokey. *Smokey and the Fouke Monster*. Fouke, Arkansas: Days Creek Production Corp., 1974. Page 190.

Craighead, John. "Studying Grizzly Habitat by Satellite," *National Geographic*, July, 1976. Page 383.

Cronin, Edward W. Jr. "The Yeti," *Atlantic Monthly*, November, 1975. Page 132.

Donovan, Roberta, and Wolverton, Keith. *Mystery Stalks the Prairie*. Raynesford, Montana: T.H.A.R. Institute, 1976. Page 298.

Eberhart, Perry. *Treasure Tales of the Rockies*. Chicago: Swallow Press, 1968. Page 44.

Eliot, Agnes Louise. Article in *Told by the Pioneers*. Olympia, Washington: U.S. Work Projects Administration, 1937-38. P. 29.

Fate, September, 1971. Page 180.

Flying Saucer Review, October-November, 1966. Page 133.

Flying Saucer Review, July-August, 1968. Page 224.

Gatti, Attilio. "On Stanley's Trail in a Trailer," *Good Housekeeping*, February, 1938. Page 133.

Gebhart, Russell. "Report from Idaho: Incidents at O Mill," *Bigfoot Bulletin*, Oakland, January-February-March, 1971. Page 288.

BIBLIOGRAPHY

Goodall, Jane Van Lawick. *In the Shadow of Man.* Boston: Houghton Mifflin, 1971. Page 466.

—————. "New Discoveries Among Africa's Chimpanzees," *National Geographic*, December, 1965. Page 458.

Guenette, Robert and Frances. *Bigfoot—The Mysterious Monster.* Los Angeles: Sun Classic, 1975. Page 309.

Householder, Patrick F. *Monsters and Murders, the Story of Deadman's Hole*, unpublished, 1972. Page 371.

Hoyt, Edward Jonathan. *Buckskin Joe.* Lincoln, Nebraska: Nebraska University Press, 1966. Page 133.

Huff, J. W. "A Possible Sasquatch Sighting in Alaska," *Bigfoot Bulletin*, Oakland, November, 1969. Page 303.

Science News Letter, "Human-Like Tracks in Stone Are Riddle to Scientists," October, 29, 1938. Page 324.

Hunter, Don, and Dahinden, Rene. *Sasquatch.* Toronto: McClelland and Stewart, 1973. Page 128.

Ingalls, Albert G. "The Carboniferous Mystery," *Scientific American*, January, 1940. Page 324.

Jones, Robert E. "Bigfoot in New Jersey." *Pursuit*, Society for the Investigation of the Unexplained, July, 1975. Page 266.

Kane, Paul. *Wanderings of an Artist.* New York: Longman, Brown, Green, 1859. Page 25.

Keel, John *Strange Creatures from Time and Space.* Fawcett Gold Books. Pages 209, 223, 265, 272, 370.

Linnaeus, Carolus. *Systema Naturae.* Stockholm: 1768. Page 137.

Locke, Charles O. *The Hell Bent Kid.* New York: W. W. Norton, 1957. Page 191.

Marsh, Richard Ogelesby, *White Indians of Darien.* New York: G. P. Putnam's Sons, 1934. Page 133.

McIlwraigh, T. F. "Certain Beliefs of the Bella Coola Indians," Archaeological Report of 1924-25, Province of Ontario. Page 21.

McKinnon, John. *In Search of the Red Ape.* New York: Holt, Rinehart and Winston, 1974. Pages 134, 449.

Merrick, Elliot. *True North.* New York: Charles Scribner's Sons, 1933. Page 252.

Montana Sports Outdoors. "Snowman or Snowjob?" December, 1960. Page 293.

Montagna, William. "From the Director's Desk," *Primate News*, Beaverton, Oregon, U.S. Primate Centre, September, 1976. Pages 130, 131.

Mozino, Jose Mariano. *Noticias de nutka.* Seattle: University of Washington Press, 1970. Page 25.

BIBLIOGRAPHY

Napier, John. *Bigfoot, the Yeti and the Sasquatch in Myth and Reality.* London: Jonathon Cape Ltd., 1972. Pages 124, 125.

Noe, Allen V. "And Still the Reports Roll In," *Pursuit,* Society for the Investigation of the Unexplained, January, 1974. Page 266.

Norman, Eric. *The Abominable Snowman.* New York: Award Books, Page 197.

Papashvily, George and Helen. *Anything Can Happen.* New York: Harper & Row, 1940. Page 435.

Patterson, Roger. *Do Abominable Snowmen of America Really Exist?* Yakima: Trailblazer Research Inc., 1966. Page 114.

Porshnev, Boris F. "The Problem of Relic Paeleaoanthropus," *Soviet Ethnography,* Moscow, U.S.S.R. Academy of Sciences, 1969. Page 137.

——————. "The Troglodytidae and the Hominidae in the Taxonomy and Evolution of Higher Primates," *Current Anthropology,* University of Chicago Press, December, 1974. Page 137.

Rawicz, Slavomir. *The Long Walk.* New York: Harper & Row, 1956. Page 136.

Rayburn, Otto Ernest. *Ozark County.* New York: Duell, Sloan & Pearce, 1941. Pages 189, 369.

Reynolds, Vernon. *The Apes.* New York: E. P. Dutton, 1967. Pages 35, 449.

Roosevelt, Theordore. *The Wilderness Hunter - Outdoor Pastimes of an American Hunter.* New York: G. P. Putnam's Sons, 1893. Page 29.

Saga. January, 1961. Pages 292, 293.

Sanderson, Ivan. *Abominable Snowmen, Legend Come to Life.* Philadelphia: Chilton, 1961. Pages 45, 133, 135, 145, 221, 252.

——————. "First Photos of 'Bigfoot', California's Legendary Abominable Snowman," *Argosy,* February, 1968. Page 119.

——————. "The Strange Story of America's Abominable Snowman," *True,* December, 1959. Page 114. Also "A New Look at America's Mystery Giant," *True,* March, 1960.

——————. "Wisconsin's Abominable Snowman," *Argosy,* April, 1969. Page 198.

Schaller, George. *Year of the Gorilla.* Chicago: University of Chicago Press, 1964. Pages 133, 346, 449.

Slate, B. Ann, and Berry, Alan. *Bigfoot.* New York: Bantam Books, 1976. Pages 318, 386, 417.

Smith, Warren, "America's Terrifying Woodland Monster Men," *Saga,* June, 1969. Page 197.

BIBLIOGRAPHY

Stephens, Harold. "Abominable Snowman of Malaysia," *Argosy*, August, 1971. Page 135.

Stoyanow, Victor. Article in *Desert Magazine*, July, 1964. Page 311.

Suttles, Wayne. "On the Cultural Track of the Sasquatch," *Northwest Anthropological Research Notes, Vo. 1, No. 6,* University of Idaho, spring, 1972. Page 20. Also Sprague, Roderick and Krantz, Grover S. *The Scientist Looks at the Sasquatch.* Moscow, Idaho: University of Idaho Press, 1977.

Tchernine, Odette. *In Pursuit of the Abominable Snowman.* New York: Taplinger Publishing Co., 1971. Page 145.

Thompson, David. *Narrative of His Explorations in Western America, 1784-1812.* Westport: Greenwood Press Inc. Page 36.

Totsgi, Jorg. Article in *The Real American*, Hoquiam, Washington, July 17, 1924. Page 90.

Tulp, Nicolaas. *Observations Medicae*, Ed. 6. Amsterdam. 1700's. Page 139.

Van Lawick-Goodall, Hugo and Jane. *Innocent Killers.* Boston: Houghton Mifflin Co., 1970. Page 458.

Warth, Robert C. "A UFO-ABSM Link," *Pursuit*, Society for the Investigation of the Unexplained, April, 1975. Page 233.

Weeks, John M. Letter to *True*, March, 1960. Page 37.

Wendt, Herbert. *In Search of Adam.* Boston: Houghton Mifflin, 1956. Page 34.

Worley, Don. "U.F.O. Anthropoids in the U.S.A.," *Argosy U.F.O.*, July, 1977. Page 221.

Wright, Bruce. Wildlife Sketches Near and Far. Fredricton: University of New Brunswick Press, 1962. Page 253.

INDEX of Places and Newspapers in the United States and Canada, listed by State and Province.

Index of Places and Newspapers

Index of Places and Newspapers

Index of Places and Newspapers

Index of Places and Newspapers

Index of Places and Newspapers

Index of Places and Newspapers

Index of Places and Newspapers

Index of Places and Newspapers

Index of Names

Index of Names

Index of Names

General Index

489

General Index

Royal Canadian Mounted Police, 247
Russian Academy of Science, 138, 145
Russian Information Service, 145
Salmon runs, 430
Scandanavia, 147
Science, attitude to sasquatch, 11, 12, 113, 114, 118, 119, 128, 130, 131, 137, 152, 153, 159, 237
Sheriff's department, Cascade County, Montana, 297-299
Sheriff's department, Los Angeles County, California, 309
Shansi Province, China, 140
Siberia, 136
Sikkim, 140
Sinkiang, 140
Smithsonian Institution, 50-51, 324, 325
Society for the Investigation of the Unexplained, 146, 187, 214, 216, 228, 232, 265, 266
South Mountain Research Group, 152
Southern States, 171, 212-214, 220, 271, 369, 443
Southern Methodist University, 326
Southwest States, 171
Spain, 19, 147
Spokane Indians, 26
Stone foot, 22
Sumatra, 135
Tajikstan, 140
Talysh Mountains, U.S.S.R., 139
Tanzania, 133
Tennessee River, 169
Texas Memorial Museum, 325
Texas Christian University, 328

Tien Shan Range, U.S.S.R., 138, 140, 149, 150
Totem pole, 22
Tracking dog, 74, 75, 118, 129, 380
Tranquilizer gun, 158, 465
Transylvania, 140
Troglodyte, 137, 138
True, 352
U.F.O.'s, 206, 207, 256, 261-264, 297, 298, 318
U.F.O. investigators, 258, 260-263, 320
United Press International, 180, 199, 203, 204, 219, 305, 306
Universal Studios, 129
University of British Columbia, 85, 118, 154
University of Illinois, 203
University of New Brunswick, 252
University of Texas, 324, 325
U.S. Inter-Agency Grizzly Bear Study Team, 466
U.S.S.R., 137-150, 359, 435
Vanguard Research, 152
Vestigia, 269
Vultures, as tool users, 458
Western United States, 211, 442-448, 453
Whitman Mission, 26
Wolfmen, 176
Wolper Productions, 129, 429
Yerkes Primate Centre, 463
Yeti Newsletter, 217, 273, 278, 434
Yeti Research Society, 217, 273, 274,
Zoologists, 79, 110, 113, 139, 145

Sasquatch Index

BEHAVIOR

Agression, rarity of, 34, 333, 334, 336
Approach people: asleep, 334; in building, 13, 14, 15, 17, 31, 38, 46, 89, 94, 222, 247, 255; in tent, 57, 177; in vehicle, 230, 279, 296, 338, 339
Bluffing agression, 338, 341-346
Break things: fences, 186, 221, 376; sluice, 38; stumps, 252, 408; trees, 199, 252, 336
Carry things: animals, 39, 172, 177, 378, 407; person, 102, 103, 412; vegetable matter, 106, 176; other objects, 39, 143, 146, 233
Caution, avoid leaving tracks, 144, 359, 388; females and infants scarce, 448, 449; infant avoid male, 424
Chase: animals, 28, 39, 375; people, 272, 338, 368, 378
Damage building, 17, 94, 206, 338
Defecate, 425
Dig out rodents, 422
Family groupings, 104, 240, 421, 448
Feeding, actual observations, 19, 45, 55, 59, 106, 143, 144, 177, 194, 269, 368, 422, 424, 425
Fight dogs, 206, 267, 272
Groups, 61, 97, 175, 201, 448
Hibernation, 144, 358
Injure person, 201, 216, 227, 308, 336
Jump on vehicle, 185, 273, 306, 338

490

492

Also by John Green

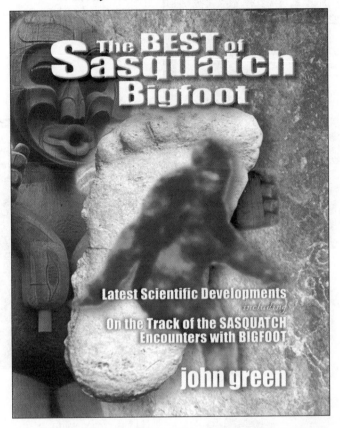

The Best of Sasquatch Bigfoot
The Latest Scientific Developments
plus all of
On the Track of the Sasquatch
and Encounters with Bigfoot

ISBN 088839-546-9
8.5 x 11 inches, softcover
144 pages

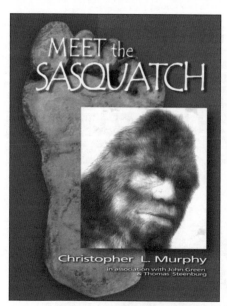

Raincoast Sasquatch
J. Robert Alley
0-88839-508-6
5½ x 8½, sc, 360 pages

Bigfoot Sasquatch Evidence
Dr. Grover S. Krantz
0-88839-447-0
5½ x 8½, sc, 348 pages

The Locals
Thom Powell
0-88839-552-3
5½ x 8½, sc, 271 pages

Sasquatch Bigfoot
Thomas N. Steenburg
0-88839-312-1
5½ x 8½, sc, 126 pages

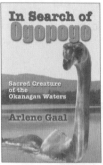

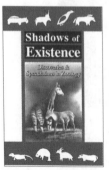